# MULTIVARIABLE AND VECTOR CALCULUS

# MULTIVARIABLE AND VECTOR CALCULUS

*An Introduction*

**Second Edition**

SARHAN M. MUSA, PHD

**MERCURY LEARNING AND INFORMATION**
Dulles, Virginia
Boston, Massachusetts
New Delhi

Publisher: David Pallai
MERCURY LEARNING AND INFORMATION
22841 Quicksilver Drive
Dulles, VA 20166
info@merclearning.com
www.merclearning.com
(800) 232-0223

S. M. Musa. *Multivariable and Vector Calculus: An Introduction, Second Edition.*
ISBN: 978-1-68392-919-2

The publisher recognizes and respects all marks used by companies, manufacturers, and developers as a means to distinguish their products. All brand names and product names mentioned in this book are trademarks or service marks of their respective companies. Any omission or misuse (of any kind) of service marks or trademarks, etc. is not an attempt to infringe on the property of others.

Library of Congress Control Number: 2022947021
232425321   This book is printed on acid-free paper in the United States of America.

Our titles are available for adoption, license, or bulk purchase by institutions, corporations, etc. For additional information, please contact the Customer Service Dept. at (800)232-0223(toll free).

All of our titles are available in digital format at authorcloudware.com and other digital vendors. Companion disc files for this title are available by contacting info@merclearning.com. The sole obligation of MERCURY LEARNING AND INFORMATION to the purchaser is to replace the disc, based on defective materials or faulty workmanship, but not based on the operation or functionality of the product.

*Dedicated to my late father Mahmoud,*
*my mother Fatmeh, and my wife Lama*

# CONTENTS

| | | |
|---|---|---|
| **Preface** | | **xiii** |
| **Acknowledgments** | | **xv** |
| **CHAPTER 1: VECTORS AND PARAMETRIC CURVES** | | **1** |
| 1.1 | Points and Vectors on the Plane | 1 |
| | Exercises 1.1 | 13 |
| 1.2 | Scalar Product on the Plane | 17 |
| | Exercises 1.2 | 25 |
| 1.3 | Linear Independence | 28 |
| | Exercises 1.3 | 31 |
| 1.4 | Geometric Transformations in Two Dimensions | 34 |
| | Exercises 1.4 | 42 |
| 1.5 | Determinants in Two Dimensions | 44 |
| | Exercises 1.5 | 51 |
| 1.6 | Parametric Curves on the Plane | 53 |
| | Exercises 1.6 | 67 |
| 1.7 | Vectors in Space | 70 |
| | Exercises 1.7 | 84 |
| 1.8 | Cross Product | 87 |
| | Exercises 1.8 | 96 |

1.9    Matrices in Three Dimensions                        98
       Exercises 1.9                                       102
1.10   Determinants in Three Dimensions                    104
       Exercises 1.10                                      109
1.11   Some Solid Geometry                                 111
       Exercises 1.11                                      116
1.12   Cavalieri and the Pappus–Guldin Rules               117
       Exercises 1.12                                      120
1.13   Dihedral Angles and Platonic Solids                 121
       Exercises 1.13                                      126
1.14   Spherical Trigonometry                              127
       Exercises 1.14                                      132
1.15   Canonical Surfaces                                  133
       Exercises 1.15                                      145
1.16   Parametric Curves in Space                          146
       Exercises 1.16                                      151
1.17   Multidimensional Vectors                            152
       Exercises 1.17                                      160

**CHAPTER 2: DIFFERENTIATION**                             **163**
2.1    Some Topology                                       163
       Exercises 2.1                                       166
2.2    Multivariable Functions                             169
       Exercises 2.2                                       174
2.3    Limits and Continuity                               176
       Exercises 2.3                                       187
2.4    Definition of the Derivative                        189
       Exercises 2.4                                       193
2.5    The Jacobi Matrix                                   193
       Exercises 2.5                                       205
2.6    Gradients and Directional Derivatives               207
       Exercises 2.6                                       216

| | | |
|---|---|---|
| 2.7 | Levi-Civitta and Einste | 218 |
| | Exercises 2.7 | 221 |
| 2.8 | Extrema | 222 |
| | Exercises 2.8 | 227 |
| 2.9 | Lagrange Multipliers | 229 |
| | Exercises 2.9 | 233 |

**CHAPTER 3: INTEGRATION** ..................................................... **235**
| | | |
|---|---|---|
| 3.1 | Differential Forms | 235 |
| | Exercises 3.1 | 240 |
| 3.2 | Zero-Manifolds | 241 |
| | Exercises 3.2 | 242 |
| 3.3 | One Manifold | 243 |
| | Exercises 3.3 | 248 |
| 3.4 | Closed and Exact Forms | 250 |
| | Exercises 3.4 | 256 |
| 3.5 | Two-Manifolds | 257 |
| | Exercises 3.5 | 267 |
| 3.6 | Change of Variables in Double Integrals | 273 |
| | Exercises 3.6 | 280 |
| 3.7 | Change to Polar Coordinates | 283 |
| | Exercises 3.7 | 287 |
| 3.8 | Three-Manifolds | 290 |
| | Exercises 3.8 | 295 |
| 3.9 | Change of Variables in Triple Integrals | 297 |
| | Exercises 3.9 | 301 |
| 3.10 | Surface Integrals | 303 |
| | Exercises 3.10 | 306 |
| 3.11 | Green's, Stokes', and Gauss' Theorems | 307 |
| | Exercises 3.11 | 313 |

**APPENDIX A: MAPLE**     **319**

A.1 Getting Started and Windows of Maple     319

A.2 Arithmetic     321

A.3 Symbolic Computation     322

A.4 Assignments     323

A.5 Working with Output     323

A.6 Solving Equations     323

A.7 Plots with Maple     324

A.8 Limits and Derivatives     326

A.9 Integration     326

A.10 Matrix     326

**APPENDIX B: MATLAB**     **327**

B.1 Getting Started and Windows of MATLAB     327

    B.1.1 Using MATLAB in Calculations     329

B.2 Plotting     336

    B.2.1 Two-dimensional Plotting     336

    B.2.2 Three-Dimensional Plotting     347

B.3 Programming in MATLAB     353

    B.3.1 For Loops     355

    B.3.2 While Loops     356

    B.3.3 If, Else, and Elseif     357

    3.3.4 Switch     359

B.4 Symbolic Computation     361

    B.4.1 Simplifying Symbolic Expressions     361

    B.4.2 Differentiating Symbolic Expressions     364

    B.4.3 Integrating Symbolic Expressions     365

    B.4.4 Limits Symbolic Expressions     366

    B.4.5 Taylor Series Symbolic Expressions     366

    B.4.6 Sums Symbolic Expressions     367

    B.4.7 Solving Equations as Symbolic Expressions     367

**APPENDIX C: ANSWERS TO ODD-NUMBERED EXERCISES**    **369**

    Chapter 1    369

    Chapter 2    386

    Chapter 3    408

**APPENDIX D: FORMULAS**    **417**

    D.1  Trigonometric Identities    417

    D.2  Hyperbolic Functions    419

    D.3  Table of Derivatives    420

    D.4  Table of Integrals    422

    D.5  Summations (Series)    424

        D.5.1  Finite Element of Terms    424

        D.5.2  Infinite Element of Terms    424

    D.6  Logarithmic Identities    425

    D.7  Exponential Identities    425

    D.8  Approximations for Small Quantities    425

    D.9  Vectors    426

        D.9.1  Vector Derivatives    426

        D.9.2  Vector Identity    427

        D.9.3  Fundamental Theorems    428

**BIBLIOGRAPHY**    **431**

**INDEX**    **435**

# PREFACE

Multivariable and vector calculus is an essential subject of the mathematical education for scientists and engineers. This book is aimed primarily at undergraduates in mathematics, engineering, and the physical sciences. Rather than concentrating on technical skills, it focuses on a deeper understanding of the subject by providing many unusual and challenging examples. Furthermore, this book can be used to impower the mathematical knowledge for Artificial Intelligence (AI) concepts. The topics of vector geometry, differentiation, and integration in several variables are explored. It also provides numerous computer illustrations and tutorials using *Maple*® and *MATLAB*®. The software applications allow the students to bridge the gap between analysis and computation. Mainly, this book compromises 3 chapters, 4 appendices and companion files.

Chapter 1 provides vectors and parametric curves. It contains points and vectors on the plane, scalar products on the plane, linear independence, geometric transformations in two dimensions, determinants in two dimensions, parametric curves on the plane, vectors in space, cross products, matrices in three dimensions, determinants in three dimensions, some solid geometry, Cavalieri and the Pappus-Guldin rules, dihedral angles and platonic solids, spherical trigonometry, canonical surfaces, parametric curves in space, and multidimensional vectors.

Chapter 2 provides differentiation of functions of several variables. This chapter mainly discusses some topology, multivariable functions, limits and continuity, definition of the derivative, the Jacobi matrix, gradients and directional derivatives, Levi-Civita and Einstein, extrema, and Lagrange multipliers.

Chapter 3 provides integrations of functions of several variables. It contains differentiation forms, zero-manifolds, one-manifolds, closed and exact forms, two-manifolds, change of variables in double integrals, change to polar coordinates, three-manifolds, change of variables in triple integrals, surface integrals, and Green's, Stokes', and Gauss' Theorems.

Finally, the book concludes with four appendices: Appendix A covers a basic tutorial on *Maple* software; Appendix B includes a basic tutorial on *MATLAB*; Appendix C provides the answers to odd-numbered exercises; Appendix D reviews the common, useful mathematical formulas.

The companion files are available from the publisher. Contact info@merclearning.com.

**Sarhan M. Musa**
Houston, Texas
January 2023

# ACKNOWLEDGMENTS

It is my pleasure to acknowledge the outstanding help and support of the team at Mercury Learning and Information in preparing this book, especially from David Pallai and Jennifer Blaney. Finally, the book is written based on the constant support, love and patience of my family.

..

# 1

# VECTORS AND PARAMETRIC CURVES

In science and engineering, certain quantities such as force and acceleration possess both a magnitude and a direction. They are represented as vectors and geometrically drawn as an arrow. For example, in dealing with systems of linear equations, the solution can be pointed in the plane if the equations have two variables, points in three space if they are equations in three variables, points in four space if they have four variables, and so on. The solutions make up the subset of large spaces and the constructed spaces are called *vector spaces*, which are used in many areas of mathematics.

We start this chapter with an introduction to some linear algebra necessary for the course.

We mainly discuss points and vectors on the plane, scalar product on the plane, linear independence, geometric transformations in two dimensions, determinants in two dimensions, parametric curves on the plane, vectors in space, cross product, matrices in three dimensions, determinants in three dimensions, some solid geometry, Cavalieri, the Pappus–Guldin rules, dihedral angles and platonic solids, spherical trigonometry, canonical surfaces, parametric curves in space, and multidimensional vectors.

## 1.1 POINTS AND VECTORS ON THE PLANE

**Definition 1.1.1** A *scalar* $a \in \mathbb{R}$ is simply a real number.

**Definition 1.1.2** A *point* $r \in \mathbb{R}^2$ is an ordered pair of real numbers, $r \in (x,y)$ with $x \in \mathbb{R}$ and $y \in \mathbb{R}$. Here the first coordinate $x$ stipulates the

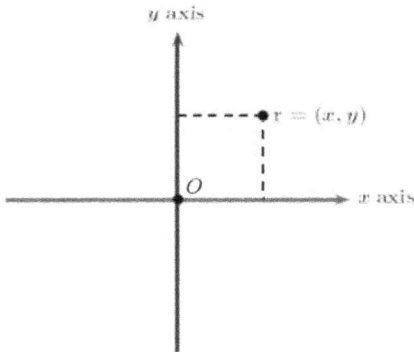

*y* axis

$r = (x, y)$

*O*

*x* axis

**FIGURE 1.1.1** A point in $\mathbb{R}^2$.

location on the horizontal axis and the second coordinate $y$ stipulates the location on the vertical axis. See Fig. 1.1.1.

We will always denote the origin by $O$, that is, the point $O = (0,0) = \begin{pmatrix} 0 \\ 0 \end{pmatrix}$. $\mathbb{R}^2$ is the set of all points $a = (a_1, a_2) = \begin{pmatrix} a_1 \\ a_2 \end{pmatrix}$ with real number coordinates on the plane.

Given $a = \begin{pmatrix} a_1 \\ a_2 \end{pmatrix} \in \mathbb{R}^2$ and $b = \begin{pmatrix} b_1 \\ b_2 \end{pmatrix} \in \mathbb{R}^2$, we define their addition as $a + b = \begin{pmatrix} a_1 \\ a_2 \end{pmatrix} + \begin{pmatrix} b_1 \\ b_2 \end{pmatrix} = \begin{pmatrix} a_1 + b_1 \\ a_2 + b_2 \end{pmatrix}$. Similarly, we define the *scalar* multiplication of a point of $\mathbb{R}^2$ by the scalar $a \in \mathbb{R}$ as $a\,a = a \begin{pmatrix} a_1 \\ a_2 \end{pmatrix} = \begin{pmatrix} a\,a_1 \\ a\,a_2 \end{pmatrix}$.

**TIP!** *The following sets have special symbols.*

$\mathbb{N} = \{0, 1, 2, \dots\}$ *denotes the set of natural numbers.*

$\mathbb{Z} = \{\dots, -2, -1, 0, 1, 2, \dots\}$ *denotes the set of integers.*

$\mathbb{Q} = \left\{ x \mid x = \dfrac{a}{b}, a, b \in \mathbb{Z}, b \neq 0 \right\}$ *denotes the set of rational numbers.*

$\mathbb{R} = \{ x \mid -\infty < x < \infty \}$ *denotes the set of real numbers.*

$\mathbb{C} = \{ a + b : a, b \in \mathbb{R} \text{ and } i = \sqrt{-1} \}$ *denotes the set of complex numbers.*

$\phi = \{\ \}$ *denotes the empty set.*

**TIP!** *The symbol $\Rightarrow$ is read " implies", and the symbol $\Leftrightarrow$ is read "if and only if".*

**Definition 1.1.3** Given two points r and r′ in $\mathbb{R}^2$ the directed line segment with departure point r and the arrival point r′ is called the *bi-point (fixed vector)* r, r′, and is denoted by [r, r′]. See Fig. 1.1.2 for an example.

The bi-point [r, r′] can be thus inter-preted as an arrow starting at $r$ and fin-ishing, with the arrow tip, at r′. We say that r is the *tail* of the bi-point [r, r′] and that r′ is its *head*.

*Let A be a set. If a belongs to the set A, then we write $a \in A$, read "a is an ele-ment of A." If a does not belong to the set A, then we write $a \notin A$, read "a is not an element of A."*

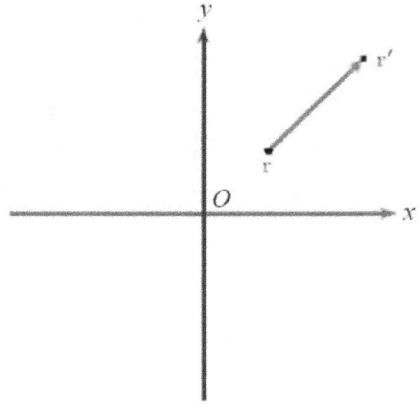

**Definition 1.1.4** A *Vector* $\vec{a} \in \mathbb{R}^2$ is a codification of the movement of a bi-point. Given the bi-point [r, r′], we associate to it the vector $\overrightarrow{rr'} = \begin{bmatrix} x' - x \\ y' - y \end{bmatrix}$

*FIGURE 1.1.2* A bi-point in $\mathbb{R}^2$.

stipulating a movement of $x' - x$ units from $(x, y)$ in the horizontal axis and of $y' - y$ units from the current position in the vertical axis. The *zero vector* $\vec{0} = \begin{bmatrix} 0 \\ 0 \end{bmatrix}$ indicates no movement in either direction. Notice that infinitely many different choices of departure and arrival points may give the same vector.

*Consider* $a \in \mathbb{R}^2$, *then* $[a, a] = O = \begin{pmatrix} 0 \\ 0 \end{pmatrix}$.

**EXAMPLE 1.1.1**

The vector into which the bi-point with tail $a = \begin{pmatrix} -1 \\ 2 \end{pmatrix}$ and head at $b = \begin{pmatrix} 3 \\ 4 \end{pmatrix}$ falls is $\overrightarrow{ab} = \begin{bmatrix} 3 - (-1) \\ 4 - 2 \end{bmatrix} = \begin{bmatrix} 4 \\ 2 \end{bmatrix}$.

### EXAMPLE 1.1.2

Consider the following points:

$a_1 = (1, 2)$, $b_1 = (3, -4)$, $a_2 = (3, 5)$, $b_2 = (5, -1)$, $O = (0, 0)$, $b = (2, -6)$. Though the bi-points $[a_1, b_1]$, $[a_2, b_2]$, $[O, b]$ are in different locations on the plane, they represent the same vector, as

$$\begin{bmatrix} 3-1 \\ -4-2 \end{bmatrix} = \begin{bmatrix} 5-3 \\ -1-5 \end{bmatrix} = \begin{bmatrix} 2-0 \\ -6-0 \end{bmatrix} = \begin{bmatrix} 2 \\ -6 \end{bmatrix}.$$

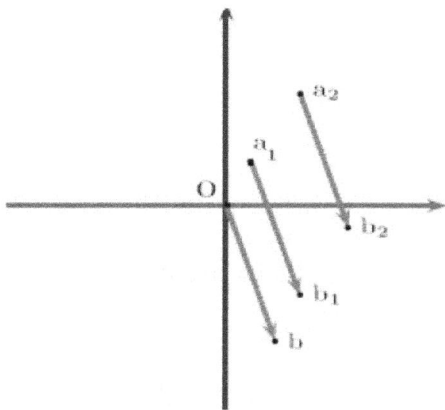

The instructions given by the vector are all the same: start at the point, go two units right and six units down. See Fig. 1.1.3.

In more technical language, a vector is an *equivalence* class of bi-points, that is, all bi-points that have the same length, have the same direction. In this sense, points are equivalent and the name of this equivalence is a *vector*. As a simple example of an equivalence class, consider the set of integers $\mathbb{Z}$. According to their remainder upon division by 3, each integer belongs to one of the three sets:

**FIGURE 1.1.3** Example 1.1.1.

$$3\mathbb{Z} = \{\ldots, -6, -3, 0, 3, 6, \ldots\}, \quad 3\mathbb{Z} + 1 = \{\ldots, -5, -2, 1, 4, 7, \ldots\},$$
$$3\mathbb{Z} + 2 = \{\ldots, -4, -1, 2, 5, 8, \ldots\}.$$

The equivalence class $3\mathbb{Z}$ comprises the integers divisible by 3, and for example, $-18 \in 3\mathbb{Z}$. Analogously, in Example 1.1.2, the bi-point $[a_1, b_1]$ belongs to the equivalence class $\begin{bmatrix} 2 \\ -6 \end{bmatrix}$, that is, $[a_1, b_1] \in \begin{bmatrix} 2 \\ -6 \end{bmatrix}$.

**Definition 1.1.5** The Vector $\overrightarrow{Oa}$ that corresponds to the point $a \in \mathbb{R}^2$ is called the *position vector* of the point a.

**Definition 1.1.6** Let $a \neq b$ be the point on the plane and let $\overrightarrow{ab}$ be the line passing through a and b. The *direction* of the bi-point [a, b] is the direction of

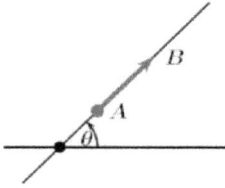

FIGURE 1.1.4 Direction of a bi-point.

the line $L$, that is, the angle $\theta \in [0;\pi]$ that the line $\overleftrightarrow{ab}$ makes with the positive $x$-axis (horizontal axis), when measured counterclockwise. The direction of a vector $\vec{v} \neq \vec{0}$ is the direction of any of its bi-point representatives. See Fig. 1.1.4.

**Definition 1.1.7**   We say that [a, b] has the same direction as [z, w] if $\overrightarrow{ab} = \overrightarrow{zw}$ .

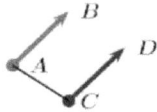

FIGURE 1.1.5 Bi-points with the same sense.

**Definition 1.1.8**   We say that the bi-points [a, b] and [z, w] have the *same sense* if they have the same direction, if when translating one so its tail is over the other's tail, and both their heads lie on the same half-plane made by the line perpendicular to their tails. See Fig. 1.1.5.

FIGURE 1.1.6 Bi-points with opposite sense.

**Definition 1.1.9**   We say that the bi-points [a, b] and [z, w] have the *opposite sense* if they have the same direction, if when translating one so its tail is over the other's tail, and their heads lie on different half-planes made by the line perpendicular to their tails. See Fig. 1.1.6.

The *sense* of a vector is the sense of any of its bi-point representatives. Two bi-points are *parallel* if the lines containing them are parallel. Two vectors are parallel if bi-point representatives of them are parallel.

**TIP!**   *Bi-point* [b, a] *has the opposite sense of* [a, b], *so we write* $[b, a] = -[a, b]$. *Similarly, we write* $\overrightarrow{ab} = -\overrightarrow{ba}$ .

**Definition 1.1.10**   The *Euclidean length (norm or magnitude)* of bi-point [a, b] is simply the distance between a and b, and it is denoted by

$$\|[a,b]\| = \sqrt{(a_1 - b_1)^2 + (a_2 - b_2)^2} \ .$$

A bi-point is said to have *unit length* if it has norm 1. The *norm of a vector* is the norm of any of its bi-point representatives.

**NOTE**   *A vector is completely determined by three things: (i) its norm, (ii) its direction, and (iii) its sense. It is clear that the norm of a vector satisfies the following properties:*

1. $\|\vec{a}\| \geq 0$

2. $\|\vec{a}\| = 0 \Leftrightarrow \vec{a} = \vec{0}$

**Definition 1.1.11**  A *unit vector* is a vector whose norm is 1. If $\vec{v}$ is a nonzero vector ($\vec{v} \neq 0$), then the vector $\vec{u} = \dfrac{\vec{v}}{\|\vec{v}\|}$ is a *unit vector* in direction of $\vec{v}$. The procedure of constructing a unit vector in the same direction as a given vector is called *normalizing* the vector.

**EXAMPLE 1.1.3**

Find the norm of the vector $\vec{v} = \begin{bmatrix} 1 \\ \sqrt{2} \end{bmatrix}$ and normalize this vector.

***Solution:***

$$\|\vec{v}\| = \sqrt{(1)^2 + \left(\sqrt{2}\right)^2} = \sqrt{3},$$

which is called *the magnitude of the given vector*.

The normalized vector is $\dfrac{1}{\sqrt{3}} \begin{bmatrix} 1 \\ \sqrt{2} \end{bmatrix} = \begin{bmatrix} \dfrac{1}{\sqrt{3}} \\ \dfrac{\sqrt{2}}{\sqrt{3}} \end{bmatrix}.$

We may use the software *MATLAB* in order to compute the norm of vectors.

$>> v = [1\ \text{sqrt}(2)];$

$>> \text{norm}(v)$

  ans =

   1.7321

$>> u = v/\text{norm}(v)$

u =

   0.5774   0.8165

Also, we may use the software *Maple*$^{TM}$ in order to compute the norm of vectors.

> $with(linalg):$

> $v := vector([1, sqrt(2)]);$

$$v := \begin{bmatrix} 1 & \sqrt{2} \end{bmatrix}$$

> $norm(v, 2);$

$$\sqrt{3}$$

> $normalize(v);$

$$\begin{bmatrix} \dfrac{1}{3}\sqrt{3} & \dfrac{1}{3}\sqrt{2}\sqrt{3} \end{bmatrix}$$

**Definition 1.1.12** If $\vec{u}$ and $\vec{v}$ are two vectors in $\mathbb{R}^2$, their *vector sum* $\vec{u} + \vec{v}$ is defined by the coordinate-wise addition:

$$\vec{u} + \vec{v} = \begin{bmatrix} u_1 \\ u_2 \end{bmatrix} + \begin{bmatrix} v_1 \\ v_2 \end{bmatrix} = \begin{bmatrix} u_1 + v_1 \\ u_2 + v_2 \end{bmatrix}. \tag{1.1}$$

It is easy to see that vector addition is commutative and associative, that the vector $\vec{0}$ acts as an additive identity, and that the additive inverse of $\vec{a}$ is $-\vec{a}$.

To add two vectors geometrically, proceed as follows. Draw a bi-point representative of $\vec{u}$. Find a bi-point representative of $\vec{v}$ having its tail at the tip of $\vec{u}$. The sum $\vec{u} + \vec{v}$ is the vector whose tail is that of the bi-point for $\vec{u}$ and whose tip is that of the bi-point for $\vec{v}$. In particular, if $\vec{u} = \overrightarrow{AB}$ and $\vec{v} = \overrightarrow{BC}$, then we have *Chasles' Rule*:

$$\overrightarrow{AB} + \overrightarrow{BC} = \overrightarrow{AC}. \tag{1.2}$$

See Figs. 1.1.7, 1.1.8, 1.1.9, and 1.1.10.

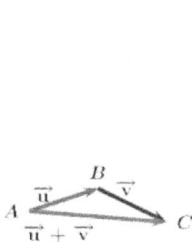

**FIGURE 1.1.7** Addition of vectors.

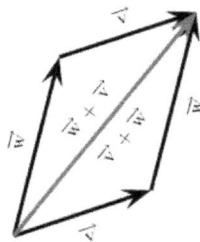

**FIGURE 1.1.8** Commutative of vector addition.

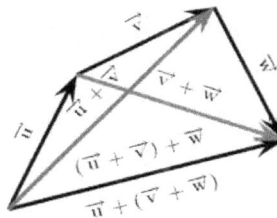

**FIGURE 1.1.9** Associative of vector addition.

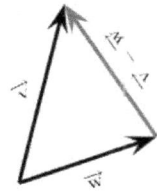

**FIGURE 1.1.10** Difference of vectors.

**Definition 1.1.13** If $a \in \mathbb{R}$ and $\vec{a} \in \mathbb{R}^2$, we define *scalar multiplication* of a vector and a scalar by the coordinatewise multiplication:

$$a \ \vec{a} = a \begin{bmatrix} a_1 \\ a_2 \end{bmatrix} = \begin{bmatrix} a \, a_1 \\ a \, a_2 \end{bmatrix}. \tag{1.3}$$

See Fig. 1.1.11.

It is easy to see that vector addition and scalar multiplication satisfies the following properties:

**1.** $a \ \left( \vec{a} + \vec{b} \right) = a \ \vec{a} + a \ \vec{b}$

**2.** $\left( a + \beta \right) \vec{a} = a \ \vec{a} + \beta \ \vec{a}$

**3.** $1 \, \vec{a} = \vec{a}$

**4.** $\left( a \ \beta \right) \vec{a} = a \left( \beta \ \vec{a} \right)$

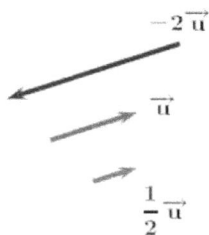

**FIGURE 1.1.11** Scalar multiplication of vectors.

We may use *MATLAB* in order to compute the sum of vectors and scalar multiplication of vectors.

>> u = [2 5];

>> v = [3 4];

>> u + v

ans =

   5   9

>> 6*u

ans =

   12   30

We may use the software *Maple*$^{TM}$ in order to compute the sum of vectors and scalar multiplication of vectors.

> $with(linalg)$ :

> $u := Vector([2,5])$;

$$u := \begin{bmatrix} 2 \\ 5 \end{bmatrix}$$

> $v := Vector([3,4])$;

$$v := \begin{bmatrix} 3 \\ 4 \end{bmatrix}$$

▷ $u + v$;

$$\begin{bmatrix} 5 \\ 9 \end{bmatrix}$$

> $6 * u$;

$$\begin{bmatrix} 12 \\ 30 \end{bmatrix}$$

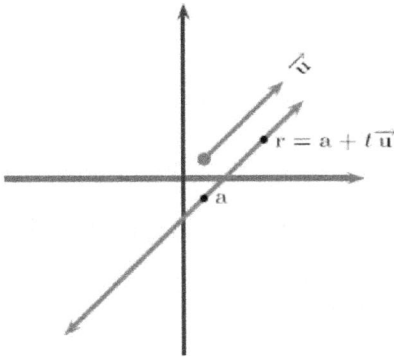

**FIGURE 1.1.12** Parametric equation of a line on the plane.

**Definition 1.1.14** Let $\vec{u} \neq \vec{0}$. Put $\mathbb{R}\,\vec{u} = \{\lambda\vec{u} : \lambda \in \mathbb{R}\}$ and let $a \in \mathbb{R}^2$. The *affine line* with direction vector $\vec{u} = \begin{bmatrix} u_1 \\ u_2 \end{bmatrix}$ and passing through a is the set off points on the plane $a + \mathbb{R}\,\vec{u} = \left\{\begin{pmatrix} x \\ y \end{pmatrix} \in \mathbb{R}^2 : x = a_1 + tu_1,\ y = a_2 + tu_2,\ t \in \mathbb{R}\right\}$. See Fig. 1.1.12.

If $u_1 = 0$, the affine line previously defined is vertical, as $x$ is constant. If $u_1 \neq 0$, then

$$\frac{x - a_1}{u_1} = t \Rightarrow y = a_2 + \frac{(x - a_1)}{u_1}u_2 = \frac{u_2}{u_1}x + a_2 - a_1\frac{u_2}{u_1},$$

that is, the affine line is the Cartesian line with slope $\frac{u_2}{u_1}$. Conversely, if $y = mx + k$ is the equation of a Cartesian line, then

$$\begin{pmatrix} x \\ y \end{pmatrix} = \begin{bmatrix} 1 \\ m \end{bmatrix}t + \begin{pmatrix} 0 \\ k \end{pmatrix},$$

that is, every Cartesian line is also an affine line and one may take the vector $\begin{bmatrix} 1 \\ m \end{bmatrix}$ as its direction vector. It also follows that two vectors $\vec{u}$ and $\vec{v}$ are parallel if and only if the affine lines $\mathbb{R}\,\vec{u}$ and $\mathbb{R}\,\vec{v}$ are parallel. Hence, $\vec{u} \parallel \vec{v}$ if there exists a scalar $\lambda \in \mathbb{R}$ such that $\vec{u} = \lambda\,\vec{v}$.

**NOTE**  *Because $\vec{0} = 0\,\vec{v}$ for any vector $\vec{v}$, the $\vec{0}$ is parallel to every vector.*

### EXAMPLE 1.1.4

Let $a \in \mathbb{R}$, $a > 0$, and let $\vec{u} \neq \vec{0}$. Find a vector with norm $a$ and parallel to $\vec{u}$.

***Solution:***

We know that $\dfrac{\vec{u}}{\|\vec{u}\|}$ has norm 1 as $\left\| \dfrac{\vec{u}}{\|\vec{u}\|} \right\| = \dfrac{\|\vec{u}\|}{\|\vec{u}\|} = 1$. Hence, the vector $a\,\dfrac{\vec{u}}{\|\vec{u}\|}$ has norm $a$ and it is in the direction of $\vec{u}$. One may also take $-a\,\dfrac{\vec{u}}{\|\vec{u}\|}$.

### EXAMPLE 1.1.5

Find a vector of length 3, parallel to $\vec{v} = \begin{bmatrix} 1 \\ \sqrt{2} \end{bmatrix}$, but in the opposite sense.

***Solution:***

Since $\|\vec{v}\| = \sqrt{(1)^2 + \left(\sqrt{2}\right)^2} = \sqrt{3}$, the vector $\dfrac{\vec{v}}{\|\vec{v}\|}$ has unit norm, and has the same direction and sense as $\vec{v}$, so the vector sought is

$$-3\,\frac{\vec{v}}{\|\vec{v}\|} = -\frac{3}{\sqrt{3}}\begin{bmatrix} 1 \\ \sqrt{2} \end{bmatrix} = \begin{bmatrix} -\sqrt{3} \\ -\sqrt{6} \end{bmatrix}.$$

### EXAMPLE 1.1.6

Find the parametric equation of the line passing through $\begin{pmatrix} 1 \\ -1 \end{pmatrix}$ and in the direction of the vector $\begin{bmatrix} 2 \\ -3 \end{bmatrix}$.

***Solution:***

The desired equation is plainly:

$$\begin{pmatrix} x \\ y \end{pmatrix} = \begin{pmatrix} 1 \\ -1 \end{pmatrix} + t\begin{bmatrix} 2 \\ -3 \end{bmatrix} \Rightarrow x = 1 + 2t, \; y = -1 - 3t, \; t \in \mathbb{R}.$$

Some plane geometry results can be easily proved by means of vectors. Here are some examples.

## EXAMPLE 1.17

Given a pentagon $ABCDE$, determine the vector sum $\overrightarrow{AB} + \overrightarrow{BC} + \overrightarrow{CD} + \overrightarrow{DE} + \overrightarrow{EA}$.

### *Solution:*

Utilizing *Chasles' Rule* several times:

$$\vec{0} = \overrightarrow{AA} = \overrightarrow{AB} + \overrightarrow{BC} + \overrightarrow{CD} + \overrightarrow{DE} + \overrightarrow{EA}.$$

## EXAMPLE 1.1.8

Consider a $\triangle ABC$. Demonstrate that the line segment joining the midpoints of two sides is parallel to the third side and it is in fact, half its length.

### *Solution:*

Let the midpoints of [A, B] and [C, A], be $M_C$ and $M_B$, respectively. We will demonstrate that $\overrightarrow{BC} = 2\,\overrightarrow{M_C M_B}$. We have, $2\,\overrightarrow{AM_C} = \overrightarrow{AB}$ and $2\,\overrightarrow{AM_B} = \overrightarrow{AC}$. Therefore,

$$\begin{aligned}
\overrightarrow{BC} &= \overrightarrow{BA} + \overrightarrow{AC} \\
&= -\overrightarrow{AB} + \overrightarrow{AC} \\
&= -2\overrightarrow{AM_C} + 2\overrightarrow{AM_B} \\
&= 2\overrightarrow{M_C A} + 2\overrightarrow{AM_B} \\
&= 2\left(\overrightarrow{M_C A} + \overrightarrow{AM_B}\right) \\
&= 2\overrightarrow{M_C M_B}
\end{aligned}$$

as we were to show.

## EXAMPLE 1.1.9

In $\triangle ABC$, let $M_C$ be the midpoint of [A, B]. Demonstrate that

$$\overrightarrow{CM_C} = \frac{1}{2}\left(\overrightarrow{CA} + \overrightarrow{CB}\right).$$

***Solution:***

As $\overrightarrow{AM_C} = \overrightarrow{M_CB}$, we have,

$$\overrightarrow{CA} + \overrightarrow{CB} = \overrightarrow{CM_C} + \overrightarrow{M_CA} + \overrightarrow{CM_C} + \overrightarrow{M_CB}$$

$$= 2\overrightarrow{CM_C} - \overrightarrow{AM_C} + \overrightarrow{M_CB}$$

$$= -2\overrightarrow{CM_C}$$

from where the results follow.

## EXAMPLE 1.1.10

If the medians $[A, M_A]$ and $[B, M_B]$ of the non-degenerate $\triangle ABC$ intersect at the point G, demonstrate that

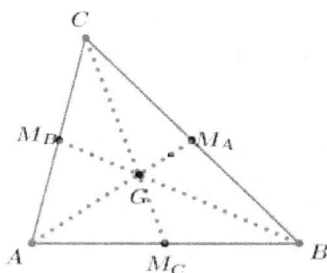

$$\overrightarrow{AG} = 2\overrightarrow{GM_A}; \ \overrightarrow{BG} = 2\overrightarrow{GM_B}.$$

See Fig. 1.1.13.

***Solution:***

Since the triangle is non-degenerate, the lines $\overrightarrow{AM_A}$ and $\overrightarrow{BM_B}$ are not parallel, and thus meet at a point G. Therefore, $\overrightarrow{AG}$ and $\overrightarrow{GM_A}$ are parallel and hence there is a scalar $a$ such that $\overrightarrow{AG} = a\overrightarrow{GM_A}$. In the same fashion, there is a scalar $b$ such that $\overrightarrow{BG} = b\overrightarrow{GM_B}$. From Example 1.1.6,

**FIGURE 1.1.13** Example 1.1.8.

$$2\overrightarrow{M_AM_B} = \overrightarrow{BA}$$

$$= \overrightarrow{BG} + \overrightarrow{GA}$$

$$= b\overrightarrow{GM_B} - a\overrightarrow{GM_A}$$

$$= b\overrightarrow{GM_A} + b\overrightarrow{M_AM_B} - a\overrightarrow{GM_A},$$

and thus

$$(2 - b)\overrightarrow{M_AM_B} = (b - a)\overrightarrow{GM_A}.$$

Since $\triangle ABC$ is non-degenerate, $\overrightarrow{M_AM_B}$ and $\overrightarrow{GM_A}$ are not parallel, where

$$(2 - b) = 0, \ (b - a) = 0 \Rightarrow a = b = 2.$$

## EXAMPLE 1.1.11

The medians of a non-degenerate triangle $\triangle ABC$ are concurrent. The point of concurrency G is called the barycenter or centroid of the triangle. See Fig. 1.1.14.

### *Solution:*

Let G be as in Example 1.1.7. We must show that the line $\overrightarrow{CM_C}$ also passes through G. Let the line $\overrightarrow{CM_C}$ and $\overrightarrow{BM_B}$ meet in $G'$. By the aforementioned example,

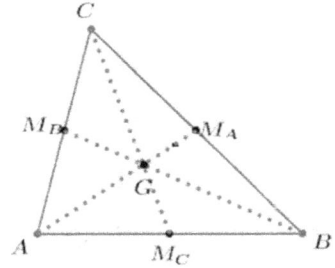

**FIGURE 1.1.14** Example 1.1.9.

$$\overrightarrow{AG} = 2\overrightarrow{GM_A} \; ; \; \overrightarrow{BG} = 2\overrightarrow{GM_B} \; ; \; \overrightarrow{BG'} = 2\overrightarrow{G'M_B} \; ;$$
$$\overrightarrow{CG'} = 2\overrightarrow{G'M_C} \; .$$

It follows that

$$\overrightarrow{GG'} = \overrightarrow{GB} + \overrightarrow{BG'}$$
$$= -2\overrightarrow{GM_B} + 2\overrightarrow{G'M_B}$$
$$= 2\left( \overrightarrow{M_BG} + \overrightarrow{G'M_B} \right)$$
$$= 2\overrightarrow{G'G} \; .$$

Therefore,

$$\overrightarrow{GG'} = -2\overrightarrow{GG'} \Rightarrow 3\overrightarrow{GG'} = \vec{0} \Rightarrow \overrightarrow{GG'} = \vec{0} \Rightarrow G = G', \text{ demonstrating the result.}$$

## EXERCISES 1.1

**1.1.1** Identify the following physical quantities as scalars or vectors:

1. time

2. pressure

3. acceleration

4. velocity

5. temperature

**6.** gravity

**7.** force

**8.** displacement

**9.** frequency

**10.** grade of a motor oil

**11.** sound

**12.** current in a river

**13.** speed

**14.** energy

**1.1.2**   Is there is any truth to the statement "a vector is that which has magnitude and direction"?

**1.1.3**   Let $\vec{u} = \begin{bmatrix} -1 \\ 5 \end{bmatrix}$, and $\vec{v} = \begin{bmatrix} -2 \\ -4 \end{bmatrix}$, be vectors in $\mathbb{R}^2$. Find $\vec{u} + \vec{v}$, $\vec{u} - \vec{v}$, $2\vec{u}$, and normalization of vector $\vec{u}$.

**1.1.4**   Name all the equal vectors in the parallelogram shown.

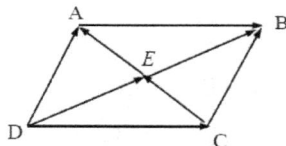

FIGURE 1.1.15 Exercise 1.1.4.

**1.1.5**   Copy the vectors in the figure and use them to draw the following vectors.

FIGURE 1.1.16 Exercise 1.1.5.

**1.** $\vec{a} + \vec{b}$

**2.** $\vec{a} - \vec{b}$

**3.** $\vec{b} + \vec{c}$

**4.** $\vec{a} + \vec{b} + \vec{c}$

**1.1.6** Let $\vec{u} = \begin{bmatrix} 1 \\ -3 \end{bmatrix}$ and $\vec{v} = \begin{bmatrix} 1 \\ 1 \end{bmatrix}$. Find:

**1.** $\|\vec{u}\| + \|\vec{v}\|$

**2.** $\|\vec{u} + \vec{v}\|$

**3.** $3\|\vec{u}\|$

**1.1.7** Let $a$ be a real number. Find the distance between $\begin{bmatrix} 1 \\ a \end{bmatrix}$ and $\begin{bmatrix} 1-a \\ 1 \end{bmatrix}$.

**1.1.8** Find all scales $\lambda$ for which $\|\lambda \vec{u}\| = \dfrac{1}{2}$, where $\vec{u} = \begin{bmatrix} 1 \\ -1 \end{bmatrix}$.

**1.1.9** For which values of $a$ will the vectors $\vec{a} = \begin{bmatrix} a+1 \\ a^2 - 1 \end{bmatrix}$, $\vec{b} = \begin{bmatrix} 2a+5 \\ a^2 - 4a + 3 \end{bmatrix}$ will be parallel?

**1.1.10** In $\triangle ABC$ let the midpoints of $[A, B]$ and $[A, C]$ be $M_C$ and $M_B$, respectively. Put $\overrightarrow{M_C B} = \vec{x}$, $\overrightarrow{M_B C} = \vec{y}$, and $\overrightarrow{CA} = \vec{z}$. Express:

**1.** $\overrightarrow{AB} + \overrightarrow{BC} + \overrightarrow{M_C M_B}$,

**2.** $\overrightarrow{AM_C} + \overrightarrow{M_C M_B} + \overrightarrow{M_B C}$,

**3.** $\overrightarrow{AC} + \overrightarrow{M_C A} - \overrightarrow{BM_B}$

In terms of $\vec{x}$, $\vec{y}$, and $\vec{z}$.

**1.1.11** ABCD is a parallelogram. E is the midpoint of [B,C] and F is the midpoint of [D,C]. Prove that $\overrightarrow{AC} + \overrightarrow{BD} = 2\overrightarrow{BC}$ .

**1.1.12** **(Varignon's Theorem)** Use vector algebra in order to prove that in any quadrilateral ABCD, whose sides do not intersect, the quadrilateral formed by the midpoints of the sides is a parallelogram.

**1.1.13** Let A,B be two points on the plane. Construct two points I and J such that $\overrightarrow{IA} = -3\overrightarrow{IB}$ , $\overrightarrow{JA} = -\frac{1}{3}\overrightarrow{JB}$ , and then demonstrate that for any arbitrary point M on the plane $\overrightarrow{MA} + 3\overrightarrow{MB} = 4\overrightarrow{MI}$ and $3\overrightarrow{MA} + \overrightarrow{MB} = 4\overrightarrow{MJ}$ .

**1.1.14** Find the Cartesian equation corresponding to the line with parametric equation $x = -1 + t$ , $y = 2 - t$ .

**1.1.15** Let x, y, z be points on the plane with $x \neq y$ and consider $\triangle xyz$ . Let Q be a point on side [x, z] such that $\|[x,Q]\| : \|[Q,z]\| = 3 : 4$ and let P be a point on [y, z] such that $\|[y,P]\| : \|[P,Q]\| = 7 : 2$ . Let T be an arbitrary point on the plane.

1.  Find rational numbers $\alpha$ and $\beta$ such that $\overrightarrow{TQ} = \alpha \overrightarrow{Tx} + \beta \overrightarrow{Tz}$ .

2.  Find rational numbers $l, m, n$ such that $\overrightarrow{TP} = l\overrightarrow{Tx} + m\overrightarrow{Ty} + n\overrightarrow{Tz}$ .

**1.1.16** Prove that if $\vec{u}$ and $\vec{v}$ are non-collinear then $x\vec{u} + y\vec{v} = \vec{0}$ implies $x = y = 0$ .

**1.1.17** Prove that the diagonals of a parallelogram bisect each other as in Fig. 1.1.17.

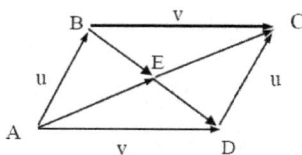

**FIGURE 1.1.17** Exercise 1.1.13.

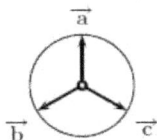

**FIGURE 1.1.18** Exercise 1.1.14.

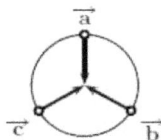

**FIGURE 1.1.19** Exercise 1.1.14.

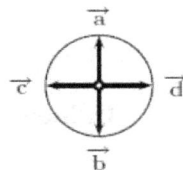

**FIGURE 1.1.20** Exercise 1.1.14.

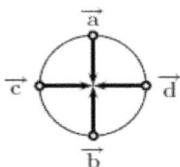

**FIGURE 1.1.21** Exercise 1.1.14.

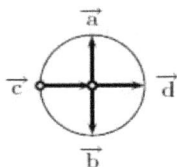

**FIGURE 1.1.22** Exercise 1.1.14.

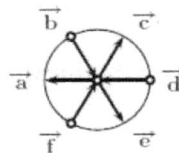

**FIGURE 1.1.23** Exercise 1.1.14.

**1.1.18**   A circle is divided into three, four, or six equal parts (Figs. 1.1.18 through 1.1.23). Find the sum of the vectors. Assume that the divisions start or stop at the center of the circle, as suggested in the figures.

## 1.2   SCALAR PRODUCT ON THE PLANE

We will now define an operation between two plane vectors, which provides a further tool to examine the geometry on the plane.

**Definition 1.2.1**   Let $\vec{x} \in \mathbb{R}^2$ and $\vec{y} \in \mathbb{R}^2$. Their *scalar product (dot product or inner product)* is defined and denoted by $\vec{x} \cdot \vec{y} = \begin{bmatrix} x_1 \\ x_2 \end{bmatrix} \cdot \begin{bmatrix} y_1 \\ y_2 \end{bmatrix} = x_1 y_1 + x_2 y_2$.

**EXAMPLE 1.2.1**

If $\vec{a} = \begin{bmatrix} 1 \\ 2 \end{bmatrix}$ and $\vec{b} = \begin{bmatrix} 3 \\ 4 \end{bmatrix}$, then $\vec{a} \cdot \vec{b} = 1 \times 3 + 2 \times 4 = 11$.

**EXAMPLE 1.2.2**

If $\vec{x} = \begin{bmatrix} 1 \\ 0 \end{bmatrix}$ and $\vec{y} = \begin{bmatrix} 0 \\ 1 \end{bmatrix}$, then $\vec{x} \cdot \vec{y} = 0$ and $\| \vec{x} \| = \| \vec{y} \| = 1$.

The following properties of the scalar product are easy to deduce from the definition.

1. **Bilinearity**

$$\left( \vec{x} + \vec{y} \right) \bullet \vec{z} = \vec{x} \bullet \vec{z} + \vec{y} \bullet \vec{z}, \ \vec{x} \bullet \left( \vec{y} + \vec{z} \right) = \vec{x} \bullet \vec{y} + \vec{x} \bullet \vec{z} \tag{1.4}$$

2. **Scalar Homogeneity**

$$\left( a \ \vec{x} \right) \bullet \vec{y} = \vec{x} \bullet \left( a \ \vec{y} \right) = a \ \left( \vec{x} \bullet \vec{y} \right), \ a \in \mathbb{R} \tag{1.5}$$

3. **Commutativity**

$$\vec{x} \bullet \vec{y} = \vec{y} \bullet \vec{x} \tag{1.6}$$

4. $\vec{x} \bullet \vec{x} \geq 0$ $\qquad\qquad\qquad\qquad\qquad\qquad\qquad$ (1.7)

5. $\vec{x} \bullet \vec{x} = 0 \Leftrightarrow \vec{x} = \vec{0}$ $\qquad\qquad\qquad\qquad\qquad$ (1.8)

6. $\left\| \vec{x} \right\| = \sqrt{\vec{x} \bullet \vec{x}}$ $\qquad\qquad\qquad\qquad\qquad\qquad$ (1.9)

The dot product of two vectors can be obtained using *MATLAB*.

```
>> a = [1; 2];
>> b = [3: 4];
>> dot (a, b)

ans =

    11
```

The dot product of two vectors also can be obtained using *Maple*[TM].

```
>   with(linalg) :
>   a := [1, 2];
```

$$a := [1, 2]$$

```
>   b := [3, 4];
```

$$b := [3, 4]$$

```
>   dotprod(a, b);
```

**Definition 1.2.2**  Given vectors $\vec{a}$ and $\vec{b}$, we define the (convex) angle between them, denoted by $\left(\widehat{\vec{a},\vec{b}}\right) \in [0;\pi]$, as the angle between the affine lines $\mathbb{R}\,\vec{a}$ and $\mathbb{R}\,\vec{b}$.

**Theorem 1.2.1**  Let $\vec{a}$ and $\vec{b}$ be vectors in $\mathbb{R}^2$.   Then,   $\vec{a}\bullet\vec{b} = \|\vec{a}\|\|\vec{b}\|\cos\left(\widehat{\vec{a},\vec{b}}\right)$.   See Fig. 1.2.1.

**FIGURE 1.2.1** Theorem 1.2.1.

**Proof:** From Fig. 1.2.1, using *Al-Kashi's Law* of Cosines on the length of the vectors and (1.4) through (1.9), we have:

$$\left\|\vec{b}-\vec{a}\right\|^2 = \|\vec{a}\|^2\|\vec{b}\|^2 - 2\|\vec{a}\|\|\vec{b}\|\cos\left(\widehat{\vec{a},\vec{b}}\right)$$

$$\left(\vec{b}-\vec{a}\right)\bullet\left(\vec{b}-\vec{a}\right) = \|\vec{a}\|^2\|\vec{b}\|^2 - 2\|\vec{a}\|\|\vec{b}\|\cos\left(\widehat{\vec{a},\vec{b}}\right)$$

$$\vec{b}\bullet\vec{b} - 2\vec{a}\bullet\vec{b} + \vec{a}\bullet\vec{a} = \|\vec{a}\|^2\|\vec{b}\|^2 - 2\|\vec{a}\|\|\vec{b}\|\cos\left(\widehat{\vec{a},\vec{b}}\right)$$

$$\left\|\vec{b}\right\|^2 - 2\vec{a}\bullet\vec{b} + \|\vec{a}\|^2 = \|\vec{a}\|^2\|\vec{b}\|^2 - 2\|\vec{a}\|\|\vec{b}\|\cos\left(\widehat{\vec{a},\vec{b}}\right)$$

$$\vec{a}\bullet\vec{b} = \|\vec{a}\|\|\vec{b}\|\cos\left(\widehat{\vec{a},\vec{b}}\right)$$

as what we wanted to show.

## EXAMPLE 1.2.3

If the vectors $\vec{a}$ and $\vec{b}$ have lengths 6 and 8, and the angle between them is $\left(\widehat{\vec{a},\vec{b}}\right) = \pi\,/\,3$, find $\vec{a}\bullet\vec{b}$ .

*Solution:*

Using Theorem 1.2.1, we have

$$\vec{a}\bullet\vec{b} = \|\vec{a}\|\|\vec{b}\|\cos\left(\pi\,/\,3\right) = (6)(8)\left(\frac{1}{2}\right) = 24\,.$$

The angle between two vectors can be obtained using *MATLAB*.

>> a = [4; 3];

>> b = [2;5];

>> c= acos (dot(a,b) / (norm(a)*norm(b)));

>> (360*c)/(2*pi)

ans =

   31.3287

The angle between two vectors can be obtained using *Maple*$^{TM}$.

> *with(linalg)* :

> *a* := [4, 3];
$$a := [4, 3]$$

> *b* := [2, 5];
$$b := [2, 5]$$

> *x* := *dotprod(a, b)*;
$$x := 23$$

> *y* := *norm(a, 2) * norm(b, 2)*;
$$y := 5\sqrt{29}$$

> $\dfrac{x}{y}$

$$\frac{23}{145}\sqrt{29}$$

> $\arccos\left(\dfrac{23}{145}\sqrt{29}\right)$

$$\arccos\left(\frac{23}{145}\sqrt{29}\right)$$

Putting $\widehat{\left(\vec{a}, \vec{b}\right)} = \dfrac{\pi}{2}$ in Theorem 1.2.1, we obtain the following corollary.

**Corollary 1.2.1**   Two vectors in $\mathbb{R}^2$ are perpendicular if and only if their dot product is 0.

NOTE   *It follows that the vector $\vec{0}$ is simultaneously parallel and perpendicular to any vector!*

**Definition 1.2.3** Two vectors are said to be *orthogonal* if they are perpendicular. If $\vec{a}$ is orthogonal to $\vec{b}$, we write $\vec{a} \perp \vec{b}$.

**EXAMPLE 1.2.4**

Show that the vectors $\vec{a} = \begin{bmatrix} -2 \\ 3 \end{bmatrix}$ and $\vec{b} = \begin{bmatrix} 3 \\ 2 \end{bmatrix}$ are orthogonal.

**Solution:**

Since $\vec{a} \cdot \vec{b} = (-2) \times (3) + (3) \times (2) = 0$, $\vec{a}$ and $\vec{b}$ are orthogonal.

**Definition 1.2.4** If $\vec{a} \perp \vec{b}$ and $\|\vec{a}\| = \|\vec{b}\| = 1$, we say that $\vec{a}$ and $\vec{b}$ are *orthonormal*.

**Definition 1.2.5** Let $\vec{a} \in \mathbb{R}^2$ be fixed. Then the orthogonal space to $\vec{a}$ is defined and denoted by $\vec{a}^{\perp} = \{\vec{x} \in \mathbb{R}^2 : \vec{x} \perp \vec{a}\}$.

Since $|\cos\theta| \le 1$, we also have the following corollary.

**Corollary 1.2.2 Cauchy–Bunyakovsky–Schwarz Inequality (CBS Inequality)**

$$|\vec{a} \cdot \vec{b}| \le \|\vec{a}\|\|\vec{b}\|.$$

Equality occurs if and only if $\vec{a} \parallel \vec{b}$.

If $\vec{a} = \begin{bmatrix} a_1 \\ a_2 \end{bmatrix}$ and $\vec{b} = \begin{bmatrix} b_1 \\ b_2 \end{bmatrix}$, the CBS Inequality takes the form

$$|a_1 b_1 + a_2 b_2| \le \left(a_1^2 + a_2^2\right)^{1/2} \left(b_1^2 + b_2^2\right)^{1/2}. \tag{1.10}$$

**Example 1.2.5**

Let $a, b$ be positive real numbers. Minimize $a^2 + b^2$ subject to the constraint $a + b = 1$.

### Solution:

By the CBS Inequality,

$$1 = \left| a \bullet 1 + b \bullet 1 \right| \le \left( a^2 + b^2 \right)^{1/2} \left( 1^2 + 1^2 \right)^{1/2} \Rightarrow a^2 + b^2 \ge \frac{1}{2}.$$

Equality occurs if and only if $\begin{bmatrix} a \\ b \end{bmatrix} = \lambda \begin{bmatrix} 1 \\ 1 \end{bmatrix}$. In this case, $a = b = \lambda$, so equality is achieved for $a = b = \frac{1}{2}$.

### Corollary 1.2.3   Triangle Inequality

$$\left\| \vec{a} + \vec{b} \right\| \le \left\| \vec{a} \right\| + \left\| \vec{b} \right\|.$$

### Proof:

$$\left| \vec{a} + \vec{b} \right|^2 = \left( \vec{a} + \vec{b} \right) \bullet \left( \vec{a} + \vec{b} \right)$$
$$= \vec{a} \bullet \vec{a} + 2 \vec{a} \bullet \vec{b} + \vec{b} \bullet \vec{b}$$
$$\le \left\| \vec{a} \right\|^2 + 2 \left\| \vec{a} \right\| \left\| \vec{b} \right\| + \left\| \vec{b} \right\|^2$$
$$= \left( \left\| \vec{a} \right\| + \left\| \vec{b} \right\| \right)^2,$$

from where the desired result follows.

### Corollary 1.2.4   Pythagorean Theorem

If $\vec{a} \perp \vec{b}$, then $\left\| \vec{a} + \vec{b} \right\|^2 = \left\| \vec{a} \right\|^2 + \left\| \vec{b} \right\|^2$.

### Proof:

Since $\vec{a} \bullet \vec{b} = 0$, we have

$$\left\| \vec{a} + \vec{b} \right\|^2 = \left( \vec{a} + \vec{b} \right) \bullet \left( \vec{a} + \vec{b} \right)$$
$$= \vec{a} \bullet \vec{a} + 2 \vec{a} \bullet \vec{b} + \vec{b} \bullet \vec{b}$$
$$= \vec{a} \bullet \vec{a} + 0 + \vec{b} \bullet \vec{b}$$
$$= \left\| \vec{a} \right\|^2 + \left\| \vec{b} \right\|^2$$

from where the desired result follows.

**EXAMPLE 1.2.6**

Let $a$, $b$, $z$ be positive real numbers. Prove that $\sqrt{2}\left(x+y+z\right)\leq\sqrt{x^2+y^2}+\sqrt{y^2+z^2}+\sqrt{z^2+x^2}$.

***Solution:***

Put $\vec{a}=\begin{bmatrix}x\\y\end{bmatrix}$, $\vec{b}=\begin{bmatrix}y\\z\end{bmatrix}$, $\vec{c}=\begin{bmatrix}z\\x\end{bmatrix}$. Then,

$$\left\|\vec{a}+\vec{b}+\vec{c}\right\|=\left\|\begin{bmatrix}x+y+z\\x+y+z\end{bmatrix}\right\|=\sqrt{2}\left(x+y+z\right).$$

Also,

$$\left\|\vec{a}\right\|+\left\|\vec{b}\right\|+\left\|\vec{c}\right\|=\sqrt{x^2+y^2}+\sqrt{y^2+z^2}+\sqrt{z^2+x^2},$$

and the assertion follows by the triangle inequality

$$\left\|\vec{a}+\vec{b}+\vec{c}\right\|\leq\left\|\vec{a}\right\|+\left\|\vec{b}\right\|+\left\|\vec{c}\right\|.$$

We now use vectors to prove a classical theorem of Euclidean geometry.

**Definition 1.2.6** Let A and B be points on the plane and let $\vec{u}$ be a unit vector. If $\overrightarrow{AB}=\lambda\,\vec{u}$, then $\lambda$ is the *directed distance or algebraic measure of the line segment* $[AB]$ *with respect to the vector* $\vec{u}$. We will denote this distance by $\overline{AB}_{\vec{u}}$, or more routinely, if the vector $\vec{u}$ is patent, by $\overline{AB}$. Observe that $\overline{AB}=-\overline{BA}$.

**Theorem 1.2.2   Thales' Theorem**

Let $\overrightarrow{D}\,y\,\overrightarrow{D'}$ be two distinct lines on the plane. Let A, B, C be distinct points of $\overrightarrow{D}$, and A′, B′, C′ be distinct points of $\overrightarrow{D'}$, $A\neq A'$, $B\neq B'$, $C\neq C'$, $A\neq B$, $A'\neq B'$. Let $\overrightarrow{AA'}\,\|\,\overrightarrow{BB'}$. Then, $\overrightarrow{AA'}\,\|\,\overrightarrow{CC'}\Leftrightarrow\dfrac{\overline{AC}}{\overline{AB}}=\dfrac{\overline{A'C'}}{\overline{A'B'}}$.

See Fig. 1.2.2.

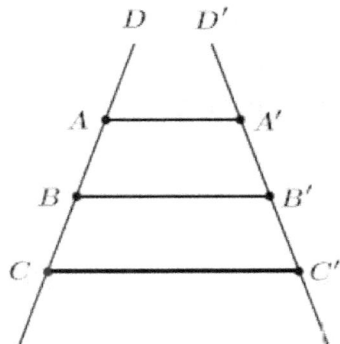

*FIGURE 1.2.2* Thales' theorem.

**Proof:**

Refer to Fig. 1.2.2. On the one hand, because they are unit vectors in the same direction,

$$\frac{\overrightarrow{AB}}{AB} = \frac{\overrightarrow{AC}}{AC}; \quad \frac{\overrightarrow{A'B'}}{A'B'} = \frac{\overrightarrow{A'C'}}{A'C'}.$$

On the other hand, by Chasles' rule,

$$\overrightarrow{BB'} = \overrightarrow{BA} + \overrightarrow{AA'} + \overrightarrow{A'B'} = \left(\overrightarrow{A'B'} - \overrightarrow{AB}\right) + \overrightarrow{AA'}.$$

Since $\overrightarrow{A'B'} = \overrightarrow{AB} + \lambda\,\overrightarrow{AA'}$.

Assembling these results,

$$\overrightarrow{CC'} = \overrightarrow{CA} + \overrightarrow{AA'} + \overrightarrow{A'C'}$$

$$= -\frac{\overrightarrow{AC}}{AB} \cdot \overrightarrow{AB} + \overrightarrow{AA'} + \frac{\overrightarrow{A'C'}}{A'B'}\left(\overrightarrow{AB} + \lambda\,\overrightarrow{AA'}\right)$$

$$= \left(\frac{\overrightarrow{A'C'}}{A'B'} - \frac{\overrightarrow{AC}}{AB}\right)\overrightarrow{AB} + \left(1 + \lambda\,\frac{\overrightarrow{A'C'}}{A'B'}\right)\overrightarrow{AA'}.$$

As the line $\overrightarrow{AA'}$ is not parallel to the line $\overrightarrow{AB}$, the preceding equality reveals that

$$\overrightarrow{AA'} \,\|\, \overrightarrow{CC'} \Leftrightarrow \frac{\overrightarrow{AC}}{AB} - \frac{\overrightarrow{A'C'}}{A'B'} = 0,$$

proving the Theorem 1.2.2.

From the preceding theorem, we immediately gather the following corollary (see Fig. 1.2.3).

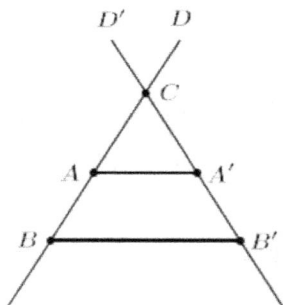

**FIGURE 1.2.3** Corollary to Thales' theorem.

**Corollary 1.2.5** Let $\overrightarrow{D}$ and $\overrightarrow{D'}$ be distinct lines, intersecting in the unique point C. Let A, B, be points on line $\overrightarrow{D}$ and $A'$, $B'$ be points on line $\overrightarrow{D'}$. Then,

$$\overrightarrow{AA'} \,\|\, \overrightarrow{BB'} \Leftrightarrow \frac{\overrightarrow{CB}}{CA} = \frac{\overrightarrow{CB'}}{CA'}.$$

**EXAMPLE 1.2.7**

Prove that the altitudes of a triangle $\triangle ABC$ on the plane are concurrent. This point is called the *orthocentre* of the triangle.

**Solution:**

Put $\vec{a} = \overrightarrow{OA}$, $\vec{b} = \overrightarrow{OB}$, $\vec{c} = \overrightarrow{OC}$. First observe that for any $\vec{x}$, we have, upon expanding,

$$(\vec{x} - \vec{a}) \cdot (\vec{b} - \vec{c}) + (\vec{x} - \vec{b}) \cdot (\vec{c} - \vec{a}) + (\vec{x} - \vec{c}) \cdot (\vec{a} - \vec{b}) = 0 \qquad (1.10)$$

Let $H$ be the point of intersection of the altitude from $A$ and the altitude from $B$. Then

$$0 = (\overrightarrow{AH}) \cdot (\overrightarrow{CB}) = (\overrightarrow{OH} - \overrightarrow{OA}) \cdot (\overrightarrow{OB} - \overrightarrow{OC}) = (\overrightarrow{OH} - \vec{a}) \cdot (\vec{b} - \vec{c}), \qquad (1.11)$$

and

$$0 = (\overrightarrow{BH}) \cdot (\overrightarrow{AC}) = (\overrightarrow{OH} - \overrightarrow{OB}) \cdot (\overrightarrow{OC} - \overrightarrow{OA}) = (\overrightarrow{OH} - \vec{b}) \cdot (\vec{c} - \vec{a}). \qquad (1.12)$$

Putting $\vec{x} = \overrightarrow{OH}$ in equation (1.10) and subtracting from it (1.11) and (1.12), we gather that

$$0 = (\overrightarrow{OH} - \vec{c}) \cdot (\vec{a} - \vec{b}) = (\overrightarrow{CH}) \cdot (\overrightarrow{AB}),$$

Which gives the results.

## EXERCISES 1.2

**1.2.1** Find $\vec{a} \cdot \vec{b}$, if:

**1.** $\vec{a} = \begin{bmatrix} 4 \\ -1 \end{bmatrix}$ and $\vec{b} = \begin{bmatrix} 3 \\ 5 \end{bmatrix}$

**2.** $\vec{a} = \begin{bmatrix} 5 \\ 0 \end{bmatrix}$ and $\vec{b} = \begin{bmatrix} 2 \\ -1 \end{bmatrix}$

**3.** $\vec{a} = \begin{bmatrix} 2 \\ 3 \end{bmatrix}$ and $\vec{b} = \begin{bmatrix} 5 \\ 6 \end{bmatrix}$

**4.** $\vec{a} = \begin{bmatrix} -5 \\ -2 \end{bmatrix}$ and $\vec{b} = \begin{bmatrix} 4 \\ 0 \end{bmatrix}$

**1.2.2** Find $\vec{a} \cdot \vec{b}$, if:

**1.** $\|\vec{a}\| = 5$, $\|\vec{b}\| = 8$, and the angle between $\vec{a}$ and $\vec{b}$ is $\pi / 3$.

**2.** $\|\vec{a}\| = \sqrt{3}$, $\|\vec{b}\| = 4$, and the angle between $\vec{a}$ and $\vec{b}$ is $\pi / 6$.

**1.2.3** Find the angle between the vectors.

**1.** $\vec{a} = \begin{bmatrix} 1 \\ \sqrt{3} \end{bmatrix}$, $\vec{b} = \begin{bmatrix} 2 \\ 0 \end{bmatrix}$

**2.** $\vec{a} = \begin{bmatrix} 0 \\ 1 \end{bmatrix}$, $\vec{b} = \begin{bmatrix} -1 \\ 1 \end{bmatrix}$

**3.** $\vec{a} = \begin{bmatrix} 4 \\ -2 \end{bmatrix}$, $\vec{b} = \begin{bmatrix} 1 \\ 2 \end{bmatrix}$

**4.** $\vec{a} = \begin{bmatrix} 0 \\ 2 \end{bmatrix}$, $\vec{b} = \begin{bmatrix} 1 \\ -1 \end{bmatrix}$

**1.2.4** Prove that $\|\alpha\,\vec{a} + \beta\,\vec{b}\|^2 = \alpha^2 \|\vec{a}\|^2 + 2\alpha\beta\,\vec{a} \cdot \vec{b} + \beta^2 \|\vec{b}\|$.

**1.2.5** Find the value of $a$ so that $\begin{bmatrix} a \\ 1-a \end{bmatrix}$ be perpendicular to $\begin{bmatrix} 1 \\ -1 \end{bmatrix}$.

**1.2.6** Let $\vec{a} \neq \vec{0}$ and $\vec{b} \neq \vec{0}$ be vectors in $\mathbb{R}^2$ such that $\vec{a} \cdot \vec{b} = 0$. Prove that

$$\alpha\,\vec{a} + \beta\,\vec{b} = \vec{0} \Rightarrow \alpha = \beta = 0.$$

**1.2.7** Prove that the diagonals of a rhombus (a parallelogram whose sides have equal length) are perpendicular.

**1.2.8** What can we say about the relative magnitudes of vectors $\vec{a}$ and $\vec{b}$, if $\vec{a} + \vec{b}$ is perpendicular to $\vec{a} - \vec{b}$?

**1.2.9** Show that the vectors $\vec{a} = \begin{bmatrix} -5 \\ \sqrt{3} \end{bmatrix}$ and $\vec{b} = \begin{bmatrix} \sqrt{27} \\ 15 \end{bmatrix}$ are perpendicular (orthogonal).

**1.2.10** If $\vec{a} = \begin{bmatrix} k \\ 3 \end{bmatrix}$ and $\vec{b} = \begin{bmatrix} k \\ -4 \end{bmatrix}$ are orthogonal, find $k$.

**1.2.11** Prove that $\left\| \vec{a} + \vec{b} \right\| = \left\| \vec{a} - \vec{b} \right\|$ if and only if $\vec{a} \cdot \vec{b} = 0$.

**1.2.12** Prove that if $\vec{a}$ and $\vec{b}$ are perpendicular, then $\left\| \vec{a} + \vec{b} \right\|^2 = \left\| \vec{a} \right\|^2 + \left\| \vec{b} \right\|^2$.

**1.2.13** Prove that $\vec{c} = \vec{a} - \left( \dfrac{\left( \vec{a} \cdot \vec{b} \right)}{\left\| \vec{b} \right\|^2} \right) \vec{b}$ is orthogonal to $\vec{b}$, where $\vec{b}$ is a nonzero vector.

**1.2.14** Find all vectors $\vec{a} \in \mathbb{R}^2$ such that $\vec{a} \perp \begin{bmatrix} -3 \\ 2 \end{bmatrix}$ and $\|a\| = \sqrt{13}$.

**1.2.15** **(Pythagorean Theorem)** If $\vec{a} \perp \vec{b}$, prove that
$$\left\| \vec{a} + \vec{b} \right\| = \left\| \vec{a} \right\|^2 + \left\| \vec{b} \right\|^2.$$

**1.2.16** Let $a, b$ be arbitrary real numbers. Prove that
$$\left( a^2 + b^2 \right) \le 2 \left( a^4 + b^4 \right).$$

**1.2.17** Let $\vec{a}$, $\vec{b}$, be fixed vectors in $\mathbb{R}^2$. Prove that if
$$\forall \, \vec{v} \in \mathbb{R}^2, \, \vec{v} \cdot \vec{a} = \vec{v} \cdot \vec{b}, \quad \vec{a} = \vec{b}.$$

**1.2.18** **(Polarization Identity)** Let $\vec{u}$, $\vec{v}$ be vectors in $\mathbb{R}^2$. Prove that
$$\vec{u} \cdot \vec{v} = \frac{1}{4} \left( \left\| \vec{u} + \vec{v} \right\|^2 - \left\| \vec{u} - \vec{v} \right\|^2 \right).$$

**1.2.19** Consider two lines on the plane $L_1$ and $L_2$ with Cartesian equations $L_1 : y = m_1 x + b_1$ and $L_2 : y = m_2 x + b_1$, where $m_1 \neq 0$, $m_2 \neq 0$. Using Corollary 1.2.1 from Section 1.2, prove that $L_1 \perp L_2 \Leftrightarrow m_1 m_2 = -1$.

**1.2.20**  Find the Cartesian equation of all lines $L'$ passing through $\begin{pmatrix} -1 \\ 2 \end{pmatrix}$, making an angle of $\dfrac{\pi}{6}$ radians with the Cartesian line $L : x + y = 1$.

**1.2.21**  Let $\vec{v}$, $\vec{w}$ be vectors on the plane, with $\vec{w} \neq \vec{0}$. Prove that the vector

$$\vec{a} = \vec{v} - \frac{\vec{v} \bullet \vec{w}}{\left\| \vec{w} \right\|^2} \vec{w} \text{ is perpendicular to } \vec{w}.$$

## 1.3   LINEAR INDEPENDENCE

Consider now two arbitrary vectors in $\mathbb{R}^2$, $\vec{x}$ and $\vec{y}$. Under which conditions can we write an arbitrary vector $\vec{v}$ on the plane as a *linear combination* of $\vec{x}$ and $\vec{y}$, that is, when can we find scalars $a, b$ such that $\vec{v} = a\,\vec{x} + b\,\vec{y}$?

The answer can be promptly obtained algebraically. Operating formally,

$$\vec{v} = a\,\vec{x} + b\,\vec{y} \Leftrightarrow v_1 = ax_1 + by_1, \ v_2 = ax_2 + by_2,$$

$$\Leftrightarrow \ a = \frac{v_1 y_2 - v_2 y_1}{x_1 y_2 - x_2 y_1}, \ b = \frac{x_1 v_2 - x_2 v_1}{x_1 y_2 - x_2 y_1}.$$

The previous expressions for $a$ and $b$ make sense only if $x_1 y_2 \neq x_2 y_1$. But, what does it mean for $x_1 y_2 = x_2 y_1$? If none of these are zero, then $\dfrac{x_1}{y_1} = \dfrac{x_2}{y_2} = \lambda$ and $\begin{bmatrix} x_1 \\ x_2 \end{bmatrix} = \lambda \begin{bmatrix} y_1 \\ y_2 \end{bmatrix} \Leftrightarrow \vec{x} \,\|\, \vec{y}$.

If $x_1 = 0$, then either $x_2 = 0$ or $y_1 = 0$. In the first case, $\vec{x} = \vec{0}$, and a fortiori $\vec{x} \,\|\, \vec{y}$, since all vectors are parallel to the zero vector. In the second case, we have $\vec{x} = x_2\,\vec{j}, \ \ \vec{y} = y_2\,\vec{j}$, so both vectors are parallel to $\vec{j}$ and hence $\vec{x} \,\|\, \vec{y}$. We have demonstrated the following theorem.

**Theorem 1.3.1** Given two vectors in $\mathbb{R}^2$, $\vec{x}$, and $\vec{y}$, an arbitrary vector $\vec{v}$ can be written as the *linear combination*

$$\vec{v} = a\vec{x} + b\vec{y}, \ a \in \mathbb{R}, \ b \in \mathbb{R}$$

if and only if $\vec{x}$ is not parallel to $\vec{y}$. In this last case, we say that $\vec{x}$ is *linearly independent* from vector $\vec{y}$. If two vectors are not linearly independent, then we say that they are *linearly dependent*.

**EXAMPLE 1.3.1**

The vectors $\begin{bmatrix} 1 \\ 0 \end{bmatrix}$ and $\begin{bmatrix} 1 \\ 1 \end{bmatrix}$ are clearly linearly independent, since one is not a scalar multiple of the other. Given an arbitrary vector $\begin{bmatrix} a \\ b \end{bmatrix}$, we can express it as a linear combination of these vectors as follows:

$$\begin{bmatrix} a \\ b \end{bmatrix} = (a-b)\begin{bmatrix} 1 \\ 0 \end{bmatrix} + b\begin{bmatrix} 1 \\ 1 \end{bmatrix}.$$

Consider now two linearly independent vectors $\vec{x}$ and $\vec{y}$. For $a \in [0;1]$, $a\,\vec{x}$ is parallel to $\vec{x}$ and traverses the whole length of $\vec{x}$: from its tip (when $a = 1$) to its tail (when $a = 0$). In the same manner, for $b \in [0;1]$, $b\,\vec{y}$ is parallel to $\vec{y}$ and traverses the whole length of $\vec{y}$. The linear combination $a\,\vec{x} + b\,\vec{y}$ is also a vector on the plane.

**EXAMPLE 1.3.2**

The vector $5\begin{bmatrix} 1 \\ 3 \end{bmatrix}$ near combination of the vector $\begin{bmatrix} 1 \\ 3 \end{bmatrix}$ of $\mathbb{R}^2$ with $a = 5$.

**EXAMPLE 1.3.3**

The vector $2\begin{bmatrix} 0 \\ 1 \end{bmatrix} + 3\begin{bmatrix} 1 \\ 2 \end{bmatrix}$ is a linear combination of the vectors $\begin{bmatrix} 0 \\ 1 \end{bmatrix}$ and $\begin{bmatrix} 1 \\ 2 \end{bmatrix}$ of

$\mathbb{R}^2$ with $a = 2$, $b = 3$, and $\vec{x} = \begin{bmatrix} 0 \\ 1 \end{bmatrix}$ and $\vec{y} = \begin{bmatrix} 1 \\ 2 \end{bmatrix}$. By applying addition and sca-

lar multiplication defined on $\mathbb{R}^2$, we get $2\begin{bmatrix} 0 \\ 1 \end{bmatrix} + 3\begin{bmatrix} 1 \\ 2 \end{bmatrix} = \begin{bmatrix} 3 \\ 8 \end{bmatrix}$. Thus, we may

also say that the vector $\vec{v} = \begin{bmatrix} 3 \\ 8 \end{bmatrix}$ is a linear combination of $\vec{x} = \begin{bmatrix} 0 \\ 1 \end{bmatrix}$ and $\vec{y} = \begin{bmatrix} 1 \\ 2 \end{bmatrix}$

because there exists $a = 2$, $b = 3$ such that $2\begin{bmatrix} 0 \\ 1 \end{bmatrix} + 3\begin{bmatrix} 1 \\ 2 \end{bmatrix} = \begin{bmatrix} 3 \\ 8 \end{bmatrix}$. We can express

this relation by saying that the vector $\vec{v} = \begin{bmatrix} 3 \\ 8 \end{bmatrix}$ is *generated by* the vectors

$\vec{x} = \begin{bmatrix} 0 \\ 1 \end{bmatrix}$ and $\vec{y} = \begin{bmatrix} 1 \\ 2 \end{bmatrix}$.

**Definition 1.3.1 (Fundamental parallelogram)**   Given two linearly independent vectors $\vec{x}$ and $\vec{y}$, consider bi-point representatives of them with the tails at the origin. The *fundamental parallelogram* of the vectors $\vec{x}$ and $\vec{y}$ is the set $\{a\,\vec{x} + b\,\vec{y} : a \in [0;1], b \in [0;1]\}$.

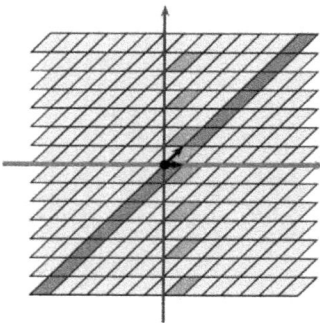

Fig. 1.3.1 shows the fundamental parallelogram of $\left\{ \begin{bmatrix} 1 \\ 0 \end{bmatrix}, \begin{bmatrix} 1 \\ 1 \end{bmatrix} \right\}$, colored in brown, and the respective tiling of the plane by various translations of it. Observe that the vertices of this parallelogram are $\left\{ \begin{pmatrix} 0 \\ 0 \end{pmatrix}, \begin{pmatrix} 1 \\ 0 \end{pmatrix}, \begin{pmatrix} 1 \\ 1 \end{pmatrix}, \begin{pmatrix} 2 \\ 1 \end{pmatrix} \right\}$. In essence then, linear independ-

**FIGURE 1.3.1** Tiling and the fundamental parallelogram.

ence of two vectors on the plane means that we may obtain every vector on the plane as a linear combination of these two vectors and hence cover the whole plane with all these linear combinations.

## EXERCISES 1.3

**1.3.1**   Show that $\begin{bmatrix} 7 \\ 3 \end{bmatrix}$ is a linear combination of the vectors $\begin{bmatrix} 1 \\ 0 \end{bmatrix}$ and $\begin{bmatrix} 3 \\ 1 \end{bmatrix}$ of $\mathbb{R}^2$.

**1.3.2**   Determine whether the first vector of the set of vectors is a linear combination of the other vectors.

**1.** $\begin{bmatrix} 7 \\ -1 \end{bmatrix}; \begin{bmatrix} 4 \\ 2 \end{bmatrix}, \begin{bmatrix} -1 \\ 1 \end{bmatrix}$

**2.** $\begin{bmatrix} 15 \\ -1 \end{bmatrix}; \begin{bmatrix} 4 \\ -1 \end{bmatrix}, \begin{bmatrix} -8 \\ 2 \end{bmatrix}$

**3.** $\begin{bmatrix} 0 \\ 4 \end{bmatrix}; \begin{bmatrix} 2 \\ -1 \end{bmatrix}, \begin{bmatrix} 4 \\ 6 \end{bmatrix}, \begin{bmatrix} 2 \\ 3 \end{bmatrix}$

**4.** $\begin{bmatrix} 5 \\ 7 \end{bmatrix}; \begin{bmatrix} -1 \\ 1 \end{bmatrix}, \begin{bmatrix} 1 \\ 5 \end{bmatrix}$

**1.3.3**   Write an arbitrary vector $\begin{bmatrix} a \\ b \end{bmatrix}$ on the plane, as a linear combination of the vectors $\begin{bmatrix} 1 \\ 1 \end{bmatrix}$ and $\begin{bmatrix} -1 \\ 1 \end{bmatrix}$.

**1.3.4**   Explain why the following are linearly dependent sets of vectors in $\mathbb{R}^2$.

**1.** $\vec{x} = \begin{bmatrix} 1 \\ 2 \end{bmatrix}$ and $\vec{y} = \begin{bmatrix} -3 \\ -6 \end{bmatrix}$

**2.** $\vec{x} = \begin{bmatrix} 2 \\ 3 \end{bmatrix}, \vec{y} = \begin{bmatrix} 6 \\ 1 \end{bmatrix}$ and $\vec{v} = \begin{bmatrix} -5 \\ 8 \end{bmatrix}$

**1.3.5**   Show that the vectors $\vec{x} = \begin{bmatrix} 1 \\ 0 \end{bmatrix}$ and $\vec{y} = \begin{bmatrix} 0 \\ 1 \end{bmatrix}$ are linearly independent.

**1.3.6**  Show that the following sets of vectors in $\mathbb{R}^2$ are linearly independent or linearly dependent.

1.  $\left\{ \begin{bmatrix} -1 \\ 2 \end{bmatrix}, \begin{bmatrix} 2 \\ 3 \end{bmatrix}, \begin{bmatrix} 1 \\ 4 \end{bmatrix} \right\}$

2.  $\left\{ \begin{bmatrix} 1 \\ 3 \end{bmatrix}, \begin{bmatrix} 3 \\ -1 \end{bmatrix} \right\}$

3.  $\left\{ \begin{bmatrix} 10 \\ 12 \end{bmatrix}, \begin{bmatrix} 5 \\ 6 \end{bmatrix} \right\}$

4.  $\left\{ \begin{bmatrix} 3 \\ 4 \end{bmatrix}, \begin{bmatrix} 1 \\ -1 \end{bmatrix} \right\}$

**1.3.7**  Consider the vectors $\begin{bmatrix} 2 \\ 1 \end{bmatrix}$, $\begin{bmatrix} 1 \\ 2 \end{bmatrix}$, and $\begin{bmatrix} 8 \\ 7 \end{bmatrix}$ in $\mathbb{R}^2$. Show that the set $\left\{ \begin{bmatrix} 2 \\ 1 \end{bmatrix}, \begin{bmatrix} 1 \\ 2 \end{bmatrix}, \begin{bmatrix} 8 \\ 7 \end{bmatrix} \right\}$ are linearly dependent.

**1.3.8**  Are vectors $\vec{x}$ and $\vec{y}$ in Figs. 1.3.2 and 1.3.3 linearly independent or dependent? Explain the reasoning.

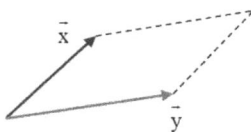

**FIGURE 1.3.2** Exercise 1.3.          **FIGURE 1.3.3** Exercise 1.3.

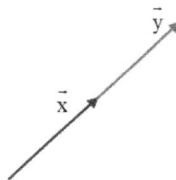

**1.3.9**  Prove that for any two vectors $\vec{x}$ and $\vec{y}$ form a linearly dependent set if and only if they are parallel.

**1.3.10**  Prove the following sets of vectors are linearly dependent or linearly independent in $\mathbb{R}^2$.

1.  $\left\{ \begin{bmatrix} 3 \\ -2 \end{bmatrix}, \begin{bmatrix} -9 \\ 6 \end{bmatrix} \right\}$

**2.** $\left\{ \begin{bmatrix} 3 \\ -1 \end{bmatrix}, \begin{bmatrix} 5 \\ 2 \end{bmatrix} \right\}$

**3.** $\left\{ \begin{bmatrix} -4 \\ 2 \end{bmatrix}, \begin{bmatrix} 2 \\ -1 \end{bmatrix} \right\}$

**4.** $\left\{ \begin{bmatrix} 0 \\ 1 \end{bmatrix}, \begin{bmatrix} 1 \\ 0 \end{bmatrix} \right\}$

**5.** $\left\{ \begin{bmatrix} 1 \\ 1 \end{bmatrix}, \begin{bmatrix} -1 \\ 1 \end{bmatrix} \right\}$

**1.3.11** Find the values of $k$ for which the following sets of vectors are linearly dependent.

**1.** $\left\{ \begin{bmatrix} 4k \\ 2k + 6 \end{bmatrix}, \begin{bmatrix} -k \\ 2 \end{bmatrix} \right\}$

**2.** $\left\{ \begin{bmatrix} -4 \\ k \end{bmatrix}, \begin{bmatrix} 2 \\ -1 \end{bmatrix} \right\}$

**1.3.12** Prove that two non-zero perpendicular vectors in $\mathbb{R}^2$ must be linearly independent.

**1.3.13** Let $ABCD$ be a parallelogram.

    **1.** Let $E$ and $F$ be points such that $\overrightarrow{AE} = \dfrac{1}{4}\overrightarrow{AC}$ and $\overrightarrow{AF} = \dfrac{3}{4}\overrightarrow{AC}$.

    Show that the lines $(BE)$ and $(DF)$ are parallel.

    **2.** Let $I$ be the midpoint of $[A, D]$ and $J$ be the midpoint of $[B, C]$. Show that the lines $(AB)$ and $(IJ)$ are parallel.

**1.3.14** $ABCD$ is a parallelogram; point $I$ is the midpoint of $[A, B]$. Point $E$ is defined by the relation $\overrightarrow{IE} = \dfrac{1}{3}\overrightarrow{ID}$. Prove that

$$\overrightarrow{AE} = \frac{1}{3}\left( \overrightarrow{AB} + \overrightarrow{AD} \right),$$ and prove that the points $A$, $C$, $E$ are collinear.

## 1.4   GEOMETRIC TRANSFORMATIONS IN TWO DIMENSIONS

We are now interested in the following fundamental functions of sets (figures) on the plane: translation, scaling (stretching or shrinking) reflection about the axes, and rotation about the origin. A handy tool for investigating all of these (with the exception of translation) is a certain construct called matrices, which we will study in the next section.

First, observe what is meant by a function $F : \mathbb{R}^2 \to \mathbb{R}^2$. This means that the input of the function is a point of the plane and the output is also a point on the plane.

The following is a rather uninteresting example, but nevertheless an important one.

### EXAMPLE 1.4.1

The function $I : \mathbb{R}^2 \to \mathbb{R}^2$, $I(x) = x$ is called the *identity transformation*. Observe that the identity transformation leaves a point untouched.

We start with the simplest of these functions.

**Definition 1.4.1**   A function $T_{\vec{v}} : \mathbb{R}^2 \to \mathbb{R}^2$ is said to be a *translation* if it is of the form $T_{\vec{v}}(x) = x + \vec{v}$, where $\vec{v}$ is a fixed vector on the plane.

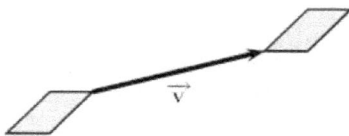

A translation simply shifts an object on the plane rigidly by a given amount of units from where it was originally to form a copy of itself (that is, it does not distort its shape or re-orient it). See Fig. 1.4.1 for an example.

FIGURE 1.4.1 A translation.

It is clear that the composition of any two translations commutes, that is, if $T_{\vec{v}_1}$, $T_{\vec{v}_2} : \mathbb{R}^2 \to \mathbb{R}^2$ are translations, then $T_{\vec{v}_1} \circ T_{\vec{v}_2} = T_{\vec{v}_2} \circ T_{\vec{v}_1}$. Let $T_{\vec{v}_1}(a) = a + \vec{v}_1$ and $T_{\vec{v}_2}(a) = a + \vec{v}_2$. Then,

$$\left( T_{\vec{v}_1} \circ T_{\vec{v}_2} \right)(a) = T_{\vec{v}_1}\left( T_{\vec{v}_2}(a) \right) = T_{\vec{v}_2}(a) + \vec{v}_1 = a + \vec{v}_2 + \vec{v}_1,$$

and

$$\left( T_{\vec{v}_2} \circ T_{\vec{v}_1} \right)(a) = T_{\vec{v}_2}\left( T_{\vec{v}_1}(a) \right) = T_{\vec{v}_1}(a) + \vec{v}_2 = a + \vec{v}_1 + \vec{v}_2,$$

from where the commutative claim is deduced.

**Definition 1.4.2** A function $S_{a,b} : \mathbb{R}^2 \to \mathbb{R}^2$ is said to be a *scaling* if it is of the form $S_{a,b}(\mathbf{r}) = \begin{pmatrix} ax \\ by \end{pmatrix}$, where $a > 0, b > 0$ are real numbers.

Fig. 1.4.2 shows the scaling $S_{2,0.5}\left( \begin{pmatrix} x \\ y \end{pmatrix} \right) = \begin{pmatrix} 2x \\ 0.5y \end{pmatrix}$.

**FIGURE 1.4.2** A scaling.

It is clear that the composition of any two scaling commute, that is, if $S_{a,b}, S_{a',b'} : \mathbb{R}^2 \to \mathbb{R}^2$ are scaling, then $S_{a,b} \circ S_{a',b'} = S_{a',b'} \circ S_{a,b}$. For

$$\left( S_{a,b} \circ S_{a',b'} \right)(\mathbf{r}) = S_{a,b}\left( S_{a',b'}(\mathbf{r}) \right) = S_{a,b}\left( \begin{pmatrix} a'x \\ b'y \end{pmatrix} \right) = \begin{pmatrix} a(a'x) \\ b(b'y) \end{pmatrix}, \text{ and}$$

$$\left( S_{a',b'} \circ S_{a,b} \right)(\mathbf{r}) = S_{a',b'}\left( S_{a,b}(\mathbf{r}) \right) = S_{a',b'}\left( \begin{pmatrix} ax \\ by \end{pmatrix} \right) = \begin{pmatrix} a'(ax) \\ b'(by) \end{pmatrix},$$

from where the commutative claim is deduced.

Translation and scaling do not necessarily commute, however. Consider the translation $T_{\vec{i}}(\mathbf{a}) = \mathbf{a} + \vec{i}$ and the scaling $S_{2,1}(\mathbf{a}) = \begin{pmatrix} 2a_1 \\ a_2 \end{pmatrix}$. Then,

$$\left( T_{\vec{i}} \circ S_{2,1} \right)\left( \begin{pmatrix} -1 \\ 0 \end{pmatrix} \right) = T_{\vec{i}}\left( S\left( \begin{pmatrix} -1 \\ 0 \end{pmatrix} \right) \right) = T_{\vec{i}}\left( \begin{pmatrix} -2 \\ 0 \end{pmatrix} \right) = \begin{pmatrix} -1 \\ 0 \end{pmatrix},$$

but

$$\left( S_{2,1} \circ T_{\vec{i}} \right)\left( \begin{pmatrix} -1 \\ 0 \end{pmatrix} \right) = S_{2,1}\left( T_{\vec{i}}\left( \begin{pmatrix} -1 \\ 0 \end{pmatrix} \right) \right) = S_{2,1}\left( \begin{pmatrix} 0 \\ 0 \end{pmatrix} \right) = \begin{pmatrix} 0 \\ 0 \end{pmatrix}.$$

**Definition 1.4.3** A function $R_H : \mathbb{R}^2 \to \mathbb{R}^2$ is said to be a *reflection about the y-axis or horizontal reflection* if it is of the form $R_H(\mathbf{r}) = \begin{pmatrix} -x \\ y \end{pmatrix}$.

A function $R_V : \mathbb{R}^2 \to \mathbb{R}^2$ is said to be a *reflection about the x-axis or vertical reflection* if it is of the form $R_V(\mathbf{r}) = \begin{pmatrix} x \\ -y \end{pmatrix}$. A function $R_O : \mathbb{R}^2 \to \mathbb{R}^2$ is said to

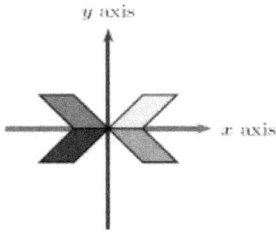

**FIGURE 1.4.3** Reflections. The original object (in the first quadrant) is yellow. Its reflection about the y-axis is magenta (on the second quadrant). Its reflection about the x-axis is cyan (on the fourth quadrant). Its reflection about the origin is blue (on the third quadrant).

be a *reflection about origin* if it is of the form

$$R_H(\mathbf{r}) = \begin{pmatrix} -x \\ -y \end{pmatrix}.$$

Some reflections appear in Fig. 1.4.3.

A few short computations establish various commutative properties among reflection, translation, and scaling. See Exercise 1.4.4.

We now define rotations. This definition will be somewhat harder than the others, so let us develop some ancillary results.

Consider a point r with polar coordinates $x = \rho \cos a$ and $y = \rho \sin a$ as in Fig. 1.4.4.

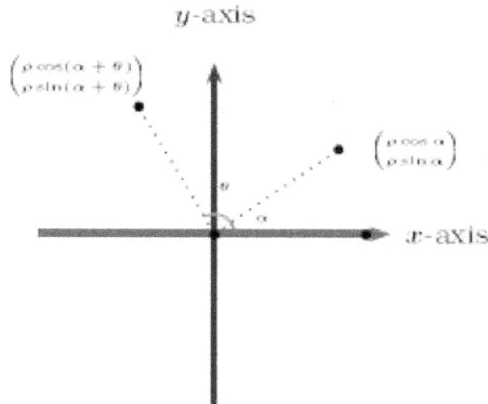

**FIGURE 1.4.4** Rotation by an angle θ in the levogyrate (counterclockwise) sense from the x-axis.

Here $\rho = \sqrt{x^2 + y^2}$ and $a \in [0; 2\pi]$. If we rotate it, in the levogyrate sense, by an angle $\theta$, we land on the new point $x'$ with $x' = \rho \cos(a + \theta)$ and $y' = \rho \sin(a + \theta)$. But

$$\rho \cos(a + \theta) = \rho \cos\theta \cos a - \rho \sin\theta \sin a = x \cos\theta - y \sin\theta,$$

and

$$\rho \sin(a + \theta) = \rho \sin a \cos\theta + \rho \sin\theta \cos a = y \cos\theta + x \sin\theta.$$

Hence, the point $\begin{pmatrix} x \\ y \end{pmatrix}$ is mapped to the point $\begin{pmatrix} x\cos\theta - y\sin\theta \\ x\sin\theta + y\cos\theta \end{pmatrix}$.

We may now formulate the definition of a rotation.

**Definition 1.4.4** A function $R_\theta : \mathbb{R}^2 \to \mathbb{R}^2$ is said to be a *levogyrate rotation about the origin by the angle* $\theta$ *measured from the positive x-axis if*

$$R_\theta(\mathrm{r}) = \begin{pmatrix} x\cos\theta - y\sin\theta \\ x\sin\theta + y\cos\theta \end{pmatrix}. \text{ Here } \rho = \sqrt{x^2 + y^2} .$$

Various properties of the composition of rotations with other plane transformations are explored in Exercises 1.4.5 and 1.4.6. We now codify some properties shared by scaling, reflection, and rotation.

**Definition 1.4.5** A function $L : \mathbb{R}^2 \to \mathbb{R}^2$ is said to be a *linear transformation* from $\mathbb{R}^2$ to $\mathbb{R}^2$ if for all points a, b on the plane and every scalar $\lambda$, it is verified that $L(a + b) = L(a) + L(b), \quad L(\lambda a) = \lambda L(a)$.

It is easy to prove that scaling, reflections, and rotations are linear transformations from $\mathbb{R}^2$ to $\mathbb{R}^2$, but not so translation.

**Definition 1.4.6** A function $L : \mathbb{R}^2 \to \mathbb{R}^2$ is said to be an *affine transformation* from $\mathbb{R}^2$ to $\mathbb{R}^2$ if there exists a linear transformation $L : \mathbb{R}^2 \to \mathbb{R}^2$ and a fixed vector $\vec{v} \in \mathbb{R}^2$ such that for all points $x \in \mathbb{R}^2$, it is verified that $A(x) = L(x) + \vec{v}$.

It is easy to see that translations are then affine transformations. In this definition, where the linear transformation is $L$, we may take $I : \mathbb{R}^2 \to \mathbb{R}^2$, then the identity transformation $I(x) = x$.

We have seen that scaling, reflection, and rotation are linear transformations. If $L : \mathbb{R}^2 \to \mathbb{R}^2$ is a linear transformation, then

$$L(r) = L(x\,\vec{i} + y\,\vec{j}) = xL(\vec{i}) + yL(\vec{j}),$$

and thus a linear transformation from R2 to R2 is solely determined by the values $L(\vec{i})$ and $L(\vec{j})$. We will now introduce a way to codify these values.

**Definition 1.4.7** Let $L:\mathbb{R}^2 \to \mathbb{R}^2$ be a linear transformation. The *matrix* $A_L$ *associated* with $L$ is the $2 \times 2$ (2 rows, 2 columns) array whose columns are (in this order) $L\left(\begin{pmatrix} 1 \\ 0 \end{pmatrix}\right)$ and $L\left(\begin{pmatrix} 0 \\ 1 \end{pmatrix}\right)$.

### EXAMPLE 1.4.2 (Scaling Matrices)

Let $a > 0$, $b > 0$ be real numbers. The matrix of the scaling transformation $S_{a,b}$ is $\begin{bmatrix} a & 0 \\ 0 & b \end{bmatrix}$. For $S_{a,b}\left(\begin{pmatrix} 1 \\ 0 \end{pmatrix}\right) = \begin{pmatrix} a \times 1 \\ b \times 0 \end{pmatrix} = \begin{pmatrix} a \\ 0 \end{pmatrix}$ and $S_{a,b}\left(\begin{pmatrix} 0 \\ 1 \end{pmatrix}\right) = \begin{pmatrix} a \times 0 \\ b \times 1 \end{pmatrix} = \begin{pmatrix} 0 \\ b \end{pmatrix}$.

### EXAMPLE 1.4.3 (Reflection Matrices)

It is easy to verify that the matrix for the transformation $R_H$ is $\begin{bmatrix} -1 & 0 \\ 0 & 1 \end{bmatrix}$, that the matrix for the transformation $R_V$ is $\begin{bmatrix} 1 & 0 \\ 0 & -1 \end{bmatrix}$, and the matrix for the transformation $R_O$ is $\begin{bmatrix} -1 & 0 \\ 0 & -1 \end{bmatrix}$.

### EXAMPLE 1.4.4 (Rotating Matrices)

It is easy to verify that the matrix for a rotation $R_\theta$ is $\begin{bmatrix} \cos\theta & -\sin\theta \\ \sin\theta & \cos\theta \end{bmatrix}$.

### EXAMPLE 1.4.5 (Identity Matrix)

The matrix for the identity linear transformation $\text{Id}:\mathbb{R}^2 \to \mathbb{R}^2$, $\text{Id}(x) = x$ is $I_2 = \begin{bmatrix} 1 & 0 \\ 0 & 1 \end{bmatrix}$.

### EXAMPLE 1.4.6 (Zero Matrix)

The matrix for the null linear transformation $N:\mathbb{R}^2 \to \mathbb{R}^2$, $N(x) = O$ is $0_2 = \begin{bmatrix} 0 & 0 \\ 0 & 0 \end{bmatrix}$.

From Example 1.4.7, we know that the composition of two linear transformations is also linear. We are now interested in how to codify the matrix of a composition of linear transformations $L_1 \circ L_2$ in terms of their individual matrices.

**Theorem 1.4.1** Let $L : \mathbb{R}^2 \to \mathbb{R}^2$ have the matrix representation $A_L = \begin{bmatrix} a & b \\ c & d \end{bmatrix}$ and let $L' : \mathbb{R}^2 \to \mathbb{R}^2$ have the matrix representation $A_{L'} = \begin{bmatrix} r & s \\ t & u \end{bmatrix}$. Then the composition $L \circ L'$ has matrix representation $\begin{bmatrix} ar + bt & as + bu \\ cr + dt & cs + du \end{bmatrix}.$

**Proof:**

We need to find $(L \circ L') \left( \begin{pmatrix} 1 \\ 0 \end{pmatrix} \right)$ and $(L \circ L') \left( \begin{pmatrix} 0 \\ 1 \end{pmatrix} \right)$.

We have

$$(L \circ L') \left( \begin{pmatrix} 1 \\ 0 \end{pmatrix} \right) = L \left( L' \left( \begin{pmatrix} 1 \\ 0 \end{pmatrix} \right) \right) = L \left( \begin{pmatrix} r \\ t \end{pmatrix} \right) = rL(\vec{i}) + tL(\vec{j}) = r \begin{pmatrix} a \\ c \end{pmatrix} + t \begin{pmatrix} b \\ d \end{pmatrix} = \begin{pmatrix} ar + bt \\ cr + dt \end{pmatrix},$$

and

$$(L \circ L') \left( \begin{pmatrix} 0 \\ 1 \end{pmatrix} \right) = L \left( L' \left( \begin{pmatrix} 0 \\ 1 \end{pmatrix} \right) \right) = L \left( \begin{pmatrix} s \\ u \end{pmatrix} \right) = sL(\vec{i}) + uL(\vec{j}) = s \begin{pmatrix} a \\ c \end{pmatrix} + u \begin{pmatrix} b \\ d \end{pmatrix} = \begin{pmatrix} as + bu \\ cs + du \end{pmatrix},$$

hence we conclude that the matrix of $L \circ L'$ is $\begin{bmatrix} ar + bt & as + bu \\ cr + dt & cs + du \end{bmatrix}$, as we wanted to show.

The preceding motivates the following definition.

**Definition 1.4.8** Let $A = \begin{bmatrix} a & b \\ c & d \end{bmatrix}$ and $B = \begin{bmatrix} r & s \\ t & u \end{bmatrix}$ be two $2 \times 2$ matrices, and $\lambda \in \mathbb{R}$ be a scalar. We define *matrix addition* as

$$A + B = \begin{bmatrix} a & b \\ c & d \end{bmatrix} + \begin{bmatrix} r & s \\ t & u \end{bmatrix} = \begin{bmatrix} a+r & b+s \\ c+t & d+u \end{bmatrix}.$$

We define *matrix multiplication* as

$$AB = \begin{bmatrix} a & b \\ c & d \end{bmatrix} \begin{bmatrix} r & s \\ t & u \end{bmatrix} = \begin{bmatrix} ar+bt & as+bu \\ cr+dt & cs+du \end{bmatrix}.$$

We define *scalar multiplication of a matrix* as

$$\lambda A = \lambda \begin{bmatrix} a & b \\ c & d \end{bmatrix} = \begin{bmatrix} \lambda a & \lambda b \\ \lambda c & \lambda d \end{bmatrix}.$$

**NOTE**  *Since the composition of functions is not necessarily commutative, neither is matrix multiplication. Since the composition of functions is associative, so is matrix multiplication.*

## EXAMPLE 1.4.7

Let $M = \begin{bmatrix} 1 & -1 \\ 0 & 1 \end{bmatrix}$, $N = \begin{bmatrix} 1 & 2 \\ -2 & 1 \end{bmatrix}$.

Then,

$$M + N = \begin{bmatrix} 1 & -1 \\ 0 & 1 \end{bmatrix} + \begin{bmatrix} 1 & 2 \\ -2 & 1 \end{bmatrix} = \begin{bmatrix} 2 & 1 \\ -2 & 2 \end{bmatrix},$$

$$2M = 2 \begin{bmatrix} 1 & -1 \\ 0 & 1 \end{bmatrix} = \begin{bmatrix} 2 & -2 \\ 0 & 2 \end{bmatrix},$$

$$MN = \begin{bmatrix} 1 & -1 \\ 0 & 1 \end{bmatrix} \begin{bmatrix} 1 & 2 \\ -2 & 1 \end{bmatrix} = \begin{bmatrix} 1\times 1 + (-1)\times(-2) & 1\times 2 + (-1)\times 1 \\ 0\times 1 + 1\times(-2) & 0\times 2 + 1\times 1 \end{bmatrix} = \begin{bmatrix} 3 & 1 \\ -2 & 1 \end{bmatrix}.$$

## EXAMPLE 1.4.8

Let $M = \begin{bmatrix} 2 & 1 \\ 0 & 3 \end{bmatrix}$, then $M^2 = \begin{bmatrix} 2 & 1 \\ 0 & 3 \end{bmatrix} \begin{bmatrix} 2 & 1 \\ 0 & 3 \end{bmatrix} = \begin{bmatrix} 4 & 5 \\ 0 & 9 \end{bmatrix}$ and $5M = 5 \begin{bmatrix} 2 & 1 \\ 0 & 3 \end{bmatrix} = \begin{bmatrix} 10 & 5 \\ 0 & 15 \end{bmatrix}$

**EXAMPLE 1.4.9**

Find a $2 \times 2$ matrix that will transform the square in Fig. 1.4.5 into the parallelogram in Fig. 1.4.6. Assume in each case that the vertices of the figures are lattice points, that is, coordinate points with integer coordinates.

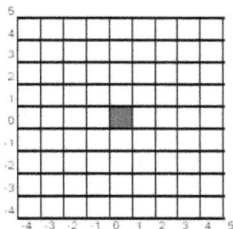

*FIGURE 1.4.5* Example 1.4.8.

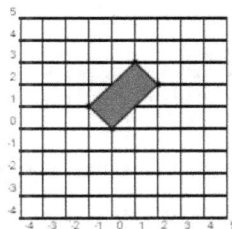

*FIGURE 1.4.6* Example 1.4.8.

*Solution:*

Let $\begin{bmatrix} a & b \\ c & d \end{bmatrix}$ be the desired matrix. Then, since

$$\begin{bmatrix} a & b \\ c & d \end{bmatrix}\begin{bmatrix} 0 \\ 0 \end{bmatrix} = \begin{bmatrix} 0 \\ 0 \end{bmatrix},$$

the point $\begin{pmatrix} 0 \\ 0 \end{pmatrix}$ is a fortiori, transformed to itself. We now assume, without loss

of generality, that each vertex of the square is transformed in the same order, counterclockwise, to each vertex of the rectangle. Then,

$$\begin{bmatrix} a & b \\ c & d \end{bmatrix}\begin{bmatrix} 1 \\ 0 \end{bmatrix} = \begin{bmatrix} 2 \\ 2 \end{bmatrix} \Rightarrow \begin{bmatrix} a \\ c \end{bmatrix} = \begin{bmatrix} 2 \\ 2 \end{bmatrix} \Rightarrow a = c = 2 \,.$$

Using these values,

$$\begin{bmatrix} a & b \\ c & d \end{bmatrix}\begin{bmatrix} 1 \\ 1 \end{bmatrix} = \begin{bmatrix} 1 \\ 3 \end{bmatrix} \Rightarrow \begin{bmatrix} a+b \\ c+d \end{bmatrix} = \begin{bmatrix} 1 \\ 3 \end{bmatrix} \Rightarrow b = -1, d = 1 \,.$$

And so the desired matrix is

$$\begin{bmatrix} 2 & -1 \\ 2 & 1 \end{bmatrix}.$$

## EXERCISES 1.4

**1.4.1** If $A = \begin{bmatrix} 1 & -1 \\ 2 & 3 \end{bmatrix}$, $B = \begin{bmatrix} a & b \\ 1 & -2 \end{bmatrix}$, and $(A+B)^2 = A^2 + 2AB + B^2$, find $a$ $b$.

**1.4.2** Let $M = \begin{bmatrix} 1 & 2 \\ 0 & -1 \end{bmatrix}$, $N = \begin{bmatrix} 1 & 2 \\ 3 & -1 \end{bmatrix}$. Find M+N, 3M, and MN.

**1.4.3** Find all matrices $A \in M_{2\times 2}(\mathbb{R})$ that $A^2 = 0_2$.

**1.4.4** Find all linear transformations from $\mathbb{R}^2$ into $\mathbb{R}^2$ that

   **1.** Carry the line $x = 0$ into the line $x = 0$.

   **2.** Carry the line $y = 0$ into the line $y = 0$.

   **3.** Carry the line $x = y$ into the line $x = y$.

   **4.** Carry the line $x = 0$ into the line $x = 0$ and carry the line $y = 0$ into the line $y = 0$.

**1.4.5** Let $L : \mathbb{R}^2 \to \mathbb{R}^2$ be defined by $L\begin{pmatrix} x \\ y \end{pmatrix} = \begin{pmatrix} 1 \\ 2 \end{pmatrix}$. Show if $L$ is linear or not linear.

**1.4.6** Let $L$ be a linear transformation by reflecting each vector $\vec{u}$ in $\mathbb{R}^2$ with respect to the line $y = -x$. Determine a matrix for $L$.

**1.4.7** Prove that the following transformation $L : \mathbb{R}^2 \to \mathbb{R}^2$ defined by $L\begin{pmatrix} x \\ y \end{pmatrix} = \begin{pmatrix} x-y \\ 3x \end{pmatrix}$ is linear.

**1.4.8** Prove that the following transformation $L : \mathbb{R}^2 \to \mathbb{R}^2$ defined by $L\begin{pmatrix} x \\ y \end{pmatrix} = \begin{pmatrix} 2x \\ x-y \end{pmatrix}$ is linear.

**1.4.9**   Consider $\triangle ABC$ with $A = \begin{pmatrix} -1 \\ 2 \end{pmatrix}$, $B = \begin{pmatrix} 0 \\ -2 \end{pmatrix}$, $C = \begin{pmatrix} 2 \\ 1 \end{pmatrix}$, as in

Fig. 1.4.7. Determine the effects of the following scaling transformations on the triangle: $S_{2,1}, S_{1,2}$, and $S_{2,2}$ .

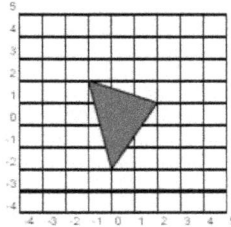

**FIGURE 1.4.7** Exercises 1.4.2, 1.4.3, and 1.4.8.

**1.4.10**   Find the equation of the image of the line $y = 3x$ under *scaling* of factor 3 in the $x$ direction and factor 5 in the $y$ direction.

**1.4.11**   Find the effects of the reflections, $R_{\frac{\pi}{2}}, R_{\frac{\pi}{4}}, R_{-\frac{\pi}{2}}$, and $R_{-\frac{\pi}{4}}$ on the triangle in Fig. 1.4.7.

**1.4.12**   Find the effects of the reflections $R_H$, $R_V$, and $R_O$ on the triangle in Fig. 1.4.7.

**1.4.13**   Determine the matrix that defines a reflection in the $y$ axis. Find the image of $\begin{bmatrix} 3 \\ 2 \end{bmatrix}$ under this transformation.

**1.4.14**   Find the equation of the image of the ellipse $\dfrac{x^2}{16} + \dfrac{y^2}{25} = 1$ under a *rotation* through an angle $\pi / 2$ .

**1.4.15**   Determine the matrix that can be used to define a rotation through $\pi / 2$ about the point $(5,1)$. Find the image of the unit square under this rotation.

**1.4.16**   Let $L : \mathbb{R}^2 \to \mathbb{R}^2$ be a counterclockwise rotation through $\pi / 4$ radians. Find the standard matrix representing $L$.

**1.4.17**   Let $L_1\left(\vec{x}\right) = A_1\vec{x}$ and $L_2\left(\vec{x}\right) = A_2\vec{x}$ be defined by the following matrices $A_1$ and $A_2$. Find $L_2 \circ L_1$, if:

**1.**   $A_1 = \begin{bmatrix} 1 & 3 \\ 2 & 0 \end{bmatrix}, A_2 = \begin{bmatrix} 1 & 4 \\ -1 & 0 \end{bmatrix}, \vec{x} = \begin{bmatrix} 2 \\ 3 \end{bmatrix}$

**2.**   $A_1 = \begin{bmatrix} 2 & -1 \\ 0 & 1 \end{bmatrix}, A_2 = \begin{bmatrix} 1 & 2 \\ 0 & 3 \\ 2 & -1 \end{bmatrix}, \vec{x} = \begin{bmatrix} -2 \\ 1 \end{bmatrix}$

**1.4.18**   Let $L_1\begin{pmatrix} x \\ y \end{pmatrix} = \begin{pmatrix} 3x \\ -y \end{pmatrix}$ and $L_2\begin{pmatrix} x \\ y \end{pmatrix} = \begin{pmatrix} 0 \\ x+y \end{pmatrix}$. Find $L_2 \circ L_1$.

## 1.5   DETERMINANTS IN TWO DIMENSIONS

We will now define a way of determining areas of plane figures on the plane. It seems reasonable to require that this area determination agrees with common formulae of areas of plane figures, in particular, the area of a parallelogram should be as we learn in elementary geometry and the area of a unit square is 1.

From Figs. 1.5.1 and 1.5.2, the area of a parallelogram spanned by $\begin{bmatrix} a \\ b \end{bmatrix}$ and $\begin{bmatrix} c \\ d \end{bmatrix}$ is

$$D\left(\begin{bmatrix} a \\ b \end{bmatrix}, \begin{bmatrix} c \\ d \end{bmatrix}\right) = ad - bc.$$

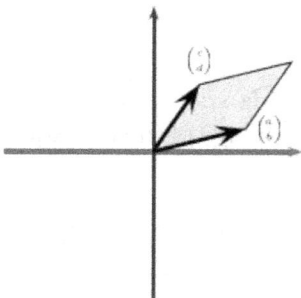

**FIGURE 1.5.1** Area of a parallelogram.

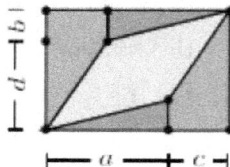

**FIGURE 1.5.2** $(a+c)(b+d) - 2\dfrac{ab}{2} - 2\dfrac{c(2b+d)}{2} = ad - bc.$

This motivates the following definition.

**Definition 1.5.1**  The *determinant* of the $2\times 2$ matrix $\begin{bmatrix} a & b \\ c & d \end{bmatrix}$ is

$$\det\begin{bmatrix} a & b \\ c & d \end{bmatrix} = ad - bc.$$

**NOTE**  *The symbol for determinant of a matrix $\begin{bmatrix} a & b \\ c & d \end{bmatrix}$ can be written as $\det\begin{bmatrix} a & b \\ c & d \end{bmatrix}$ or $\begin{vmatrix} a & b \\ c & d \end{vmatrix}$.*

**EXAMPLE 1.5.1**

Evaluate the determinants of $M = \begin{bmatrix} 3 & 1 \\ 5 & -2 \end{bmatrix}$.

**Solution:**

$$\det(M) = (3)(-2) - (1)(5) = -11 .$$

**EXAMPLE 1.5.2**

Evaluate the $\det\begin{bmatrix} -2 & 4 \\ 5 & 7 \end{bmatrix}$.

**Solution:**

$$\det\begin{bmatrix} -2 & 4 \\ 5 & 7 \end{bmatrix} = (-2)(7) - (4 - 5) = -13$$

**EXAMPLE 1.5.3**

Find the values of $x$ for which $\det(M) = 0$ when $M = \begin{bmatrix} x-2 & 1 \\ -5 & x+4 \end{bmatrix}$.

**Solution:**

$$\det(M) = \begin{vmatrix} x-2 & 1 \\ -5 & x+4 \end{vmatrix} = (x-2)(x+4) - (1)(-5)$$
$$= x^2 + 2x - 3 = (x-1)(x+3)$$

Hence, $\det(M) = 0 \Leftrightarrow x = 1$ or $x = -3$ .

### Theorem 1.5.1

If $M$ and $N$ are square matrices of same size, then $\det(MN) = \det(M)\det(N)$.

### EXAMPLE 1.5.4

Let $M = \begin{bmatrix} 1 & 2 \\ 1 & 3 \end{bmatrix}$ and $N = \begin{bmatrix} 8 & 5 \\ 3 & -1 \end{bmatrix}$, then $MN = \begin{bmatrix} 14 & 3 \\ 17 & 2 \end{bmatrix}$,

$\det(M) = 3 - 2 = 1$, $\det(N) = -8 - 15 = -23$, and $\det(MN) = 28 - 51 = -23$.
Thus, $\det(MN) = \det(M)\det(N)$ as expected.

Consider now a simple quadrilateral with vertices

$r_1 = (x_1, y_1)$, $r_2 = (x_2, y_2)$, $r_3 = (x_3, y_3)$, $r_4 = (x_4, y_4)$, listed in counterclockwise

order, as in Fig. 1.5.3. This quadrilateral is spanned by the vectors

$$\overrightarrow{r_1 r_2} = \begin{bmatrix} x_2 - x_1 \\ y_2 - y_1 \end{bmatrix}, \quad \overrightarrow{r_1 r_4} = \begin{bmatrix} x_4 - x_1 \\ y_4 - y_1 \end{bmatrix},$$

and hence, its area is given by

$$A = \det \begin{bmatrix} x_2 - x_1 & x_4 - x_1 \\ y_2 - y_1 & y_4 - y_1 \end{bmatrix} = D\left(\vec{r_2} - \vec{r_1}, \vec{r_4} - \vec{r_1}\right).$$

Similarly, noticing that the quadrilateral is also spanned by

$$\overrightarrow{r_3 r_4} = \begin{bmatrix} x_4 - x_3 \\ y_4 - y_3 \end{bmatrix}, \quad \overrightarrow{r_3 r_2} = \begin{bmatrix} x_2 - x_3 \\ y_2 - y_3 \end{bmatrix},$$

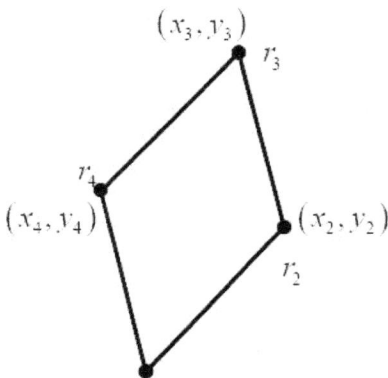

**FIGURE 1.5.3** Area of a quadrilateral.

its area is also given by

$$A = \det \begin{bmatrix} x_4 - x_3 & x_2 - x_3 \\ y_4 - y_3 & y_2 - y_3 \end{bmatrix} = D\left(\vec{r_4} - \vec{r_3}, \vec{r_2} - \vec{r_3}\right).$$

Using the properties derived in previous theorems, we see that

$$A = \frac{1}{2}\left(D\left(\vec{r_2} - \vec{r_1}, \vec{r_4} - \vec{r_1}\right) + D\left(\vec{r_4} - \vec{r_3}, \vec{r_2} - \vec{r_3}\right)\right)$$

$$A = \frac{1}{2}\left(D\left(\vec{r_2}, \vec{r_4}\right) - D\left(\vec{r_2}, \vec{r_1}\right) - D\left(\vec{r_1}, \vec{r_4}\right) + D\left(\vec{r_1}, \vec{r_1}\right)\right) + \frac{1}{2}\left(D\left(\vec{r_4}, \vec{r_2}\right) - D\left(\vec{r_3}, \vec{r_2}\right) - D\left(\vec{r_4}, \vec{r_3}\right) + D\left(\vec{r_3}, \vec{r_3}\right)\right)$$

$$A = \frac{1}{2}\left(D\left(\vec{r_2}, \vec{r_4}\right) - D\left(\vec{r_2}, \vec{r_1}\right) - D\left(\vec{r_1}, \vec{r_4}\right)\right) + \frac{1}{2}\left(D\left(\vec{r_4}, \vec{r_2}\right) - D\left(\vec{r_3}, \vec{r_2}\right) - D\left(\vec{r_4}, \vec{r_3}\right)\right)$$

$$A = \frac{1}{2}\left(D\left(\vec{r_1}, \vec{r_2}\right) + D\left(\vec{r_2}, \vec{r_3}\right) + D\left(\vec{r_3}, \vec{r_4}\right) + D\left(\vec{r_4}, \vec{r_1}\right)\right).$$

We conclude that the area of a quadrilateral with vertices $(x_1, y_1)$, $(x_2, y_2)$, $(x_3, y_3)$, $(x_4, y_4)$ listed in counterclockwise order is

$$\frac{1}{2}\left(\det\begin{bmatrix} x_1 & x_2 \\ y_1 & y_2 \end{bmatrix} + \det\begin{bmatrix} x_2 & x_3 \\ y_2 & y_3 \end{bmatrix} + \det\begin{bmatrix} x_3 & x_4 \\ y_3 & y_4 \end{bmatrix} + \det\begin{bmatrix} x_4 & x_1 \\ y_4 & y_1 \end{bmatrix}\right). \quad (1.11)$$

To find the area of a triangle of vertices, $\vec{r_1} = (x_1, y_1), \vec{r_2} = (x_2, y_2), \vec{r_3} = (x_3, y_3)$, listed in counterclockwise order as in Fig. 1.5.4, reflects about one of its sides, as in Fig. 1.5.5, creating a parallelogram.

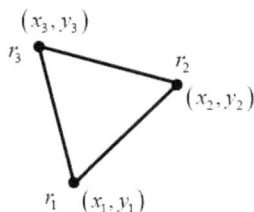

**FIGURE 1.5.4** Area of a triangle.

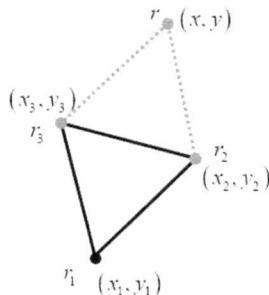

**FIGURE 1.5.5** Area of a triangle.

The area of the triangle is now half the area of the parallelogram, which, by virtue of equation (1.11), is

$$\frac{1}{4}\left(D\left(\vec{r_1},\vec{r_2}\right)+D\left(\vec{r_2},\vec{r}\right)+D\left(\vec{r},\vec{r_3}\right)+D\left(\vec{r_3},\vec{r_1}\right)\right).$$

This is equivalent to

$$\frac{1}{2}\left(D\left(\vec{r_1},\vec{r_2}\right)+D\left(\vec{r_2},\vec{r_3}\right)+D\left(\vec{r_3},\vec{r_1}\right)\right)$$

$$-\frac{1}{4}\left(D\left(\vec{r_1},\vec{r_2}\right)-D\left(\vec{r_2},\vec{r}\right)-D\left(\vec{r},\vec{r_3}\right)+D\left(\vec{r_3},\vec{r_1}\right)+2D\left(\vec{r_2},\vec{r_3}\right)\right).$$

We will prove that

$$D\left(\vec{r_1},\vec{r_2}\right)-D\left(\vec{r_2},\vec{r}\right)-D\left(\vec{r},\vec{r_3}\right)+D\left(\vec{r_3},\vec{r_1}\right)+2D\left(\vec{r_2},\vec{r_3}\right)=0\ .$$

To do this, we appeal once again to the bi-linearity properties derived in previous theorems, and observe that we have a parallelogram, $\vec{r}-\vec{r_3}=\vec{r_2}-\vec{r_1}$ , which means, $\vec{r}=\vec{r_3}+\vec{r_2}-\vec{r_1}$ . Thus,

$$D\left(\vec{r_1},\vec{r_2}\right)-D\left(\vec{r_2},\vec{r}\right)-D\left(\vec{r},\vec{r_3}\right)+D\left(\vec{r_3},\vec{r_1}\right)+2D\left(\vec{r_2},\vec{r_3}\right)=$$

$$=D\left(\vec{r_1},\vec{r_2}\right)-D\left(\vec{r_2},\vec{r_3}+\vec{r_2}-\vec{r_1}\right)+2D\left(\vec{r_2},\vec{r_3}\right)-D\left(\vec{r_3}+\vec{r_2}-\vec{r_1},\vec{r_3}\right)+D\left(\vec{r_3},\vec{r_1}\right)$$

$$=D\left(\vec{r_1},\vec{r_2}-\vec{r_3}\right)+D\left(\vec{r_3}+\vec{r_2}-\vec{r_1},\vec{r_2}-\vec{r_3}\right)+2D\left(\vec{r_2},\vec{r_3}\right)$$

$$=D\left(\vec{r_3}+\vec{r_2},\vec{r_2}-\vec{r_3}\right)+2D\left(\vec{r_2},\vec{r_3}\right)$$

$$=D\left(\vec{r_3},\vec{r_2}\right)-D\left(\vec{r_2},\vec{r_3}\right)+2D\left(\vec{r_2},\vec{r_3}\right)$$

$$=D\left(\vec{r_3},\vec{r_2}\right)-D\left(\vec{r_2},\vec{r_3}\right)$$

$$=0,$$

as claimed. We have proved then that the area of a triangle, whose vertices $(x_1,y_1),(x_2,y_2),(x_3,y_3)$ are listed in counterclockwise order, is

$$\frac{1}{2}\left(\det\begin{bmatrix}x_1 & x_2\\ y_1 & y_2\end{bmatrix}+\det\begin{bmatrix}x_2 & x_3\\ y_2 & y_3\end{bmatrix}+\det\begin{bmatrix}x_3 & x_1\\ y_3 & y_1\end{bmatrix}\right). \tag{1.12}$$

In general, we have the following theorem.

**Theorem 1.5.1 (Surveyor's Theorem)**     Let $(x_1, y_1), (x_2, y_2), \ldots, (x_n, y_n)$ be the vertices of a simple (non-crossing) polygon, listed in counterclockwise order. Then its area is given by

$$\frac{1}{2}\left( \det\begin{bmatrix} x_1 & x_2 \\ y_1 & y_2 \end{bmatrix} + \det\begin{bmatrix} x_2 & x_3 \\ y_2 & y_3 \end{bmatrix} + \cdots + \det\begin{bmatrix} x_{n-1} & x_n \\ y_{n-1} & y_n \end{bmatrix} + \det\begin{bmatrix} x_n & x_1 \\ y_n & y_1 \end{bmatrix} \right).$$

**Proof:**

The proof is by induction on $n$. We have already proved the cases $n = 3$ and $n = 4$ in (1.12) and (1.11), respectively. Consider now a simple polygon $P$ with n vertices. If $P$ is convex, then we may take any vertex and draw a line to the other vertices, triangulating the polygon, creating $n - 2$ triangles. If $P$ is not convex, then there must be a vertex that has a reflex angle. A ray produced from this vertex must hit another vertex, creating a diagonal; otherwise, the polygon would have an infinite area. This diagonal divides the polygon into two sub-polygons. These two sub-polygons are either both convex or at least one is not convex. In the latter case, we repeat the argument, finding another diagonal and creating a new sub-polygon. Eventually, since the number of vertices is infinite, we end up triangulating the polygon. Moreover, the polygon can be triangulated in such a way that all triangles inherit the positive orientation of the original polygon but each neighboring pair of triangles have opposite orientations. Applying (1.12), we obtain that the area is

$$\sum \det\begin{bmatrix} x_i & x_j \\ y_i & y_j \end{bmatrix},$$

where the sum is over each oriented edge. Since each diagonal occurs twice, but having opposite orientations, the terms

$$\det\begin{bmatrix} x_i & x_j \\ y_i & y_j \end{bmatrix} + \det\begin{bmatrix} x_j & x_i \\ y_j & y_i \end{bmatrix} = 0$$

disappear from the sum and we are simply left with

$$\frac{1}{2}\left( \det\begin{bmatrix} x_1 & x_2 \\ y_1 & y_2 \end{bmatrix} + \det\begin{bmatrix} x_2 & x_3 \\ y_2 & y_3 \end{bmatrix} + \cdots + \det\begin{bmatrix} x_{n-1} & x_n \\ y_{n-1} & y_n \end{bmatrix} + \det\begin{bmatrix} x_n & x_1 \\ y_n & y_1 \end{bmatrix} \right).$$

We may use the software *Maple*<sup>TM</sup> in order to speed up computations with vectors. Most of the commands we will need are in the *linalg* package. For example, let us define two vectors, $\vec{a} = \begin{bmatrix} 1 \\ 2 \end{bmatrix}$ and $\vec{b} = \begin{bmatrix} 2 \\ 1 \end{bmatrix}$, and a matrix $A := \begin{bmatrix} 1 & 2 \\ 3 & 4 \end{bmatrix}$. Let us compute their dot product, find a unit vector in the direction of $\vec{a}$, and the angle between the vectors, the determinant of a matrix$_A$. (There must be either a colon or a semicolon at the end of each statement. The result will not display if a colon is chosen.)

```
>  with(linalg) :
>  a := vector([1, 2]);
```
$$a := \begin{bmatrix} 1 & 2 \end{bmatrix}$$

```
>  b := vector([2, 1]);
```
$$b := \begin{bmatrix} 2 & 1 \end{bmatrix}$$

```
>  normalize(a);
```
$$\begin{bmatrix} \frac{1}{5}\sqrt{5} & \frac{2}{5}\sqrt{5} \end{bmatrix}$$

```
>  dotprod(a, b);
```
$$4$$

```
>  angle(a, b);
```
$$\arccos\left(\frac{4}{5}\right)$$

```
>  A := matrix([[1, 2], [3, 4]]);
```
$$A := \begin{bmatrix} 1 & 2 \\ 3 & 4 \end{bmatrix}$$

```
>  det(A);
```
$$-2$$

We may also use *MATLAB* in order to speed up computations with vectors.

```
>> a = [1,2];
>> b = [2,1];
>> norm(a)
```

ans =

  2.2361

>> normalized = a/norm(a)

normalized =

  0.4472   0.8944

>> CosTheta = dot(a,b)/(norm(a)*norm(b));

>> CosTheta = dot(a,b)/(norm(a)*norm(b));
>> ThetaInDegrees = acos(CosTheta)*180/pi

ThetaInDegrees =

  36.8699

>> A=[1 2;3 4]

A =

  1  2

  3  4

>> det (A)

ans =

  −2

# EXERCISES 1.5

**1.5.1**   Calculate

     **1.**  $\det \begin{bmatrix} 1 & -2 \\ 6 & 3 \end{bmatrix}$

**2.**  $\det \begin{bmatrix} 1 & 2 \\ 5 & 4 \end{bmatrix}$

**3.**  $\det \begin{bmatrix} 4 & -2 \\ 5 & 3 \end{bmatrix}$

**4.**  $\det \begin{bmatrix} -2 & -7 \\ 6 & -5 \end{bmatrix}$

**5.**  $\det \begin{bmatrix} x-2 & -3 \\ 5 & x-3 \end{bmatrix}$

**6.**  $\det \begin{bmatrix} \sqrt{3} & 4 \\ \sqrt{6} & \sqrt{2} \end{bmatrix}$

**1.5.2**  For what values of $a$, does

**1.**  $\det \begin{bmatrix} 1+a & 1 \\ 1 & 1-a \end{bmatrix} = 0$

**2.**  $\det \begin{bmatrix} a+4 & -5 \\ 1 & a-2 \end{bmatrix} = 0$

**1.5.3**  Compute the $\det \begin{bmatrix} \cos\theta & \sin\theta \\ -\sin\theta & \cos\theta \end{bmatrix}$ for any real number.

**1.5.4**  Compute the $\det \begin{bmatrix} \cos\theta & \sin\theta \\ \sin\theta & \cos\theta \end{bmatrix}$ when

**1.**  $\theta = \pi / 4$

**2.**  $\theta = \pi / 3$

**1.5.5**  Let $k$ be a number, and let $A$ be a $2 \times 2$ matrix. How does $\det(kA)$ differ from $\det(A)$?

**1.5.6**  Let matrix $B$ be formed from matrix $A$ by interchanging two rows. Prove then that $\det(A) = -\det(B)$.

**1.5.7**  Prove that if $A$ and $B$ are square matrices of the same size, then $\det(AB) = \det(BA)$.

**1.5.8** Prove that if the sum of elements of each row (or column) of a square matrix $A$ zero, then $\det(A) = 0$.

**1.5.9** Let the vectors $\vec{r_1} = \begin{bmatrix} a \\ b \end{bmatrix}$ and $\vec{r_2} = \begin{bmatrix} c \\ d \end{bmatrix}$, and that the vector $\vec{r_3}$

obtained by rotating $\vec{r_1}$ counterclockwise by $\pi / 2$ is $\vec{r_3} = \begin{bmatrix} -b \\ a \end{bmatrix}$.

Show that $\vec{r_3} \bullet \vec{r_2} = D\left(\vec{r_1}, \vec{r_2}\right)$.

**1.5.10** Prove that the area of the parallelogram spanned by the vectors

$\vec{r_1} = \begin{bmatrix} a \\ b \end{bmatrix}$ and $\vec{r_2} = \begin{bmatrix} c \\ d \end{bmatrix}$ is the $\left\| D\left(\vec{r_1}, \vec{r_2}\right) \right\|$.

## 1.6 PARAMETRIC CURVES ON THE PLANE

**Definition 1.6.1** Let $[a;b] \subseteq \mathbb{R}$. A *parametric curve* representation r of a

curve $\Gamma$ is function $r : \lfloor a;b \rfloor \to \mathbb{R}^2$, with $r(t) = \begin{pmatrix} x(t) \\ y(t) \end{pmatrix}$, and such that

$r\left([a;b]\right) = \Gamma$.

r $(a)$ is the *initial point* of the curve and r $(b)$ *its terminal point*. A curve is *closed* if its initial point and its final point coincide. The *trace* of the curve r is the set of all images of r, that is, $\Gamma$. If there exists $t_1 \neq t_2$ such that $r(t_1) = r(t_2) = p$, then p is a *multiple point* of the curve. The curve is *simple* if curve has no multiple points. A closed curve whose only multiple points are its endpoints is called a *Jordan curve*.

Graphing parametric equations is a difficult art. A theory akin to the one studied for Cartesian equations in a first Calculus course has been developed. Our interest is not in graphing curves, but in obtaining suitable parameterizations of simple Cartesian curves. However, we mention in passing that *Maple*$^{TM}$ has excellent capabilities for graphing parametric equations. For example, the commands to graph the various curves in Figs. 1.6.1 through 1.6.4 follow.

```
> with( plots ) :
> plot( [ sin( 2 * t ), cos( 6 * t ), t = 0 ..2 * Pi ], x = −5 ..5, y = −5 ..5 );
```

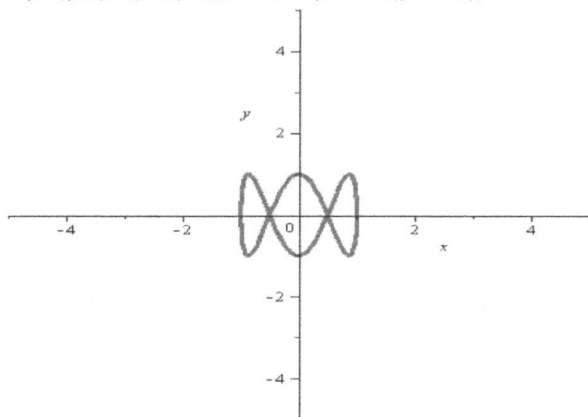

**FIGURE 1.6.1** $x = \sin 2t,\ y = \cos 6t$.

```
> plot( [ 2^( t/ 10 ) * cos( t ), 2^( t/ 10 ) * sin( t ), t = −20 ..10 ], x = −5 ..5, y = −5
  ..5 );
```

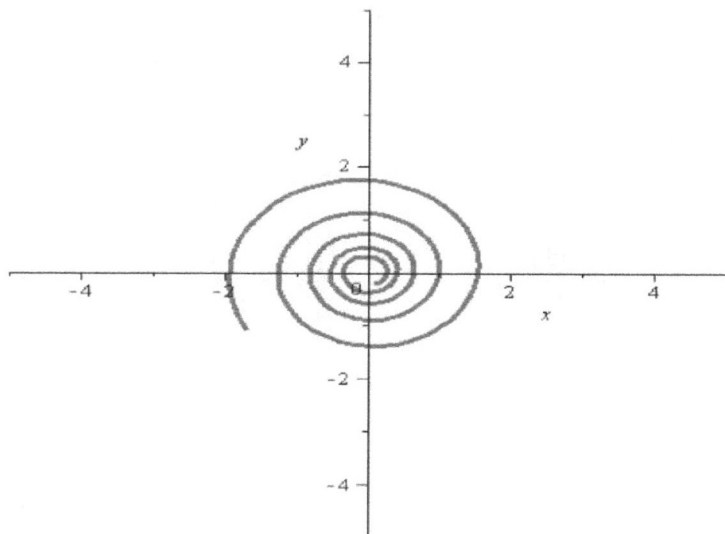

**FIGURE 1.6.2** $x = 2^{t/10} \cos t,\ y = 2^{t/10} \sin t$.

```
> plot( [ ( 1 − t^2 )/ ( 1 + t^2 ), ( t − t^3 )/ ( 1 + t^2 ), t = −2 ..2 ], x = −5 ..5, y =
  −5 ..5 );
```

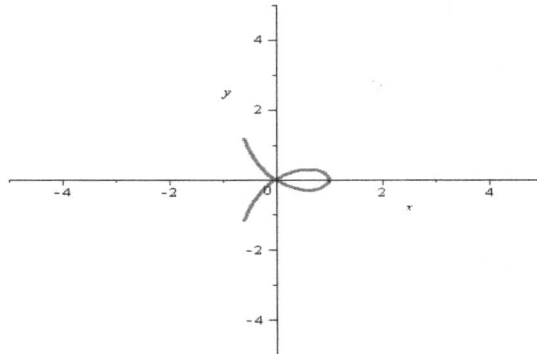

FIGURE 1.6.3  $x = \dfrac{1-t^2}{1+t^2}, \; y = \dfrac{t-t^3}{1+t^2}.$

> $plot([(1 + \cos(t))/2, \sin(t) * (1 + \cos(t))/2, t = 0 .. 2 * \text{Pi}], x = -5 .. 5, y = -5 .. 5);$

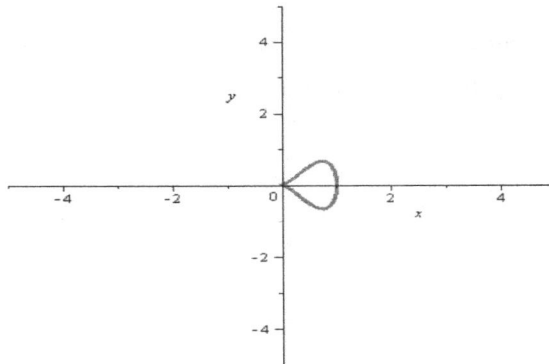

FIGURE 1.6.4  $x = \dfrac{(1+\cos t)}{2}, \; y = \dfrac{(\sin t)(1+\cos t)}{2}.$

Our main focus of attention is the following. Given a Cartesian curve with equation $f(x,y) = 0$, we wish to find suitable parameterizations for them. That is, we want to find functions $x : t \to a(t)$, $y : t \to b(t)$ and an interval $I$ such that the graphs of $f(x,y) = 0$ and $f(a(t), b(t)) = 0$, $t \in I$ coincide. These parameterizations may differ in features, according to the choice of functions and the choice of intervals.

**EXAMPLE 1.6.1**

Suppose that $m \in \mathbb{R}^2$, $b \in \mathbb{R}^2$. Write the Cartesian equation of the line $y = mt + b$ in parametric form.

**Solution:**

We put $x = t$, then $y = mt + b$ and so the desired parametric form is

$$\begin{bmatrix} x \\ y - b \end{bmatrix} = t \begin{bmatrix} 1 \\ m \end{bmatrix}.$$

**EXAMPLE 1.6.2**

Consider the parabola with the Cartesian equation $y = x^2$. We will give various parameterizations for portions of this curve.

1. If $x = t$ and $y = t^2$, then clearly $y = t^2 = x^2$. This works for every $t \in \mathbb{R}$, and hence the parameterization

$$x = t, \ y = t^2, \ t \in \mathbb{R},$$

works for the whole curve. Notice that as t increases, the curve is traversed from left to right.

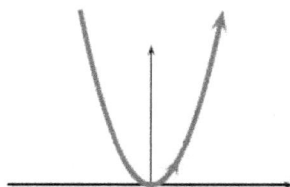

**FIGURE 1.6.5** $x = t, y = t^2, t \in \mathbb{R}$

2. If $x = \sqrt{t}$ and $y = t$, then again $y = t = \left(\sqrt{t}\right)^2 = x^2$. This works only for $t \geq 0$, and hence the parameterization

$$x = \sqrt{t}, \ y = t, \ t \in [0; +\infty[,$$

gives the half of the curve for which $x \geq 0$. As t increases, the curve is traversed from left to right.

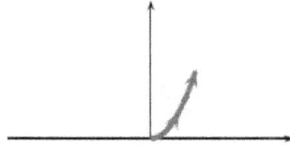

**FIGURE 1.6.6** $x = \sqrt{t},\, y = t, t \in [0; +\infty[$.

3. Similarly, if $x = -\sqrt{t}$ and $y = t$, then again $y = t = \left(\sqrt{t}\right)^2 = x^2$. This works only for $t \geq 0$, and hence the parameterization

$$x = -\sqrt{t},\, y = t,\ \ t \in [0; +\infty[,$$

gives the half of the curve for which $x \geq 0$. As $t$ increases, $x$ decreases, and so the curve is traversed from right to left.

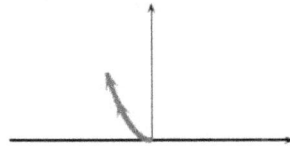

**FIGURE 1.6.7** $x = -\sqrt{t},\, y = t, t \in [0; +\infty[$.

4. If $x = \cos t$ and $y = \cos^2 t = \left(\cos t\right)^2 = x^2$. Both $x$ and $y$ are periodic with period $2\pi$, and so this parameterization only agrees with the curve $y = x^2$ when $-1 \leq x \leq 1$. For $t \in [0; \pi]$, the cosine decreases from 1 to $-1$ and so the curve is traversed from right to left in this interval.

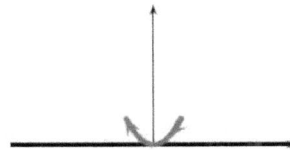

**FIGURE 1.6.8** $x = \cos t,\, y = \cos^2 t,\ t \in [0; \pi]$.

The identities

$$\cos^2 \theta + \sin^2 = 1,\ \ \tan^2 \theta - \sec^2 \theta = 1,\ \ \cosh^2 \theta - \sinh^2 \theta = 1,$$ are often useful when parametrizing quadratic curves.

## EXAMPLE 1.6.3

Give two distinct parameterizations of the ellipse $\dfrac{(x-1)^2}{4}+\dfrac{(y+2)^2}{9}=1$.

1. The first parameterization must satisfy that as $t$ traverses the values in the interval $[0;2\pi]$, one starts at the point $(3, -2)$, traverses the ellipse once counterclockwise, finishing at $(3, -2)$.

2. The second parameterization must satisfy that as $t$ traverses the interval $[0; 1]$, one starts at the point $(3, -2)$, traverses the ellipse twice clockwise, and returns to $(3, -2)$.

***Solution:***

What formula do we know where a sum of two squares equals 1? We use a trigonometric substitution, a sort of "polar coordinates." Observe that for $t \in [0;2\pi]$, the point $(\cos t, \sin t)$ traverses the unit circle once, starting at $(1, 0)$ and ending there. Put

$$\frac{x-1}{2} = \cos t \Rightarrow x = 1 + 2\cos t,$$

and

$$\frac{y+2}{3} = \sin t \Rightarrow y = -2 + 3\sin t.$$

Then

$$x = 1 + 2\cos t, \quad y = -2 + 3\sin t, \quad t \in [0;2\pi]$$

is the desired first parameterization.

For the second parameterization, notice that as $t$ traverses the interval $[0; 1]$, $(\sin 4\pi t, \cos 4\pi t)$ traverses the unit interval twice, clockwise, but begins and ends at the point $(0, 1)$. To begin at the point $(1, 0)$ we must make a shift: $\left(\sin\left(4\pi t + \dfrac{\pi}{2}\right), \cos\left(4\pi t + \dfrac{\pi}{2}\right)\right)$ will start at $(1, 0)$ and travel clockwise twice, as $t$ traverses $[0; 1]$. Hence we may take

$$x = 1 + 2\sin\left(4\pi t + \frac{\pi}{2}\right), \ \ y = -2 + 3\cos\left(4\pi t + \frac{\pi}{2}\right), \ \ t \in [0;1]$$

as our parameterization.

Some classic curves can be described by mechanical means, as the curves drawn by a spirograph. We will consider one such curve.

### EXAMPLE 1.6.4

A *hypocycloid* is a curve traced out by a fixed point $P$ on a circle $C$ of radius $\rho$ as $C$ rolls on the inside of a circle with center at $O$ and radius $R$. If the initial position of $P$ is $\begin{pmatrix} R \\ 0 \end{pmatrix}$, and $\theta$ is the angle, measured counterclockwise, with a ray starting at $O$ and passing through the center of $C$, which makes the $x$-axis, show that a parameterization of the hypocycloid is

$$x = (R - \rho)\cos\theta + \rho\cos\left(\frac{(R - \rho)}{\rho}\right),$$

$$y = (R - \rho)\sin\theta + \rho\sin\left(\frac{(R - \rho)}{\rho}\right).$$

**Solution:**

Suppose that starting from $\theta = 0$, the centre $O'$ of the small circle moves counterclockwise inside the larger circle by an angle $\theta$, and the point $P = (x, y)$ moves clockwise an angle $\phi$. The arc length traveled by the center of the small circle is $(R - \rho)\theta$ radians. At the same time, the point $P$ has rotated $\rho\phi$ radians, and so $(R - \rho)\theta = \rho\phi$. See Fig. 1.6.9, where $O'B$ is parallel to the $x$-axis.

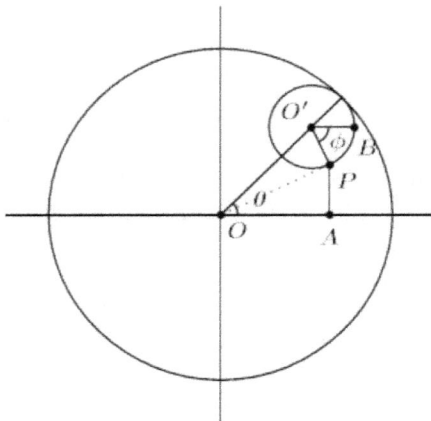

**FIGURE 1.6.9** Construction of the hypocycloid.

Let $A$ be the projection of $P$ on the $x$-axis. Then $\angle OAP = \angle OPO' = \dfrac{\pi}{2}$,

$\angle OO'P = \pi - \phi - \theta$, $\angle POA = \dfrac{\pi}{2} - \phi$, and $OP = (R - \rho)\sin(\pi - \phi - \theta)$.

$$x = (OP)\cos\angle POA = (R - \rho)\sin(\pi - \phi - \theta)\cos\left(\frac{\pi}{2} - \phi\right),$$

$$y = (R - \rho)\sin(\pi - \phi - \theta)\sin\left(\frac{\pi}{2} - \phi\right).$$

Now,

$$x = (R - \rho)\sin(\pi - \phi - \theta)\cos\left(\frac{\pi}{2} - \phi\right)$$

$$= (R - \rho)\sin(\phi + \theta)\sin\phi$$

$$= \frac{(R - \rho)}{2}(\cos\theta - \cos(2\phi + \theta))$$

$$= (R - \rho)\cos\theta - \frac{(R - \rho)}{2}(\cos\theta + \cos(2\phi + \theta))$$

$$= (R - \rho)\cos\theta - (R - \rho)(\cos(\theta + \phi)\cos\phi).$$

Also,

$$\cos(\theta + \phi) = -\cos(\pi - \theta - \phi) = -\frac{\rho}{OO'} = -\frac{\rho}{R - \rho}$$

and

$$\cos\phi = \cos\left(\frac{(R - \rho)\theta}{\rho}\right), \text{ and so}$$

$$x = (R - \rho)\cos\theta - (R - \rho)\left(\cos(\theta + \phi)\cos\phi\right) = (R - \rho)\cos\theta + \rho\cos\left(\frac{(R - \rho)\theta}{\rho}\right),$$

as required. The identity for y is proved similarly. Particular examples appear in Figs. 1.6.10 and 1.6.11.

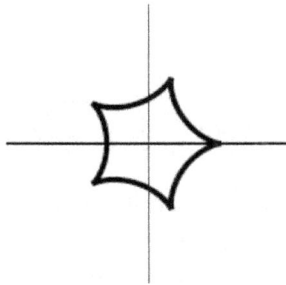

**FIGURE 1.6.10** Hypocycloid with $R = 5$, $\rho = 1$.

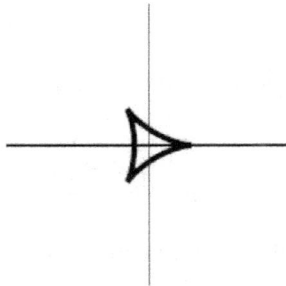

**FIGURE 1.6.11** Hypocycloid with $R = 3$, $\rho = 2$.

Given a curve $\Gamma$ how can we find its length? The idea, as seen in Fig. 1.6.12 is to consider the projections dx, dy at each point.

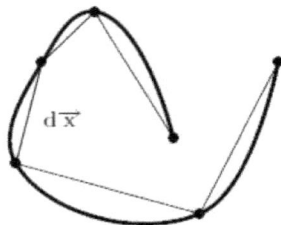

**FIGURE 1.6.12** Length of a curve.

The length of the vector

$$dr = \begin{bmatrix} dx \\ dy \end{bmatrix}$$

is

$$\|dr\| = \sqrt{(dx)^2 + (dy)^2}.$$

Hence the length of $\Gamma$ is given by

$$\int_\Gamma \|dr\| = \int_\Gamma \sqrt{(dx)^2 + (dy)^2}. \tag{1.13}$$

Similarly, suppose that $\Gamma$ is a simple closed curve in $\mathbb{R}^2$. How do we find the (oriented) area of the region it encloses? The idea, as seen in Fig. 1.6.13, borrowed from finding areas polygons, is to split the region into triangles, each area

$$\frac{1}{2}\det\begin{bmatrix} x & x+dx \\ y & y+dy \end{bmatrix} = \frac{1}{2}\det\begin{bmatrix} x & dx \\ y & dy \end{bmatrix} = \frac{1}{2}(x\,dy - y\,dx),$$

and to sum over the closed curve, obtaining a total oriented area of

$$\frac{1}{2}\oint_\Gamma \det\begin{bmatrix} x & dx \\ y & dy \end{bmatrix} = \frac{1}{2}\oint_\Gamma (x\,dy - y\,dx). \tag{1.14}$$

Hence $\oint_\Gamma$ denotes integration around the closed curve.

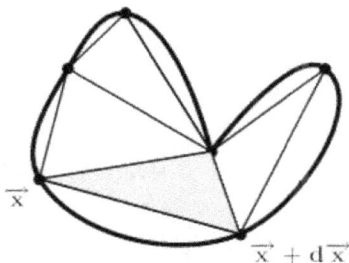

**FIGURE 1.6.13** Area enclosed by a simple closed curve.

## EXAMPLE 1.6.5

Let $(A, B) \in \mathbb{R}^2, A > 0, B > 0$. Find a parameterization of the ellipse

$$\Gamma : \left\{ (x, y) \in \mathbb{R}^2 : \frac{x^2}{A^2} + \frac{y^2}{B^2} = 1 \right\},$$

in Fig. 1.6.14. Furthermore, find an integral expression for the perimeter of this ellipse and find the area it encloses.

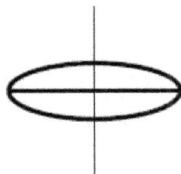

**FIGURE 1.6.14** Example 1.6.4.

### *Solution:*

Consider the parameterization $\Gamma : [0, 2\pi] \to \mathbb{R}^2$, with

$$\begin{bmatrix} x \\ y \end{bmatrix} = \begin{bmatrix} A \cos t \\ B \sin t \end{bmatrix}.$$

This is a parameterization of the ellipse, for

$$\frac{x^2}{A^2} + \frac{y^2}{B^2} = \frac{A^2 \cos^2 t}{A^2} + \frac{B^2 \sin^2 t}{B^2} = \cos^2 t + \sin^2 t = 1.$$

Notice that this parameterization goes around once the ellipse is counterclockwise.

The perimeter of the ellipse is given by

$$\int_{\Gamma} \|d\vec{r}\| = \int_0^{2\pi} \sqrt{A^2 \sin^2 t + B^2 \cos^2 t} \; dt$$

The above integral is an elliptic integral, and we do not have a closed form for it (in terms of the elementary functions studied in Calculus I). We will have better luck with the area of the ellipse, which is given by

$$\frac{1}{2}\oint_{\Gamma}(x dy - y dx) = \frac{1}{2}\oint_{\Gamma}\left(A\cos t \; d(B\sin t) - B\sin t \; d(A\cos t)\right)$$
$$= \frac{1}{2}\int_0^{2\pi}\left(AB\cos^2 t + AB\sin^2 t\right) dt$$
$$= \frac{1}{2}\int_0^{2\pi} AB \; dt$$
$$= \pi AB.$$

## EXAMPLE 1.6.6

Find a parametric representation for the astroid

$$\Gamma : \left\{(x,y) \in \mathbb{R}^2 : x^{2/3} + y^{2/3} = 1\right\},$$

in Fig. 1.6.15.

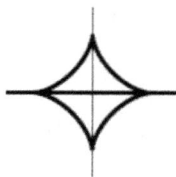

**FIGURE 1.6.15** Example 1.6.5.

Find the perimeter of the astroid and the area it encloses.

**Solution:**

Take

$$\begin{bmatrix} x \\ y \end{bmatrix} = \begin{bmatrix} \cos^3 t \\ \sin^3 t \end{bmatrix}$$

with $t \in [0; 2\pi]$. Then

$$x^{2/3} + y^{2/3} = \cos^2 t + \sin^2 t = 1$$

The perimeter of the astroid is

$$\int_\Gamma \|dr\| = \int_0^{2\pi} \sqrt{9\cos^4 t \sin^2 t + 9\sin^4 t \cos^2 t} \ dt$$

$$= \int_0^{2\pi} 3|\sin t \cos t| \ dt$$

$$= \frac{3}{2} \int_0^{2\pi} |\sin 2t| \ dt$$

$$= 6\int_0^{\pi/2} \sin 2t \ dt$$

$$= 6.$$

The area of the astroid is given by

$$\frac{1}{2} \oint_\Gamma (x\,dy - y\,dx) = \frac{1}{2} \oint \left( \cos^3 t \ d(\sin^3 t) - \sin^3 t \ d(\cos^3 t) \right)$$

$$= \frac{1}{2} \int_0^{2\pi} \left( 3\cos^4 t \sin^2 t + 3\sin^4 t \cos^2 t \right) dt$$

$$= \frac{3}{2} \int_0^{2\pi} \left( \sin t \cos t \right)^2 dt$$

$$= \frac{3}{8} \int_0^{2\pi} \left( \sin 2t \right)^2 dt$$

$$= \frac{3}{16} \int_0^{2\pi} \left( 1 - \cos 4t \right) dt$$

$$= \frac{3\pi}{8}.$$

We can use $Maple^{TM}$ to calculate the above integrals. For example, if $(x,y) = (\cos^3 t, \sin^3 t)$, to compute the arc length we use the path integral command and to compute the area, we use the line integral command with the vector field $\begin{bmatrix} -y/2 \\ x/2 \end{bmatrix}$.

> $with(Student[VectorCalculus])$ :
> $PathInt(1, [x, y] = Path(\langle(\cos(t))^3, (\sin(t))^3\rangle, 0..2*Pi));$

$$6$$

> $LineInt(VectorField(\langle -y/2, x/2 \rangle), Path(\langle(\cos(t))^3, (\sin(t))^3\rangle, 0..2*Pi));$

$$\frac{3}{8}\pi$$

We include here for convenience, some $Maple^{TM}$ commands to compute various arc lengths and areas.

## EXAMPLE 1.6.7

To obtain the arc length of the path in Fig. 1.6.16, we type

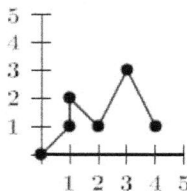

**FIGURE 1.6.16** Line path.

> $with(Student[VectorCalculus])$ :
> $PathInt(1, [x, y] = LineSegments(\langle 0, 0\rangle, \langle 1, 1\rangle, \langle 1, 2\rangle, \langle 2, 1\rangle, \langle 3, 3\rangle, \langle 4, 1\rangle));$

$$1 + 2\sqrt{2} + 2\sqrt{5}$$

To obtain the arc length of the path in Fig. 1.6.17, we type

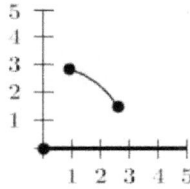

**FIGURE 1.6.17** Arc of the circle of radius 3, angle $\dfrac{\pi}{6} \leq \theta \leq \dfrac{\pi}{5}$ .

```
> with( Student[ VectorCalculus ]) :
> PathInt( 1, [x, y] = Arc( Circle( ⟨0, 0⟩, 3), Pi/6, Pi/5 ) );
```

$$\frac{1}{10}\pi$$

To obtain the area inside the curve in Fig. 1.6.18, we type

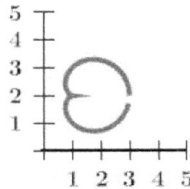

**FIGURE 1.6.18** Area inside the curve $x = 1 + (1 + \cos t)(\cos t)$, $y = 2 + (1 + \cos t)(\sin t)$ **.**

```
> LineInt( VectorField( ⟨−y/2, x/2⟩ ), Path( ⟨(1 + cos(t)) * (cos(t))
     + 1, (1 + cos(t)) * (sin(t)) + 2⟩, 0 ..2 * Pi) );
```

$$\frac{3}{2}\pi$$

## EXERCISES 1.6

**1.6.1**  A curve is represented parametrically by $x(t) = t^3 - 2t$, $y(t) = t^3 + 2t$ .

Find its Cartesian equation.

**1.6.2**  Give an implicit Cartesian equation for the parametric representation

$$x(t) = \frac{t^2}{1 + t^5}, \ y = \frac{t^3}{1 + t^5}.$$

**1.6.3**   Let a, b, c, and d be strictly positive real constants. In each case give an implicit Cartesian equation for the parametric representation and describe the trace of the parametric curve.

1.  $x = at + b, y = ct + d$

2.  $x = \cos t, y = 0$

3.  $x = a \cosh t, y = b \sinh t$

4.  $x = a \sec t, y = b \tan t, \ t \in \left] -\dfrac{\pi}{2}; \dfrac{\pi}{2} \right[$

**1.6.4**   Parameterize the curve $y = \log \cos x$ for $0 \le x \le \dfrac{\pi}{3}$. Then find its arc length.

**1.6.5**   Describe the trace of the parametric curve

$$\begin{bmatrix} x \\ y \end{bmatrix} = \begin{bmatrix} \sin t \\ 2\sin t + 1 \end{bmatrix}, \ t \in [0; 4\pi].$$

**1.6.6**   Consider the plane curve defined implicitly by $\sqrt{x} + \sqrt{y} = 1$. Give a suitable parameterization of this curve, and find its length. The graph of the curve appears in Fig. 1.6.19.

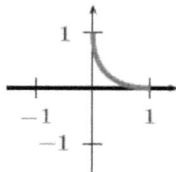

FIGURE 1.6.19 Exercise 1.6.6.

**1.6.7**   Consider the graph given parametrically by $x(t) = t^3 + 1$, $y(t) = 1 - t^2$. Find the area under the graph, over the $x$-axis, and between the lines $x = 1$ and $x = 2$.

**1.6.8**   Find the arc length of the curve given parametrically by $x(t) = 3t^2$, $y(t) = 2t^3$ for $0 \le x \le 1$.

**1.6.9**   Let $C$ be the curve in $\mathbb{R}^2$ defined by

$$x(t) = \frac{t^2}{2}, \ y(t) = \frac{(2t+1)^{3/2}}{3}, \ t \in \left[-\frac{1}{2}; +\frac{1}{2}\right].$$

Find the length of this curve.

**1.6.10** Find the area enclosed by the curve
$x(t) = \sin^3 t, \ y(t) = (\cos t)(1 + \sin^2 t)$. The curve appears in
Fig. 1.6.20.

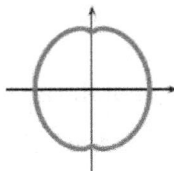

**FIGURE 1.6.20** Exercise 1.6.10.

**1.6.11** Let $C$ be the curve in $\mathbb{R}^2$ defined by

$$x(t) = \frac{3t}{1+t^3}, \ y(t) = \frac{3t^2}{1+t^3}, \ t \in \mathbb{R} \setminus \{-1\},$$

which you may see in Fig. 1.6.21. Find the area enclosed by the
loop of this curve.

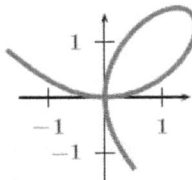

**FIGURE 1.6.21** Exercise 1.6.11.

**1.6.12** Let $P$ be a point at a distance $d$ from the center of a circle of radius
$\rho$. The curve traced out $P$ as the circle rolls along a straight line,
without slipping, is called a *cycloid*, as shown in Fig. 1.6.22. Find a
parameterization of the cycloid.

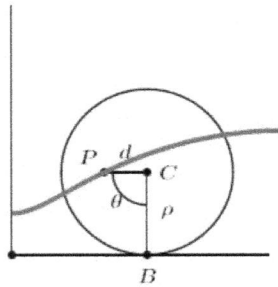

**FIGURE 1.6.22** Exercise 1.6.12, Cycloid.

**1.6.13** Find the arc length of the arc of the cycloid
$$x = \rho\left(t - \sin t\right), \; y = \rho\left(1 - \cos t\right), \; t \in [0; 2\pi].$$

**1.6.14** Find the length of the parametric curve given by
$$x = e^t \cos t, \; y = e^t \sin t, \; t \in [0; \pi].$$

**1.6.15** A shell strikes an airplane flying at a height h above the ground. It is known that the shell was fired from a gun on the ground with a muzzle velocity of magnitude V, but the position of the gun and its angle of elevation are both unknown. Deduce that the gun is situated within a circle whose center lies directly below the airplane and whose radius is

$$\frac{V\sqrt{V^2 - 2gh}}{g}.$$

**1.6.16** The parabola $y^2 = -4px$ rolls without slipping around the parabola $y^2 = 4px$ Find the equation of the locus of the vertex of the rolling parabola.

## 1.7 VECTORS IN SPACE

**Definition 1.7.1** The three-dimensional Cartesian Space is defined and denoted by $\mathbb{R}^3 = \left\{ r = \left(x, y, z\right) : x \in \mathbb{R}, y \in \mathbb{R}, z \in \mathbb{R} \right\}$

In Fig. 1.7.1, we have pictured the point (2, 1, 3).

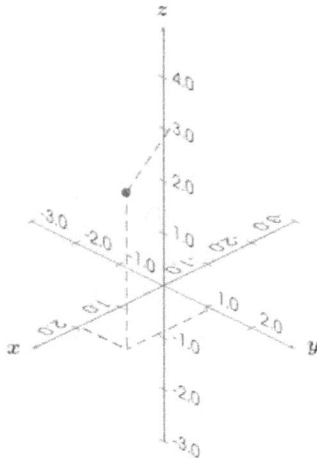

**FIGURE 1.7.1** A point in space.

**Definition 1.7.2** A system of unit vectors $\vec{i}$, $\vec{j}$, $\vec{k}$ is *right-handed* if the shortest-route rotation which brings $\vec{i}$ to coincide with $\vec{j}$ is performed in a counter-clockwise manner. It is *left-handed* if the rotation is done in a clock-wise manner.

Having oriented the $z$-axis upwards, we have a choice for the orientation of the $x$ and $y$-axis. Figs. 1.7.2 and 1.7.3 show the right-handed system and the right-hand, respectively. While Fig. 1.7.4 shows the left-handed system. Here, we adopt the convention right-*handed coordinate system*, as in Fig. 1.7.2.

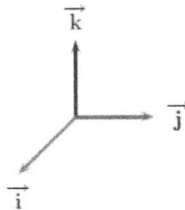

**FIGURE 1.7.2** Right-handed system.

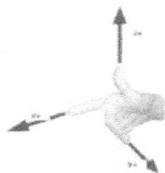

**FIGURE 1.7.3** Right-hand.

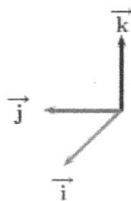

**FIGURE 1.7.4** Left-handed system.

Let us explain. In analogy to $\mathbb{R}^2$ we put

$$\vec{i} = \begin{bmatrix} 1 \\ 0 \\ 0 \end{bmatrix}, \quad \vec{j} = \begin{bmatrix} 0 \\ 1 \\ 0 \end{bmatrix}, \quad \vec{k} = \begin{bmatrix} 0 \\ 0 \\ 1 \end{bmatrix},$$

and observe that

$$r = (x, y, z) = x\vec{i} + y\vec{j} + z\vec{k}.$$

Most of what we did in $\mathbb{R}^2$ transfers to $\mathbb{R}^3$ without major complications.

**NOTE**   *The $\vec{i}$ is the unit vector in the x-axis direction, $\vec{j}$ is the unit vector in the y-axis direction, and $\vec{k}$ is the unit vector in the z-axis direction.*

The rectangular coordinate form of a vector $\vec{a}$ in $\mathbb{R}^3$ is $\vec{a} = \begin{bmatrix} a_1 \\ a_2 \\ a_3 \end{bmatrix}$, then the

unit vector form of vector $\vec{a}$ is $\vec{a} = a_1\vec{i} + a_2\vec{j} + a_3\vec{k}$. We also have

$$\left\|\vec{i}\right\| = \left\|\vec{j}\right\| = \left\|\vec{k}\right\| = 1 \text{ and } \vec{i} \cdot \vec{j} = \vec{j} \cdot \vec{k} = \vec{k} \cdot \vec{i} = 0.$$

To calculate the magnitude of vectors in $\mathbb{R}^3$, we need a distance formula for points in Euclidean space.

### Theorem 1.7.1

The distance $d\left(\vec{a}, \vec{b}\right)$ between vectors $\vec{a} = \begin{bmatrix} a_1 \\ a_2 \\ a_3 \end{bmatrix}$ and $\vec{b} = \begin{bmatrix} b_1 \\ b_2 \\ b_3 \end{bmatrix}$ in $\mathbb{R}^3$ is

$$d\left(\vec{a}, \vec{b}\right) = \left\|\vec{a} - \vec{b}\right\| = \sqrt{\left(a_1 - b_1\right)^2 + \left(a_2 - b_2\right)^2 + \left(a_3 - b_3\right)^2}$$

**Definition 1.7.3**   The dot (inner) product of two vectors $\vec{a}$ and $\vec{b}$ in $\mathbb{R}^3$ is

$$\vec{a} \cdot \vec{b} = a_1 b_1 + a_2 b_2 + a_3 b_3.$$

The norm of a vector $\vec{a} = \begin{bmatrix} a_1 \\ a_2 \\ a_3 \end{bmatrix}$ in $\mathbb{R}^3$ is

$$\left\|\vec{a}\right\| = \sqrt{\vec{a} \cdot \vec{a}} = \sqrt{\left(a_1\right)^2 + \left(a_2\right)^2 + \left(a_3\right)^2}.$$

Just as in $\mathbb{R}^2$, the dot product satisfies $\vec{a} \cdot \vec{b} = \left\|\vec{a}\right\|\left\|\vec{b}\right\|\cos\theta$, where $\theta \in [0; \pi]$ is the convex angle between the two vectors.

**NOTE**   *Two vectors $\vec{a}$ and $\vec{b}$ in $\mathbb{R}^3$ are orthogonal (perpendicular) if $\vec{a} \cdot \vec{b} = 0$.*

### Theorem 1.7.1 (Pythagorean Theorem in $\mathbb{R}^3$)

If $\vec{a}$ and $\vec{b}$ are orthogonal vectors in $\mathbb{R}^3$, then $\left\|\vec{a} + \vec{b}\right\|^2 = \left\|\vec{a}\right\|^2 + \left\|\vec{b}\right\|^2$.

### Proof:

Since $\vec{a}$ and $\vec{b}$ are orthogonal, $\vec{a} \cdot \vec{b} = 0$

$$\left\|\vec{a}+\vec{b}\right\|^2 = \left(\vec{a}+\vec{b}\right)\bullet\left(\vec{a}+\vec{b}\right)$$
$$= \vec{a}\bullet\vec{a} + \vec{a}\bullet\vec{b} + \vec{b}\bullet\vec{a} + \vec{b}\bullet\vec{b}$$
$$= \left\|\vec{a}\right\|^2 + \left\|\vec{b}\right\|^2$$

The Cauchy–Schwarz–Bunyakovsky Inequality takes the form

$$\left|\vec{a}\bullet\vec{b}\right| \le \left\|\vec{a}\right\|\left\|\vec{b}\right\| \Rightarrow \left|a_1 b_1 + a_2 b_2 + a_3 b_3\right| \le \left(a_1^{\ 2} + a_2^{\ 2} + a_3^{\ 2}\right)^{1/2}\left(b_1^{\ 2} + b_2^{\ 2} + b_3^{\ 2}\right)^{1/2},$$

equality holding if and only if the vectors are parallel.

### EXAMPLE 1.7.1

Find the norm of the vector $\vec{a} = \begin{bmatrix} 8 \\ 5 \\ -2 \end{bmatrix}$ in $\mathbb{R}^3$.

***Solution:***

$$\left\|\vec{a}\right\| = \sqrt{(8)^2 + (5)^2 + (-2)^2} = \sqrt{64 + 25 + 4} = \sqrt{93}.$$

### EXAMPLE 1.7.2

Normalize the vector $\vec{a} = \begin{bmatrix} \dfrac{2}{3} \\ \dfrac{1}{2} \\ -\dfrac{1}{4} \end{bmatrix}$ and generate a unit vector.

***Solution:***

$$\left\|\vec{a}\right\| = \begin{bmatrix} \dfrac{2}{3} \\ \dfrac{1}{2} \\ -\dfrac{1}{4} \end{bmatrix} = \sqrt{\left(\dfrac{2}{3}\right)^2 + \left(\dfrac{1}{2}\right)^2 + \left(-\dfrac{1}{4}\right)^2} = \sqrt{\left(\dfrac{4}{9}\right) + \left(\dfrac{1}{4}\right) + \left(\dfrac{1}{16}\right)} = \dfrac{\sqrt{109}}{12}$$

The normalized vector of $\vec{a}$ is $\vec{u} = \dfrac{\vec{a}}{\|\vec{a}\|} = \begin{bmatrix} \dfrac{8}{\sqrt{109}} \\ \dfrac{6}{\sqrt{109}} \\ -\dfrac{3}{\sqrt{109}} \end{bmatrix}$.

The normalized vector $\vec{u}$ is a unit vector since $\|\vec{u}\| = 1$.

## EXAMPLE 1.7.3

Let $\mathbb{R}^3$ have the Euclidean dot product. For which values of $a$ are $\vec{a} = \begin{bmatrix} a \\ 5 \\ 3 \end{bmatrix}$

and $\vec{b} = \begin{bmatrix} a \\ a \\ 2 \end{bmatrix}$ orthogonal.

### Solution:

$$\vec{a} \bullet \vec{b} = 0$$

$$\begin{bmatrix} a \\ 5 \\ 3 \end{bmatrix} \bullet \begin{bmatrix} a \\ a \\ 2 \end{bmatrix} = a^2 + 5a + 6 = 0$$

$$(a + 3)(a + 2) = 0$$

$$a = -3 \text{ or } -2$$

## EXAMPLE 1.7.4

Let $x, y, z$ positive real numbers such that $x^2 + 4y^2 + 9z^2 = 27$. Maximize $x + y + z$.

### Solution:

Since $x, y, z$ are positive, $|x + y + z| = x + y + z$. By Cauchy's Inequality,

$$\left| x+y+z \right| = \left| x + 2y\left(\frac{1}{2}\right) + 3z\left(\frac{1}{3}\right) \right| \le \left( x^2 + 4y^2 + 9z^2 \right)^{1/2} \left( 1 + \frac{1}{4} + \frac{1}{9} \right)^{1/2} = \sqrt{27}\left(\frac{7}{6}\right) = \frac{7\sqrt{3}}{2}.$$

Equality occurs if and only if

$$\begin{bmatrix} x \\ 2y \\ 3z \end{bmatrix} = \lambda \begin{bmatrix} 1 \\ 1/2 \\ 1/3 \end{bmatrix} \Rightarrow x = \lambda, y = \frac{\lambda}{4}, z = \frac{\lambda}{9} \Rightarrow \lambda^2 + \frac{\lambda^2}{4} + \frac{\lambda^2}{9} = 27 \Rightarrow \lambda = \pm\frac{18\sqrt{3}}{7}.$$

Therefore for a maximum, we take

$$x = \frac{18\sqrt{3}}{7}, y = \frac{9\sqrt{3}}{14}, z = \frac{2\sqrt{3}}{7}.$$

**Definition 1.7.4**   Let a be a point in $\mathbb{R}^3$ and let $\vec{v} \ne \vec{0}$ a vector in $\mathbb{R}^3$. The *parametric line passing* through a in the direction of $\vec{v}$ is the set $\left\{ r \in \mathbb{R}^3 : r = a + t\vec{v} \right\}$.

**EXAMPLE 1.7.5**

Find the parametric equation of the line passing through $\begin{pmatrix} 1 \\ 2 \\ 3 \end{pmatrix}$ and $\begin{pmatrix} -2 \\ -1 \\ 0 \end{pmatrix}$.

*Solution:*

The line follows the direction $\begin{bmatrix} 1-(-2) \\ 2-(-1) \\ 3-0 \end{bmatrix} = \begin{bmatrix} 3 \\ 3 \\ 3 \end{bmatrix}$.

The desired equation is $\begin{pmatrix} x \\ y \\ z \end{pmatrix} = \begin{pmatrix} 1 \\ 2 \\ 3 \end{pmatrix} + t\begin{bmatrix} 3 \\ 3 \\ 3 \end{bmatrix}$.

NOTE   *Given two lines in space, one of the following three situations might arise: (i) the lines intersect at a point, (ii) the lines are parallel, and (iii) the lines are skew (non-parallel, one over the other, without intersecting, lying on different planes). See Fig. 1.7.5.*

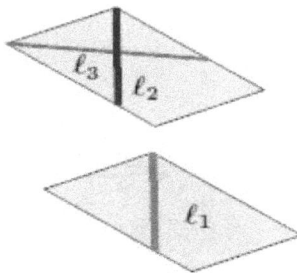

**FIGURE 1.7.5** $\ell_1 \| \ell_2$ . $\ell_1$ and $\ell_3$ are skew.

Consider now two non-zero vectors $\vec{a}$ and $\vec{b}$ in $\mathbb{R}^3$. If $\vec{a} \| \vec{b}$, then the set

$$\left\{ s\vec{a} + t\vec{b} : s \in \mathbb{R}, t \in \mathbb{R} \right\} = \left\{ \lambda \vec{a} : \lambda \in \mathbb{R} \right\},$$

which is a line through the origin. Suppose now that $\vec{a}$ and $\vec{b}$ are not parallel. We saw in the preceding sections that if the vectors were on the plane, they would span the whole plane $\mathbb{R}^2$. In the case at hand, the vectors are in space, they still span a plane, passing through the origin. Thus

$$\left\{ s\vec{a} + t\vec{b} : s \in \mathbb{R}, t \in \mathbb{R}, \vec{a} \nparallel \vec{b} \right\}$$

is a plane passing through the origin. We will say, abusing language that two vectors are *coplanar* if there exist bi-point representatives of the vector that lie on the same plane. We will say, again abusing language, that a vector is *parallel to a specific plane* or that it *lies on a specific plane* if there exists a bi-point representative of the vector that lies on the particular plane. All the above gives the following result.

**Theorem 1.7.2**: Let $\vec{v}$, $\vec{w}$ in $\mathbb{R}^3$ be non-parallel vectors. Then every vector $\vec{u}$ of the form $\vec{u} = a\vec{v} + b\vec{w}$, where $a, b$ are arbitrary scalars, is coplanar with both $\vec{v}$ and $\vec{w}$. Conversely, any vector $\vec{t}$ coplanar with both $\vec{v}$ and $\vec{w}$ can be uniquely expressed in the form $\vec{t} = p\vec{v} + q\vec{w}$. See Fig. 1.7.5.

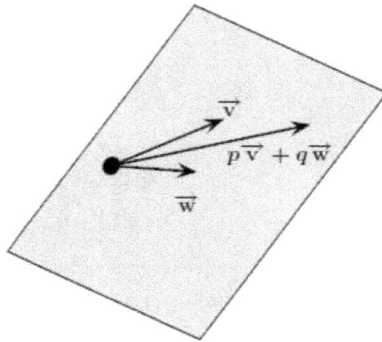

**FIGURE 1.7.6** Theorem 1.7.1.

From the above theorem, if a vector $\vec{a}$ is not a linear combination of two other vectors $\vec{b}$, $\vec{c}$, then linear combinations of these three vectors may lie outside the plane containing $\vec{b}$, $\vec{c}$. This prompts the following theorem.

**Theorem 1.7.3:** Three vectors $\vec{a}$, $\vec{b}$, $\vec{c}$ in $\mathbb{R}^3$ said to be *linearly independent* if

$$\alpha\,\vec{a} + \beta\,\vec{b} + \gamma\,\vec{c} = \vec{0} \Rightarrow \alpha = \beta = \gamma = 0.$$

Any vector in $\mathbb{R}^3$ can be written as a linear combination of three linearly independent vectors in $\mathbb{R}^3$.

A plane is determined by three non-collinear points. Suppose that a, b, and c are non-collinear points on the same plane and that $r = \begin{pmatrix} x \\ y \\ z \end{pmatrix}$ is another arbitrary point on this plane. Since a, b, and c are non-collinear, $\overrightarrow{ab}$ and $\overrightarrow{ac}$, which are coplanar, are non-parallel. Since $\overrightarrow{ax}$ also lies on the plane, we have by theorem 1, that there exist real numbers $p$, $q$ with

$$\overrightarrow{ar} = p\overrightarrow{ab} + q\overrightarrow{ac}.$$

By Chasles's rule,

$$\vec{r} = \vec{a} + p\left(\vec{b} - \vec{a}\right) + q\left(\vec{c} - \vec{a}\right),$$

is the equation of a plane containing the three non-collinear points a, b, and c, where $\vec{a}$, $\vec{b}$, and $\vec{c}$ are the position vectors of these points. Thus we have the following theorem.

**Theorem 1.7.4:** Let $\vec{u}$ and $\vec{v}$ be linearly independent vectors. The *parametric equation* of a plane containing the point a, and parallel to the vectors $\vec{u}$ and $\vec{v}$ is given by

$$\vec{r} - \vec{a} = p\vec{u} + q\vec{v}.$$

Component wise this takes the for

$$x - a_1 = pu_1 + qv_1,$$
$$y - a_2 = pu_2 + qv_2,$$
$$z - a_3 = pu_3 + qv_3.$$

Multiplying the first equation by $u_2 v_3 - u_3 v_2$, the second by $u_3 v_1 - u_1 v_3$, and the third by $u_1 v_2 - u_2 v_1$, we obtain,

$$\left(u_2 v_3 - u_3 v_2\right)\left(x - a_1\right) = \left(u_2 v_3 - u_3 v_2\right)\left(pu_1 + qv_1\right),$$
$$\left(u_3 v_1 - u_1 v_3\right)\left(y - a_2\right) = \left(u_3 v_1 - u_1 v_3\right)\left(pu_2 + qv_2\right),$$
$$\left(u_1 v_2 - u_2 v_1\right)\left(z - a_3\right) = \left(u_1 v_2 - u_2 v_1\right)\left(pu_3 + qv_3\right).$$

Adding gives,

$$\left(u_2 v_3 - u_3 v_2\right)\left(x - a_1\right) + \left(u_3 v_1 - u_1 v_3\right)\left(y - a_2\right) + \left(u_1 v_2 - u_2 v_1\right)\left(z - a_3\right) = 0.$$

Put

$$a = u_2 v_3 - u_3 v_2, \; b = u_3 v_1 - u_1 v_3, \; c = u_1 v_2 - u_2 v_1,$$

and

$$d = a_1 \left(u_2 v_3 - u_3 v_2\right) + a_2 \left(u_3 v_1 - u_1 v_3\right) + a_3 \left(u_1 v_2 - u_2 v_1\right).$$

Since $\vec{v}$ is linearly independent $\vec{u}$, not all of $a, b, c$ are zero. This gives the following theorem.

**Theorem 1.7.5:** The equation of the plane in space can be written in the form

$$ax + by + cz = d,$$

which is the *Cartesian equation* of the plane. Here $a^2 + b^2 + c^2 \neq 0$, that is, at least one of the coefficients is non-zero. Moreover, the vector $\vec{n} = \begin{bmatrix} a \\ b \\ c \end{bmatrix}$ is normal to the plane with the Cartesian equation $ax + by + cz = d$. See Fig. 1.7.7.

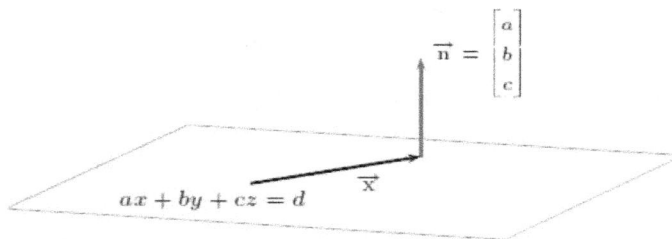

**FIGURE 1.7.7** Theorem 1.7.4.

**Proof:**

We have already proved the first statement. For the second statement, observe that if $\vec{u}$ and $\vec{v}$ are non-parallel vectors and $\vec{r} - \vec{a} = p\vec{u} + q\vec{v}$ is the equation of the plane containing the point a and parallel to the vectors $\vec{u}$ and $\vec{v}$, then if $\vec{n}$ is simultaneously perpendicular to $\vec{u}$ and $\vec{v}$ then $(\vec{r} - \vec{a}) \cdot \vec{n} = 0$ for $\vec{u} \cdot \vec{n} = 0 = \vec{v} \cdot \vec{n}$. Now, since at least one of $a, b, c$ is non-zero, we may assume $a \neq 0$. The argument is similar if one of the other letters is non-zero and $a = 0$. In this case, we can see that

$$x = \frac{d}{a} - \frac{b}{a}y - \frac{c}{a}z.$$

Put $y = s$ and $z = t$. Then

$$\begin{pmatrix} x - \dfrac{d}{a} \\ y \\ z \end{pmatrix} = s\begin{bmatrix} -\dfrac{b}{a} \\ 1 \\ 0 \end{bmatrix} + t\begin{bmatrix} -\dfrac{c}{a} \\ 0 \\ 1 \end{bmatrix},$$

is a parametric equation for the plane. We have

$$a\left(-\frac{b}{a}\right)+b(1)+c(0)=0, \quad a\left(-\frac{c}{a}\right)+b(0)+c(1)=0,$$

and so $\begin{bmatrix} a \\ b \\ c \end{bmatrix}$ is simultaneously perpendicular to $\begin{bmatrix} -\dfrac{b}{a} \\ 1 \\ 0 \end{bmatrix}$ and $\begin{bmatrix} -\dfrac{c}{a} \\ 0 \\ 1 \end{bmatrix}$, proving the

second statement.

## EXAMPLE 1.7.5

The equation of the plane passing through the point $\begin{pmatrix} 1 \\ -1 \\ 2 \end{pmatrix}$ and normal to

vector $\begin{bmatrix} -3 \\ 2 \\ 4 \end{bmatrix}$ is

$$-3(x-1)+2(y+1)+4(z-2)=0 \implies -3x+2y+4z=3.$$

## EXAMPLE 1.7.6

Find both the parametric equation and the Cartesian equation of the plane

parallel to the vectors $\begin{bmatrix} 1 \\ 1 \\ 1 \end{bmatrix}$ and $\begin{bmatrix} 1 \\ 1 \\ 0 \end{bmatrix}$, and passing through the point $\begin{pmatrix} 0 \\ -1 \\ 2 \end{pmatrix}$.

*Solution:*

The desired parametric equation is

$$\begin{pmatrix} x \\ y+1 \\ z-2 \end{pmatrix} = s\begin{bmatrix} 1 \\ 1 \\ 1 \end{bmatrix} + t\begin{bmatrix} 1 \\ 1 \\ 0 \end{bmatrix}.$$

This gives

$$s = z - 2, \quad t = y + 1 - s = y + 1 - z + 2 = y - z + 3$$

and

$$x = s + t = z - 2 + y - z + 3 = y + 1.$$

Hence the Cartesian equation is $x - y = 1$.

**Definition 1.7.5** If $\vec{n}$ is perpendicular to plane $\Pi_1$ and $\vec{n'}$ is perpendicular to plane $\Pi_2$, the *angle between the two planes* is the angle between the two vectors $\vec{n}$ and $\vec{n'}$.

## EXAMPLE 1.7.7

1. Draw the intersection of the plane $z = 1 - x$ with the first octant.
2. Draw the intersection of the plane $z = 1 - y$ with the first octant.
3. Find the angle between the planes $z = 1 - x$ and $z = 1 - y$.
4. Draw the solid S which results from the intersection of the planes $z = 1 - x$ and $z = 1 - y$ with the first octant.
5. Find the volume of the solid S.

*Solution:*

1. This appears in Fig. 1.7.8.

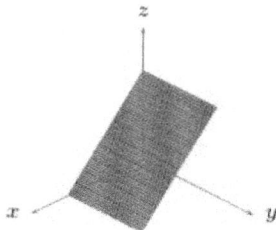

*FIGURE 1.7.8* The plane $z = 1 - x$.

**2.** This appears in Fig. 1.7.9.

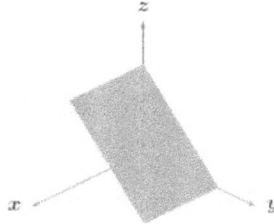

**FIGURE 1.7.9** The plane $z = 1 - y$.

**3.** The vector $\begin{bmatrix} 1 \\ 0 \\ 1 \end{bmatrix}$ is normal to the plane $x + z = 1$, and the vector $\begin{bmatrix} 0 \\ 1 \\ 1 \end{bmatrix}$ is nor-

mal to the plane $y + z = 1$. If $\theta$ is the angle between these two vectors, then

$$\cos\theta = \frac{1 \cdot 0 + 0 \cdot 1 + 1 \cdot 1}{\sqrt{1^2 + 1^2} \cdot \sqrt{1^2 + 1^2}} \Rightarrow \cos\theta = \frac{1}{2} \Rightarrow \theta = \frac{\pi}{3}.$$

**4.** This appears in Fig. 1.7.10.

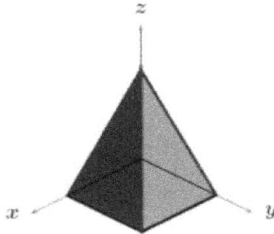

**FIGURE 1.7.10** Solid bounded by the planes $z = 1 - x$ and $z = 1 - y$ in the first octant.

**5.** The resulting solid is a pyramid with square base of area $A = 1 \cdot 1 = 1$. Recall that the volume of a pyramid is given by the formula $V = \dfrac{Ah}{3}$, where $A$ is area of the base of the pyramid and $h$ is its height. Now, the height of this pyramid is clearly 1, and hence the volume required is $\dfrac{1}{3}$.

## EXERCISES 1.7

**1.7.1**   Vectors $\vec{a}$, $\vec{b}$ satisfy $\|\vec{a}\| = 13, \|\vec{b}\| = 19, \|\vec{a} + \vec{b}\| = 24$. Find $\|\vec{a} - \vec{b}\|$.

**1.7.2**   Find the equation of the line passing through $\begin{pmatrix} 1 \\ 2 \\ 3 \end{pmatrix}$ in the direction of $\begin{bmatrix} -2 \\ -1 \\ 0 \end{bmatrix}$.

**1.7.3**   Find the value of $a$ when the vectors $\vec{a} = \begin{bmatrix} 6 \\ 1 \\ 5 \end{bmatrix}$ and $\vec{b} = \begin{bmatrix} 1 \\ 4 \\ k \end{bmatrix}$ orthogonal.

**1.7.4**   Find the Euclidean distance between $\vec{a} = \begin{bmatrix} -5 \\ 4 \\ 3 \end{bmatrix}$ and $\vec{b} = \begin{bmatrix} 2 \\ -1 \\ 6 \end{bmatrix}$.

**1.7.5**   Find the equation of plane containing the point $\begin{pmatrix} 1 \\ 1 \\ 1 \end{pmatrix}$ and perpendicular to the line $x = 1 + t, y = -2t, z = 1 - t$.

**1.7.6**   Find the equation of plane containing the point $\begin{pmatrix} 1 \\ -1 \\ -1 \end{pmatrix}$ and containing the line $x = 2y = 3z$.

**1.7.7**   **(Putnam Exam 1984)** Let $A$ be a solid $a \times b \times c$ rectangular brick in three dimensions, where $a > 0, b > 0, c > 0$. Let $B$ be the set of all points which are at distance at most 1 from some point of $A$ (in particular, $A \subseteq B$). Express the volume of $B$ as a polynomial in $a, b, c$.

**1.7.8** It is known that $\|\vec{a}\| = 3, \|\vec{b}\| = 4, \|\vec{c}\| = 5$ and that $\vec{a} + \vec{b} + \vec{c} = \vec{0}$. Find $\vec{a} \bullet \vec{b} + \vec{b} \bullet \vec{c} + \vec{c} \bullet \vec{a}$.

**1.7.9** Find the equation of the line perpendicular to the plane

$$ax + a^2 y + a^3 z = 0, a \neq 0 \text{ and passing through the point } \begin{pmatrix} 0 \\ 0 \\ 1 \end{pmatrix}.$$

**1.7.10** Find the equation of the plane perpendicular to the line

$$ax = by = cz, abc \neq 0 \text{ and passing through the point } \begin{pmatrix} 1 \\ 1 \\ 1 \end{pmatrix} \text{ in } \mathbb{R}^3.$$

**1.7.11** Find the (shortest) distance from the point $(1,2,3)$ to the plane $x - y + z = 1$.

**1.7.12** Determine whether the lines

$$L_1 : \begin{pmatrix} x \\ y \\ z \end{pmatrix} = \begin{pmatrix} 1 \\ 1 \\ 1 \end{pmatrix} + t \begin{bmatrix} 2 \\ 1 \\ 1 \end{bmatrix},$$

$$L_2 : \begin{pmatrix} x \\ y \\ z \end{pmatrix} = \begin{pmatrix} 0 \\ 0 \\ 1 \end{pmatrix} + t \begin{bmatrix} 2 \\ -1 \\ 1 \end{bmatrix},$$

Intersect. Find the angle between them.

**1.7.13** Let $a, b, c$ be arbitrary real numbers. Prove that

$$\left(a^2 + b^2 + c^2\right)^2 \leq 3\left(a^4 + b^4 + c^4\right).$$

**1.7.14** Let $a > 0, b > 0, c > 0$ be the lengths of the sides of $\triangle ABC$. (Vertex $A$ is opposite to the side measuring $a$, etc.) Recall that by Heron's Formula, the area of this triangle is

$$S(a,b,c) = \sqrt{s(s-a)(s-b)(s-c)}, \text{ where } s = \frac{a+b+c}{2} \text{ is the}$$

semiperimeter of the triangle. Prove that $f(a,b,c) = \dfrac{S(a,b,c)}{a^2 + b^2 + c^2}$ is maximized when $\triangle ABC$ is equivalent, and find this maximum.

**1.7.15** Find the Cartesian equation of the plane passing through $\begin{pmatrix} 1 \\ 0 \\ 0 \end{pmatrix}$, $\begin{pmatrix} 0 \\ 1 \\ 0 \end{pmatrix}$, and $\begin{pmatrix} 0 \\ 0 \\ 1 \end{pmatrix}$. Draw this plane and its intersection with the first octant. Find the volume of the tetrahedron with vertices at $\begin{pmatrix} 0 \\ 0 \\ 0 \end{pmatrix}$, $\begin{pmatrix} 1 \\ 0 \\ 0 \end{pmatrix}$, $\begin{pmatrix} 0 \\ 1 \\ 0 \end{pmatrix}$, and $\begin{pmatrix} 0 \\ 0 \\ 1 \end{pmatrix}$.

**1.7.16** Prove that there do not exist three unit vectors in $\mathbb{R}^3$ such that the angle between any two of them be $> \dfrac{2\pi}{3}$.

**1.7.17** Let $(\vec{r} - \vec{a}) \cdot \vec{n} = 0$ be a plane passing through the point a and perpendicular to vector $\vec{n}$. If b is not a point on the plane, then the distance from b to the plane is

$$\frac{\left| (\vec{a} - \vec{b}) \cdot \vec{n} \right|}{\left\| \vec{n} \right\|}.$$

**1.7.18** **(Putnam Exam 1980)** Let $S$ be solid in a three-dimensional space consisting of all points $\overrightarrow{AC'}$ satisfying the following system of six conditions:

$$x \geq 0, \ y \geq 0, \ z \geq 0,$$
$$x + y + z \leq 11,$$
$$2x + 4y + 3z \leq 36,$$
$$2x + 3z \leq 24.$$

Determine the number of vertices and the number of edges of $S$.

## 1.8 CROSS PRODUCT

We now define the standard cross product in $\mathbb{R}^3$ as a product satisfying the following properties.

**Definition 1.8.1**  Let $\vec{x}, \vec{y}, \vec{z}$ be vectors in $\mathbb{R}^3$, and let $a \in \mathbb{R}$ be a scalar. The cross product $(\times)$ is a closed binary operation satisfying

1. **Anti-commutativity**:

$$\vec{x} \times \vec{y} = -\left(\vec{y} \times \vec{x}\right)$$

2. **Bilinearity**:

$$\left(\vec{x} + \vec{z}\right) \times \vec{y} = \left(\vec{x} \times \vec{y}\right) + \left(\vec{z} \times \vec{y}\right), \text{ and } \vec{x} \times \left(\vec{z} + \vec{y}\right) = \left(\vec{x} \times \vec{z}\right) + \left(\vec{x} \times \vec{y}\right)$$

3. **Scalar homogeneity**:

$$\left(a\,\vec{x}\right) \times \vec{y} = \vec{x} \times \left(a\,\vec{y}\right) = a\left(\vec{x} \times \vec{y}\right)$$

4. **Cross product of a vector with itself is always zero**

$$\vec{x} \times \vec{x} = \vec{0}$$

5. **Right-hand Rule**:

$$\vec{i} \times \vec{j} = \vec{k}, \quad \vec{j} \times \vec{k} = \vec{i}, \quad \vec{k} \times \vec{i} = \vec{j}$$

It follows that the cross product is an operation that, given two non-parallel vectors on a plane, allows us to "get out" of that plane.

**NOTE**  *The cross product of vectors in $\mathbb{R}^3$ is not associative, since*

$$\vec{i} \times \left(\vec{i} \times \vec{j}\right) = \vec{i} \times \vec{k} = -\vec{j}$$

but

$$\left(\vec{i} \times \vec{i}\right) \times \vec{j} = \vec{0} \times \vec{j} = \vec{0}.$$

### EXAMPLE 1.8.1

Find $\begin{bmatrix} 1 \\ 0 \\ -3 \end{bmatrix} \times \begin{bmatrix} 0 \\ 1 \\ 2 \end{bmatrix}$.

### Solution:

We have

$$\left(\vec{i}-3\vec{k}\right)\times\left(\vec{j}+2\vec{k}\right)=\vec{i}\times\vec{j}+2\vec{i}\times\vec{k}-3\vec{k}\times\vec{j}-6\vec{k}\times\vec{k}$$
$$=\vec{k}-2\vec{j}+3\vec{i}+6\vec{0}$$
$$=3\vec{i}-2\vec{j}+\vec{k}.$$

Hence

$$\begin{bmatrix}1\\0\\-3\end{bmatrix}\times\begin{bmatrix}0\\1\\2\end{bmatrix}=\begin{bmatrix}3\\-2\\1\end{bmatrix}$$

Operating as in Example 1.8.1, we obtain

**Theorem 1.8.1:** Let $\vec{x}=\begin{bmatrix}x_1\\x_2\\x_3\end{bmatrix}$ and $\vec{y}=\begin{bmatrix}y_1\\y_2\\y_3\end{bmatrix}$ be vectors in $\mathbb{R}^3$. Then,

$$\vec{x}\times\vec{y}=\left(x_2y_3-x_3y_2\right)\vec{i}+\left(x_3y_1-x_1y_3\right)\vec{j}+\left(x_1y_2-x_2y_1\right)\vec{k}=\begin{bmatrix}\left(x_2y_3-x_3y_2\right)\\\left(x_3y_1-x_1y_3\right)\\\left(x_1y_2-x_2y_1\right)\end{bmatrix}\quad\text{or in}$$

determinant notation

$$\vec{x}\times\vec{y}=\begin{bmatrix}\begin{vmatrix}x_2&x_3\\y_2&y_3\end{vmatrix}\\-\begin{vmatrix}x_1&x_3\\y_1&y_3\end{vmatrix}\\\begin{vmatrix}x_1&x_2\\y_1&y_2\end{vmatrix}\end{bmatrix}$$

### Proof:

Since $\vec{i}\times\vec{i}=\vec{j}\times\vec{j}=\vec{k}\times\vec{k}=\vec{0}$, we only worry about the mixed products, obtaining,

$$\vec{x} \times \vec{y} = \left(x_1\vec{i} + x_2\vec{j} + x_3\vec{k}\right) \times \left(y_1\vec{i} + y_2\vec{j} + y_3\vec{k}\right)$$

$$= x_1y_2\vec{i} \times \vec{j} + x_1y_3\vec{i} \times \vec{k} + x_2y_1\vec{j} \times \vec{i} + x_2y_3\vec{j} \times \vec{k} + x_3y_1\vec{k} \times \vec{i} + x_3y_2\vec{k} \times \vec{j}$$

$$= \left(x_1y_2 - y_1x_2\right)\vec{i} \times \vec{j} + \left(x_2y_3 - x_3y_2\right)\vec{j} \times \vec{k} + \left(x_3y_1 - x_1y_3\right)\vec{k} \times \vec{i}$$

$$= \left(x_1y_2 - y_1x_2\right)\vec{k} + \left(x_2y_3 - x_3y_2\right)\vec{i} + \left(x_3y_1 - x_1y_3\right)\vec{j},$$

proving the theorem.

## EXAMPLE 1.8.2

Let $\vec{a} = \begin{bmatrix} 1 \\ 2 \\ -3 \end{bmatrix}$ and $\vec{b} = \begin{bmatrix} 5 \\ 0 \\ 1 \end{bmatrix}$, find $\vec{a} \times \vec{b}$.

## *Solution:*

We form the $2 \times 3$ matrix whose first row contains the components of $\vec{a}$ and whose second row contains the component of $\vec{b}$, that is,

$$\begin{bmatrix} a_1 & a_2 & a_3 \\ b_1 & b_2 & b_3 \end{bmatrix} \Rightarrow \begin{bmatrix} 1 & 2 & -3 \\ 5 & 0 & 1 \end{bmatrix}$$

$$\vec{a} \times \vec{b} = \begin{bmatrix} \begin{vmatrix} 2 & -3 \\ 0 & 1 \end{vmatrix} \\ -\begin{vmatrix} 1 & -3 \\ 5 & 1 \end{vmatrix} \\ \begin{vmatrix} 1 & 2 \\ 5 & 0 \end{vmatrix} \end{bmatrix} = \begin{bmatrix} 2 \\ -16 \\ -10 \end{bmatrix}.$$

Using the cross product, we may obtain a third vector simultaneously perpendicular to two other vectors in space.

**Theorem 1.8.2:** $\vec{x} \perp \left(\vec{x} \times \vec{y}\right)$ and $\vec{y} \perp \left(\vec{x} \times \vec{y}\right)$, that is, the cross product of two vectors is simultaneously perpendicular to both original vectors.

**Proof:**

We will only check the first assertion, the second verification is analogous.

$$\vec{x} \cdot \left(\vec{x} \times \vec{y}\right) = \left(x_1\vec{i} + x_2\vec{j} + x_3\vec{k}\right) \cdot \left(\left(x_2y_3 - x_3y_2\right)\vec{i} + \left(x_3y_1 - x_1y_3\right)\vec{j} + \left(x_1y_2 - x_2y_1\right)\vec{k}\right)$$

$$= x_1x_2y_3 - x_1x_3y_2 + x_2x_3y_1 - x_2x_1y_3 + x_3x_1y_2 - x_3x_2y_1$$

$$= 0$$

completing the proof.

**Theorem 1.8.3:**

$$\vec{a} \times \left(\vec{b} \times \vec{c}\right) = (\vec{a}\cdot\vec{c})\vec{b} - \left(\vec{a}\cdot\vec{b}\right)\vec{c} \ .$$

**Proof:**

$$\vec{a} \times \left(\vec{b} \times \vec{c}\right) = \left(a_1\vec{i} + a_2\vec{j} + a_3\vec{k}\right) \times \left(\left(b_2c_3 - b_3c_2\right)\vec{i} + \left(b_3c_1 - b_1c_3\right)\vec{j} + \left(b_1c_2 - b_2c_1\right)\vec{k}\right)$$

$$= a_1\left(b_3c_1 - b_1c_3\right)\vec{k} - a_1\left(b_1c_2 - b_2c_1\right)\vec{j} - a_2\left(b_2c_3 - b_3c_2\right)\vec{k} + a_2\left(b_1c_2 - b_2c_1\right)\vec{i}$$

$$+ a_3\left(b_2c_3 - b_3c_2\right)\vec{j} - a_3\left(b_3c_1 - b_1c_3\right)\vec{i}$$

$$= \left(a_1c_1 + a_2c_2 + a_3c_3\right)\left(b_1\vec{i} + b_2\vec{j} + b_3\vec{i}\right) + \left(-a_1b_1 - a_2b_2 - a_3b_3\right)\left(c_1\vec{i} + c_2\vec{j} + c_3\vec{i}\right)$$

$$= (\vec{a}\cdot\vec{c})\vec{b} - \left(\vec{a}\cdot\vec{b}\right)\vec{c} \ ,$$

completing the proof.

**Theorem 1.8.4 (Jacobi's Identity):**

$$\vec{a} \times \left(\vec{b} \times \vec{c}\right) + \vec{b} \times (\vec{c} \times \vec{a}) + \vec{c} \times \left(\vec{a} \times \vec{b}\right) = \vec{0}$$

**Proof:**

From Theorem 1.8.1, we have

$$\vec{a} \times \left(\vec{b} \times \vec{c}\right) = (\vec{a}\cdot\vec{c})\vec{b} - \left(\vec{a}\cdot\vec{b}\right)\vec{c} \ ,$$

$$\vec{b} \times (\vec{c} \times \vec{a}) = \left(\vec{b}\cdot\vec{a}\right)\vec{c} - \left(\vec{b}\cdot\vec{c}\right)\vec{a} \ ,$$

$$\vec{c} \times \left(\vec{a} \times \vec{b}\right) = \left(\vec{c}\cdot\vec{b}\right)\vec{a} - (\vec{c}\cdot\vec{a})\vec{b} \ ,$$

and adding yields the result.

**Theorem 1.8.5:**   Let $\widehat{(\vec{x},\vec{y})}\in[0;\pi]$ be the convex angle between two vectors $\vec{x}$ and $\vec{y}$. Then

$$\left\|\vec{x}\times\vec{y}\right\|=\left\|\vec{x}\right\|\ \left\|\vec{y}\right\|\ \sin\widehat{\left(\vec{x},\vec{y}\right)}.$$

See Fig. 1.8.1.

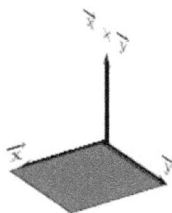

**FIGURE 1.8.1** Theorem 1.8.4.

**Proof:**

We have

$$\left\|\vec{x}\times\vec{y}\right\|=\left(x_2y_3-x_3y_2\right)^2+\left(x_3y_1-x_1y_3\right)^2+\left(x_1y_2-x_2y_1\right)^2$$
$$=y^2y_3^2-2x_2y_3x_3y_2+z^2y_2^2+z^2y_1^2-2x_3y_1x_1y_3+x^2y_3^2+x^2y_2^2-2x_1y_2x_2y_1+y^2y_1^2$$
$$=\left(x^2+y^2+z^2\right)\left(y_1^2+y_2^2+y_3^2\right)-\left(x_1y_1+x_2y_2+x_3y_3\right)^2$$
$$=\left\|\vec{x}\right\|^2\left\|\vec{y}\right\|^2-\left(\vec{x}\bullet\vec{y}\right)^2$$
$$=\left\|\vec{x}\right\|^2\left\|\vec{y}\right\|^2-\left\|\vec{x}\right\|^2\left\|\vec{y}\right\|^2\cos^2\widehat{\left(\vec{x},\vec{y}\right)}$$
$$=\left\|\vec{x}\right\|^2\left\|\vec{y}\right\|^2\sin^2\widehat{\left(\vec{x},\vec{y}\right)},$$

where the theorem follows.

Theorem 4 has the following geometric significance: $\left\|\vec{x}\times\vec{y}\right\|$ is the area of the parallelogram formed when the tails of the vectors are joined. See Fig. 1.8.2.

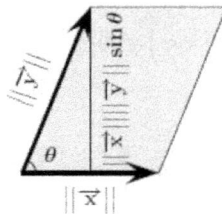

**FIGURE 1.8.2** Area of a parallelogram.

The following corollaries easily follow Theorem 1.8.5.

**Corollary 1.8.1:** Two non-zero vectors $\vec{x}$, $\vec{y}$ satisfy $\vec{x} \times \vec{y} = \vec{0}$ if and only if they are parallel.

**Corollary 1.8.2 (Lagrange's Identity):**

$$\left\| \vec{x} \times \vec{y} \right\| = \left\| \vec{x} \right\|^2 \left\| \vec{y} \right\|^2 - \left( \vec{x} \bullet \vec{y} \right)^2.$$

The following result mixes the dot and the cross product.

**Theorem 1.8.6:** Let $\vec{a}, \vec{b}, \vec{c}$, be linearly independent vectors in $\mathbb{R}^3$. The signed volume of the parallelepiped spanned by them is $\left( \vec{a} \times \vec{b} \right) \bullet \vec{c}$.

See Fig. 1.8.3.

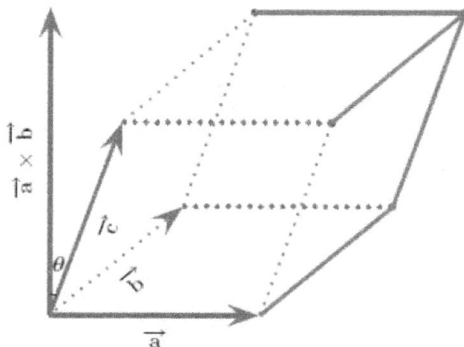

**FIGURE 1.8.3** Theorem 1.8.5.

**Proof:**

The area of the base of the parallelepiped is the area of the parallelogram determined by the vectors $\vec{a}$ and $\vec{b}$, which has area $\left\| \vec{a} \times \vec{b} \right\|$. The altitude of the parallelepiped is $\left\| \vec{c} \right\| \cos\theta$ where $\theta$ is the angle between $\vec{c}$ and $\vec{a} \times \vec{b}$. The volume of the parallelepiped is thus

$$\left\| \vec{a} \times \vec{b} \right\| \left\| \vec{c} \right\| \cos\theta = \left( \vec{a} \times \vec{b} \right) \bullet \vec{c} \,,$$

proving the theorem.

**NOTE** *Since we may have used any of the faces of the parallelepiped, it follows that*

$$\left( \vec{a} \times \vec{b} \right) \bullet \vec{c} = \left( \vec{b} \times \vec{c} \right) \bullet \vec{a} = \left( \vec{c} \times \vec{a} \right) \bullet \vec{b} \,.$$

*In particular, it is possible to "exchange" the cross and dot products:*

$$\vec{a} \bullet \left( \vec{b} \times \vec{c} \right) = \left( \vec{a} \times \vec{b} \right) \bullet \vec{c}$$

## EXAMPLE 1.8.3

Let $\vec{x} \in \mathbb{R}^3$, and $\left\| \vec{x} \right\| = 1$. Find $\left\| \vec{x} \times \vec{i} \right\|^2 + \left\| \vec{x} \times \vec{j} \right\|^2 + \left\| \vec{x} \times \vec{k} \right\|^2$.

### Solution:

By Lagrange's Identity,

$$\left\| \vec{x} \times \vec{i} \right\|^2 = \left\| \vec{x} \right\|^2 \left\| \vec{i} \right\|^2 - \left( \vec{x} \bullet \vec{i} \right)^2 = 1 - \left( \vec{x} \bullet \vec{i} \right)^2 \,,$$

$$\left\| \vec{x} \times \vec{j} \right\|^2 = \left\| \vec{x} \right\|^2 \left\| \vec{j} \right\|^2 - \left( \vec{x} \bullet \vec{j} \right)^2 = 1 - \left( \vec{x} \bullet \vec{j} \right)^2 \,,$$

$$\left\| \vec{x} \times \vec{k} \right\|^2 = \left\| \vec{x} \right\|^2 \left\| \vec{k} \right\|^2 - \left( \vec{x} \bullet \vec{k} \right)^2 = 1 - \left( \vec{x} \bullet \vec{k} \right)^2 \,,$$

and since $\left( \vec{x} \bullet \vec{i} \right)^2 + \left( \vec{x} \bullet \vec{j} \right)^2 + \left( \vec{x} \bullet \vec{k} \right)^2 = \left\| \vec{x} \right\|^2 = 1$, the desired sum equals $3 - 1 = 2$.

## EXAMPLE 1.8.4

Consider the rectangular parallelepiped $ABCDD'C'B'A'$, see Fig. 1.8.4 with vertices $A(2,0,0)$, $B(2,3,0)$, $C(0,3,0)$, $D(0,0,0)$, $D'(0,0,1)$, $C'(0,3,1)$, $B'(2,3,1)$, $A'(2,0,1)$. Let $M$ be the midpoint of the line segment joining the vertices $B$ and $C$.

1. Find the Cartesian equation of the plane containing the points $A$, $D'$, and $M$.
2. Find the area of $\triangle AD'M$.
3. Find the parametric equation of the line $\overrightarrow{AC'}$.
4. Suppose that a line through $M$ is drawn cutting the line segment $[AC']$ in $N$ and the line $\overrightarrow{DD'}$ in $P$. Find the parametric equation of $\overrightarrow{MP}$.

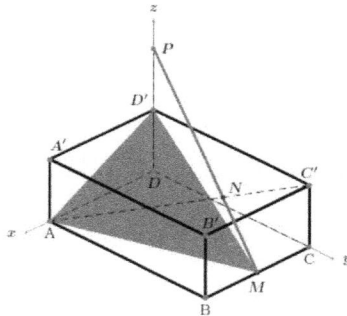

**FIGURE 1.8.4** Example 1.8.2.

## *Solution:*

1. Form the following vectors and find their cross product:

$$\overrightarrow{AD'} = \begin{bmatrix} -2 \\ 0 \\ 1 \end{bmatrix}, \quad \overrightarrow{AM} = \begin{bmatrix} -1 \\ 3 \\ 0 \end{bmatrix} \Rightarrow \overrightarrow{AD'} \times \overrightarrow{AM} = \begin{bmatrix} -3 \\ -1 \\ -6 \end{bmatrix}.$$

The equation of plane is thus

$$\begin{bmatrix} x-2 \\ y-0 \\ z-0 \end{bmatrix} \cdot \begin{bmatrix} -3 \\ -1 \\ -6 \end{bmatrix} = 0 \Rightarrow 3(x-2)+1(y)+6z = 0 \Rightarrow 3x+y+6z = 6.$$

**2.** The area of the triangle is

$$\frac{\left\|\overrightarrow{AD'} \times \overrightarrow{AM}\right\|}{2} = \frac{1}{2}\sqrt{3^2 + 1^2 + 6^2} = \frac{\sqrt{46}}{2} \ .$$

**3.** We have $\overrightarrow{AC'} = \begin{bmatrix} -2 \\ 3 \\ 1 \end{bmatrix}$, and hence the line $\overrightarrow{AC'}$ has parametric equation

$$\begin{pmatrix} x \\ y \\ z \end{pmatrix} = \begin{pmatrix} 2 \\ 0 \\ 0 \end{pmatrix} + t\begin{bmatrix} -2 \\ 3 \\ 1 \end{bmatrix} \Rightarrow x = 2 - 2t, y = 3t, z = t.$$

**4.** Since $P$ in on the $z$-axis, $P = \begin{pmatrix} 0 \\ 0 \\ z' \end{pmatrix}$ for some real number $z' > 0$. The para-

metric equation of the line $\overrightarrow{MP}$ is thus

$$\begin{pmatrix} x \\ y \\ z \end{pmatrix} = \begin{pmatrix} 1 \\ 3 \\ 0 \end{pmatrix} + s\begin{bmatrix} -1 \\ -3 \\ z' \end{bmatrix} \Rightarrow x = 1 - s, y = 3 - 3s, z = sz'.$$

Since $N$ is on both $\overrightarrow{MP}$ and $\overrightarrow{AC'}$, we must have

$$2 - 2t = 1 - s, \ 3t = 3 - 3s, \ t = sz'.$$

Solving the first two equations gives $s = \dfrac{1}{3}, t = \dfrac{2}{3}$. Putting this into the third

equation, we deduce $z' = 2$. Thus $P = \begin{pmatrix} 0 \\ 0 \\ 2 \end{pmatrix}$ and the desired equation is

$$\begin{pmatrix} x \\ y \\ z \end{pmatrix} = \begin{pmatrix} 1 \\ 3 \\ 0 \end{pmatrix} + s[2] \Rightarrow x = 1 - s, \ y = 3 - 3s, \ z = 2s.$$

## EXERCISES 1.8

**1.8.1**   Prove that

$$\left(\vec{a}-\vec{b}\right)\times\left(\vec{a}+\vec{b}\right)=2\vec{a}\times\vec{b}.$$

**1.8.2**   Prove that $\vec{x}\times\vec{x}=\vec{0}$ follows from the anti-commutativity of the cross product.

**1.8.3**   Find the area of the triangle whose vertices are at $A=\begin{pmatrix}0\\0\\1\end{pmatrix}$, $B=\begin{pmatrix}0\\1\\0\end{pmatrix}$, and $C=\begin{pmatrix}1\\0\\0\end{pmatrix}$.

**1.8.4**   Find a vector simultaneously perpendicular to $\begin{bmatrix}1\\1\\1\end{bmatrix}$ and $\begin{bmatrix}1\\1\\0\end{bmatrix}$, and having norm 3.

**1.8.5**   Prove or disprove! The cross product is associative.

**1.8.6**   Expand the product $\left(\vec{a}-\vec{b}\right)\times\left(\vec{a}+\vec{b}\right)$.

**1.8.7**   If $\vec{b}-\vec{a}$ and $\vec{c}-\vec{a}$ are parallel and it is known that $\vec{c}\times\vec{a}=\vec{i}-\vec{j}$ and $\vec{a}\times\vec{b}=\vec{j}+\vec{k}$, find $\vec{b}\times\vec{c}$.

**1.8.8**   Redo Example 1.7.6 (Section 1.7), that is, find the Cartesian equation of the plane parallel to the vectors $\begin{bmatrix}1\\1\\1\end{bmatrix}$ and $\begin{bmatrix}1\\1\\0\end{bmatrix}$, and passing through the point $(0,-1,2)$, by finding a normal to the plane.

**1.8.9**   Find the equation of the plane passing through the points $(a,0,a)$, $(a,0,a)$, $(-a,1,0)$, and $(0,1,2a)$ in $\mathbb{R}^3$.

**1.8.10**  Let $a\in\mathbb{R}$. Find a vector of unit length simultaneously perpendicular to $\vec{v}=\begin{bmatrix}0\\-a\\a\end{bmatrix}$ and $\vec{w}=\begin{bmatrix}1\\a\\0\end{bmatrix}$.

**1.8.11** **(Jacobi's Identity)** Let $\vec{a}$, $\vec{b}$, $\vec{c}$ be vectors in $\mathbb{R}^3$. Prove that
$$\vec{a} \times (\vec{b} \times \vec{c}) + \vec{b} \times (\vec{c} \times \vec{a}) + \vec{c} \times (\vec{a} \times \vec{b}) = 0.$$

**1.8.12** Let $\vec{x} \in \mathbb{R}^3, \|\vec{x}\| = 1$. Find $\left\|\vec{x} \times \vec{i}\right\|^2 + \left\|\vec{x} \times \vec{j}\right\|^2 + \left\|\vec{x} \times \vec{k}\right\|^2$.

**1.8.13** The vectors $\vec{a}, \vec{b}$ are constant vectors. Solve the equation
$$\vec{a} \times (\vec{x} \times \vec{b}) = \vec{b} \times (\vec{x} \times \vec{a}).$$

**1.8.14** Let $\vec{a}$ and $\vec{b}$ be vectors in $\mathbb{R}^3$ and $k$ be a scalar, prove that
$k(\vec{a} \times \vec{b}) = k\vec{a} \times \vec{b} = \vec{a} \times k\vec{b}$.

**1.8.15** Prove that the position vectors $\vec{a}, \vec{b}$, and $\vec{c}$ of $\mathbb{R}^3$ all lie in a plane if and only if $\vec{a} \cdot (\vec{b} \times \vec{c}) = 0$.

**1.8.16** The vectors $\vec{a}, \vec{b}, \vec{c}$ are constant vectors. Solve the system of equations
$$2\vec{x} + \vec{y} \times \vec{a} = \vec{b}, \quad 3\vec{y} + \vec{x} \times \vec{a} = \vec{c}.$$

**1.8.17** Let $\vec{a}, \vec{b}, \vec{c}, \vec{d}$, be vectors in $\mathbb{R}^3$. Prove the following vector identity,
$$(\vec{a} \times \vec{b}) \cdot (\vec{c} \times \vec{d}) = (\vec{a} \cdot \vec{c})(\vec{b} \cdot \vec{d}) - (\vec{a} \cdot \vec{d})(\vec{b} \cdot \vec{c}).$$

**1.8.18** Let $\vec{a}, \vec{b}, \vec{c}, \vec{d}$ be vectors in $\mathbb{R}^3$. Prove that
$$(\vec{b} \times \vec{c}) \cdot (\vec{a} \times \vec{d}) + (\vec{c} \times \vec{a}) \cdot (\vec{b} \times \vec{d}) + (\vec{a} \times \vec{b}) \cdot (\vec{c} \times \vec{d}) = 0.$$

**1.8.19** Consider the plane $\Pi$ passing through the points $A(6,0,0), B(0,4,0)$, and $C(0,0,3)$, as shown in Fig. 1.8.5. The plane $\Pi$ intersects a $3 \times 3 \times 3$ cube, one of whose vertices is at the origin and that has three of its edges on the coordinate axes, as in the figure. This intersection forms a pentagon $CPQRS$.

1. Find $\overrightarrow{CA} \times \overrightarrow{CB}$.

2. Find $\left\|\overrightarrow{CA} \times \overrightarrow{CB}\right\|$.

3. Find the parametric equation of the line $L_{CA}$ joining $C$ and $A$, with a parameter $t \in \mathbb{R}$.

4. Find the parametric equation of the line $L_{DE}$ joining $D$ and $E$, with a parameter $s \in \mathbb{R}$.

5. Find the intersection point between the lines $L_{CA}$ and $L_{DE}$.

6. Find the coordinates of the points $P, Q, R$, and $S$.

7. Find the area of the pentagon $CPQRS$.

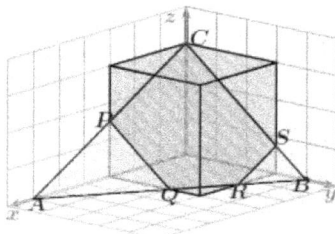

**FIGURE 1.8.5** Exercise 1.8.15.

## 1.9  MATRICES IN THREE DIMENSIONS

We will briefly introduce $3 \times 3$ matrices. Most of the material will flow like that for $2 \times 2$ matrices.

**Definition 1.9.1**  A *linear transformation* $T : \mathbb{R}^3 \to \mathbb{R}^3$ is a function such that

$$T(a + b) = T(a) + T(b), \qquad T(\lambda a) = \lambda T(a),$$

for all points $a, b$ in $\mathbb{R}^3$ and all scalars $\lambda$. Such a linear transformation has a $3 \times 3$ matrix representation whose columns are the vectors $T(i), T(j)$, and $T(k)$.

**EXAMPLE 1.9.1**

Consider $L : \mathbb{R}^3 \to \mathbb{R}^3$, with

$$L \left( \begin{pmatrix} x_1 \\ x_2 \\ x_3 \end{pmatrix} \right) = \begin{pmatrix} x_1 - x_2 - x_3 \\ x_1 + x_2 + x_3 \\ x_3 \end{pmatrix}.$$

1. Prove that $L$ is a linear transformation.

2. Find the matrix corresponding to $L$ under the standard basis.

### *Solution:*

1. Let $a \in \mathbb{R}$ and let u, v be points in $\mathbb{R}^3$. Then

$$L(u+v) = L\left(\begin{pmatrix} u_1 + v_1 \\ u_2 + v_2 \\ u_3 + v_3 \end{pmatrix}\right) = \begin{pmatrix} (u_1 + v_1) - (u_2 + v_2) - (u_3 + v_3) \\ (u_1 + v_1) + (u_2 + v_2) + (u_3 + v_3) \\ u_3 + v_3 \end{pmatrix}$$

$$= \begin{pmatrix} u_1 - u_2 - u_3 \\ u_1 + u_2 + u_3 \\ u_3 \end{pmatrix} + \begin{pmatrix} v_1 - v_2 - v_3 \\ v_1 + v_2 + v_3 \\ v_3 \end{pmatrix}$$

$$= L\left(\begin{pmatrix} u_1 \\ u_2 \\ u_3 \end{pmatrix}\right) + L\left(\begin{pmatrix} v_1 \\ v_2 \\ v_3 \end{pmatrix}\right)$$

$$= L(u) + L(v),$$

and also

$$L(a\,u) = L\left(\begin{pmatrix} a\,u_1 \\ a\,u_2 \\ a\,u_3 \end{pmatrix}\right)$$

$$= \begin{pmatrix} a\,(u_1) - a\,(u_2) - a\,(u_3) \\ a\,(u_1) + a\,(u_2) + a\,(u_3) \\ a\,u_3 \end{pmatrix}$$

$$= a \begin{pmatrix} u_1 - u_2 - u_3 \\ u_1 + u_2 + u_3 \\ u_3 \end{pmatrix}$$

$$= a\,L\left(\begin{pmatrix} u_1 \\ u_2 \\ u_3 \end{pmatrix}\right)$$

$$= a\,L(u),$$

proving that $L$ is a linear transformation.

2. We have $L\begin{pmatrix} 1 \\ 0 \\ 0 \end{pmatrix} = \begin{pmatrix} 1 \\ 1 \\ 0 \end{pmatrix}, L\begin{pmatrix} 0 \\ 1 \\ 0 \end{pmatrix} = \begin{pmatrix} -1 \\ 1 \\ 0 \end{pmatrix}$, and $L\begin{pmatrix} 0 \\ 0 \\ 1 \end{pmatrix} = \begin{pmatrix} -1 \\ 1 \\ 1 \end{pmatrix}$, where the desired

matrix is

$$\begin{bmatrix} 1 & -1 & -1 \\ 1 & 1 & 1 \\ 0 & 0 & 1 \end{bmatrix}.$$

In addition, scalar multiplication, and matrix multiplication are defined for $3 \times 3$ matrices in a manner analogous to those operations for $2 \times 2$ matrices.

**Definition 1.9.2**   Let $A, B$ be $3 \times 3$ matrices. Then we define

$$A + B = \left[ a_{ij} + b_{ij} \right], \quad a\,A = \left[ a\,a_{ij} \right], \quad AB = \left[ \sum_{k=1}^{3} a_{ik} b_{kj} \right].$$

**EXAMPLE 1.9.2**

If $A = \begin{bmatrix} 1 & 2 & 3 \\ 4 & 5 & 0 \\ 6 & 0 & 0 \end{bmatrix}$, and $B = \begin{bmatrix} a & b & c \\ a & b & 0 \\ a & 0 & 0 \end{bmatrix}$, then

$$A + B = \begin{bmatrix} 1+a & 2+a & 3+c \\ 4+a & 5+b & 0 \\ 6+a & 0 & 0 \end{bmatrix},$$

$$3A = \begin{bmatrix} 3 & 6 & 9 \\ 12 & 15 & 0 \\ 18 & 0 & 0 \end{bmatrix},$$

$$AB = \begin{bmatrix} 6a & 3b & c \\ 9a & 9b & 4c \\ 6a & 6b & 6c \end{bmatrix},$$

$$BA = \begin{bmatrix} a+4b+6c & 2a+5b & 3a \\ a+4b & 2a+5b & 3a \\ a & 2a & 3a \end{bmatrix}.$$

**Definition 1.9.3**   A *scaling matrix* is one of the form

$$S_{a,b,c} = \begin{bmatrix} a & 0 & 0 \\ 0 & b & 0 \\ 0 & 0 & c \end{bmatrix},$$

where $a > 0, b > 0, c > 0$.

It is an easy exercise to prove that the product of two scaling matrices commutes.

**Definition 1.9.4**   A *rotation matrix* about the $z$-axis by an angle $\theta$ in the counterclockwise sense is

$$R_z(\theta) = \begin{bmatrix} \cos\theta & -\sin\theta & 0 \\ \sin\theta & \cos\theta & 0 \\ 0 & 0 & 1 \end{bmatrix}.$$

A *rotation matrix* about the $y$-axis by an angle $\theta$ in the counterclockwise sense is

$$R_y(\theta) = \begin{bmatrix} \cos\theta & 0 & -\sin\theta \\ 0 & 1 & 0 \\ \sin\theta & 0 & \cos\theta \end{bmatrix}.$$

A *rotation matrix* about the $x$-axis by an angle $\theta$ in the counterclockwise sense is

$$R_x(\theta) = \begin{bmatrix} 1 & 0 & 0 \\ 0 & \cos\theta & -\sin\theta \\ 0 & \sin\theta & \cos\theta \end{bmatrix}.$$

Easy to find counterexamples should convince the reader that the product of two rotations in space does not necessarily commute.

**Definition 1.9.5**   A *reflection matrix* about $x$-axis is

$$R_x = \begin{bmatrix} -1 & 0 & 0 \\ 0 & 1 & 0 \\ 0 & 0 & 1 \end{bmatrix}.$$

A *reflection matrix* about $y$-axis is

$$R_y = \begin{bmatrix} 1 & 0 & 0 \\ 0 & -1 & 0 \\ 0 & 0 & 1 \end{bmatrix}.$$

A *reflection matrix* about $z$-axis is

$$R_z = \begin{bmatrix} 1 & 0 & 0 \\ 0 & 1 & 0 \\ 0 & 0 & -1 \end{bmatrix}.$$

## EXERCISES 1.9

**1.9.1**  If $A = \begin{bmatrix} -1 & 0 & 2 \\ 0 & 3 & 1 \\ 0 & 5 & 4 \end{bmatrix}$, and $B = \begin{bmatrix} 2 & 1 & 3 \\ -1 & 0 & 1 \\ 1 & 4 & -3 \end{bmatrix}$, find:

**1.**  $A + B$    **2.**  $2A$    **3.**  $AB$.

**1.9.2**  Let $A \in M_{3\times3}(\mathbb{R})$ be given by $A = \begin{bmatrix} 1 & 1 & 1 \\ 1 & 1 & 1 \\ 1 & 1 & 1 \end{bmatrix}$. Demonstrate, using

induction, that $A^n = 3^{n-1}A$ for $n \in \mathbb{N}, n \geq 1$.

**1.9.3**  Consider $L: \mathbb{R}^3 \to \mathbb{R}^3$, $L\begin{bmatrix} x \\ y \\ z \end{bmatrix} = \begin{bmatrix} x - y - z \\ x + y + z \\ z \end{bmatrix}$. Prove that $L$ is linear.

**1.9.4**  Let $L: \mathbb{R}^3 \to \mathbb{R}^3$ be a perpendicular projection of $\mathbb{R}^3$ onto the $xy$-plane. Show that $L \circ L = L$.

**1.9.5**  Consider the $n \times n$ matrix

$$\mathbf{A} = \begin{bmatrix} 1 & 1 & 1 & 1 & \cdots & 1 & 1 \\ 0 & 1 & 1 & 1 & \cdots & 1 & 1 \\ 0 & 0 & 1 & 1 & \cdots & 1 & 1 \\ \cdots & \cdots & \vdots & \vdots & \vdots & \cdots & \cdots \\ 0 & 0 & 0 & 0 & \cdots & 0 & 1 \end{bmatrix}.$$

Describe $A^2$ and $A^3$ in terms of $n$.

**1.9.6**   Let $x$ be a real number, and put

$$m(x) = \begin{bmatrix} 1 & 0 & x \\ -x & 1 & -\dfrac{x^2}{2} \\ 0 & 0 & 1 \end{bmatrix}.$$

If $a, b$ are real numbers, prove that

**1.** $m(a)m(b) = m(a+b)$.

**2.** $m(a)m(-a) = \mathbf{I}_3$, the $3 \times 3$ identity matrix.

**1.9.7**   Determine the standard matrix representing the linear transformation $L : \mathbb{R}^3 \to \mathbb{R}^3$ which is defined as $L\left(\begin{pmatrix} x_1 \\ x_2 \\ x_3 \end{pmatrix}\right) = \begin{pmatrix} x_1 - x_2 \\ x_1 + x_2 \\ x_2 - x_3 \end{pmatrix}$.

**1.9.8**   Find the matrix of the linear operator $L\left(\begin{pmatrix} x_1 \\ x_2 \\ x_3 \end{pmatrix}\right) = \begin{pmatrix} x_1 \\ 3x_2 \\ 5x_2 \end{pmatrix}$ on $\mathbb{R}^3$

with respect to the standard basis of $\mathbb{R}^3$, then use the matrix to find the image of the vector $\begin{pmatrix} -1 \\ 4 \\ 2 \end{pmatrix}$.

**1.9.9**   Find the matrix of the linear operator $L\left(\begin{pmatrix} x_1 \\ x_2 \\ x_3 \end{pmatrix}\right) = \begin{pmatrix} 0 \\ x_2 \\ 0 \end{pmatrix}$ on $\mathbb{R}^3$ with

respect to the standard basis of $\mathbb{R}^3$, then use the matrix to find the image of the vector $\begin{pmatrix} -1 \\ 4 \\ 2 \end{pmatrix}$.

**1.9.10**   Find the matrix of each of the following linear transformations:

**1.** $L\begin{bmatrix} x_1 \\ x_2 \\ x_3 \end{bmatrix} = \begin{bmatrix} 3x_1 - x_2 + 5x_3 \end{bmatrix}$

**2.** $L\begin{bmatrix} x_1 \\ x_2 \\ x_3 \end{bmatrix} = \begin{bmatrix} x_1 + 4x_2 - x_3 \\ x_1 + 3x_2 + 2x_3 \\ 3x_1 + 2x_2 + 5x_3 \end{bmatrix}$

**1.9.11** Determine whether the transformation $L : \mathbb{R}^3 \to \mathbb{R}^3$;

$L\begin{pmatrix} x_1 \\ x_2 \\ x_3 \end{pmatrix} = \begin{pmatrix} 1 \\ x_2 \\ x_3 \end{pmatrix}$ is linear.

**1.9.12** Let $R(\theta) = \begin{bmatrix} \cos\theta & -\sin\theta & 0 \\ \sin\theta & \cos\theta & 0 \\ 0 & 0 & 1 \end{bmatrix}$. Describe geometrically the linear

transformation $L : \mathbb{R}^3 \to \mathbb{R}^3$ given by $L\vec{x} = R(\theta)\vec{x}$.

**1.9.13** Find the standard matrix for the linear operator $L : \mathbb{R}^3 \to \mathbb{R}^3$,

which maps a vector $\vec{x} = \begin{pmatrix} x_1 \\ x_2 \\ x_3 \end{pmatrix}$ into its reflection through the

$xz$-plane.

**1.9.14** Find the standard matrix for the linear operator $L : \mathbb{R}^3 \to \mathbb{R}^3$, which rotates each vector $\pi/2$ counterclockwise about $y$-axis (looking along the positive $y$-axis toward the origin).

## 1.10 DETERMINANTS IN THREE DIMENSIONS

We now define the notion of *determinant* of a $3 \times 3$ matrix. Consider now the vectors

$$\vec{a} = \begin{bmatrix} a_1 \\ a_2 \\ a_3 \end{bmatrix}, \ \vec{b} = \begin{bmatrix} b_1 \\ b_2 \\ b_3 \end{bmatrix}, \ \vec{c} = \begin{bmatrix} c_1 \\ c_2 \\ c_3 \end{bmatrix}, \ \text{in } \mathbb{R}^3, \text{ and the } 3 \times 3 \text{ matrix } A = \begin{bmatrix} a_1 & b_1 & c_1 \\ a_2 & b_2 & c_2 \\ a_3 & b_3 & c_3 \end{bmatrix}.$$

Since, in

In Theorem 1.8.5, the volume of the parallelepiped spanned by these vectors is $\vec{a} \cdot (\vec{b} \times \vec{c})$, we *define* the determinant of $A$ ( $\det A$ ) to be

$$D(\vec{a}, \vec{b}, \vec{c}) = \det \begin{bmatrix} a_1 & b_1 & c_1 \\ a_2 & b_2 & c_2 \\ a_3 & b_3 & c_3 \end{bmatrix} = \begin{vmatrix} a_1 & b_1 & c_1 \\ a_2 & b_2 & c_2 \\ a_3 & b_3 & c_3 \end{vmatrix} = \vec{a} \cdot (\vec{b} \times \vec{c}). \tag{1.15}$$

We now establish that the properties of the determinant of a $3 \times 3$ as previously defined are analogous to those of the determinant of $2 \times 2$ matrix defined in the preceding chapter.

**Theorem 1.10.1** The determinant of a $3 \times 3$ matrix A as defined by Equation (1.15) satisfies the following properties:

1. $D$ is linear in each of its arguments.
2. If the parallelepiped is flat then the volume is 0, that is, if $\vec{a}, \vec{b}, \vec{c}$, are linearly dependent, then $D(\vec{a}, \vec{b}, \vec{c}) = 0$.
3. $D(\vec{i}, \vec{j}, \vec{k}) = 0$, and accords with the right-hand rule.

**Proof:**

1. If $D(\vec{a}, \vec{b}, \vec{c}) = \vec{a} \cdot (\vec{b} \times \vec{c})$, linearity of the first component follows by the distributive law for the dot product:

$$D(\vec{a} + \vec{a}', \vec{b}, \vec{c}) = (\vec{a} + \vec{a}') \cdot (\vec{b} \times \vec{c})$$

$$= \vec{a} \cdot (\vec{b} \times \vec{c}) + \vec{a}' \cdot (\vec{b} \times \vec{c})$$

$$= D(\vec{a}, \vec{b}, \vec{c}) + D(\vec{a}', \vec{b}, \vec{c}),$$

and if $\lambda \in \mathbb{R}$,

$$D(\lambda \vec{a}, \vec{b}, \vec{c}) = (\lambda \vec{a}) \cdot (\vec{b} \times \vec{c}) = \lambda ((\vec{a}) \cdot (\vec{b} \times \vec{c})) = \lambda D(\vec{a}, \vec{b}, \vec{c}).$$

The linearity on the second and third components can be established by using the distributive law of the cross product. For example, for the second component we have,

$$D\left(\vec{a},\vec{b}+\vec{b'},\vec{c}\right)=\vec{a}\bullet\left(\left(\vec{b}+\vec{b'}\right)\times\vec{c}\right)$$

$$=\vec{a}\bullet\left(\vec{b}\times\vec{c}+\vec{b'}\times\vec{c}\right)$$

$$=\vec{a}\bullet\left(\vec{b}\times\vec{c}\right)+\vec{a}\bullet\left(\vec{b'}\times\vec{c}\right)$$

$$=D\left(\vec{a},\vec{b},\vec{c}\right)+D\left(\vec{a},\vec{b'},\vec{c}\right),$$

and if $\lambda\in\mathbb{R}$,

$$D\left(\vec{a},\lambda\vec{b},\vec{c}\right)=\vec{a}\bullet\left(\left(\lambda\vec{b}\right)\times\vec{c}\right)=\lambda\left(\vec{a}\bullet\left(\vec{b}\times\vec{c}\right)\right)=\lambda D\left(\vec{a},\vec{b},\vec{c}\right).$$

2. If $\vec{a},\vec{b},\vec{c}$ are linearly dependent, then they lie on the same plane and the parallelepiped spanned by them is flat, hence $D\left(\vec{a},\vec{b},\vec{c}\right)=0$.

3. Since $\vec{j}\times\vec{k}=\vec{i}$, and $\vec{i}\times\vec{i}=1$, $D\left(\vec{i},\vec{j},\vec{k}\right)=\vec{i}\bullet\left(\vec{j}\times\vec{k}\right)=\vec{i}\times\vec{i}=1$.

Observe that,

$$\det\begin{bmatrix} a_1 & b_1 & c_1 \\ a_2 & b_2 & c_2 \\ a_3 & b_3 & c_3 \end{bmatrix}=\vec{a}\bullet\left(\vec{b}\times\vec{c}\right) \tag{1.16}$$

$$=\vec{a}\bullet\left(\left(b_2c_3-b_3c_2\right)\vec{i}+\left(b_3c_1-b_1c_3\right)\vec{j}+\left(b_1c_2-b_2c_1\right)\vec{k}\right) \tag{1.17}$$

$$=a_1\left(b_2c_3-b_3c_2\right)+a_2\left(b_3c_1-b_1c_3\right)+a_3\left(b_1c_2-b_2c_1\right) \tag{1.18}$$

$$=a_1\det\begin{bmatrix} b_2 & c_2 \\ b_3 & c_3 \end{bmatrix}+a_2\det\begin{bmatrix} b_1 & c_1 \\ b_3 & c_3 \end{bmatrix}+a_3\det\begin{bmatrix} b_1 & c_1 \\ b_2 & c_2 \end{bmatrix}, \tag{1.19}$$

which reduces the computation of $3\times3$ determinants to $2\times2$ determinants.

**EXAMPLE 1.10.1**

Find the $\det\begin{bmatrix} 1 & 2 & 3 \\ 4 & 5 & 6 \\ 7 & 8 & 9 \end{bmatrix}$.

*Solution:*

Using Equation (1.19), we have

$$\det A = 1\det\begin{bmatrix} 5 & 6 \\ 8 & 9 \end{bmatrix} - 4\det\begin{bmatrix} 2 & 3 \\ 8 & 9 \end{bmatrix} + 7\det\begin{bmatrix} 2 & 3 \\ 5 & 6 \end{bmatrix}$$

$$= 1(45-48) - 4(18-24) + 7(12-15)$$

$$= -3 + 24 - 21$$

$$= 0.$$

**EXAMPLE 1.10.2**

Find the $\det\begin{bmatrix} t & 1 & a \\ t^2 & 1 & b \\ t^3 & 1 & c \end{bmatrix} = t\det\begin{bmatrix} 1 & 1 & a \\ t & 1 & b \\ t^2 & 1 & c \end{bmatrix}$.

**EXAMPLE 1.10.3**

Find the area of parallelogram with edges $\vec{a} = \begin{bmatrix} 1 \\ 2 \\ 3 \end{bmatrix}$ and $\vec{b} = \begin{bmatrix} 3 \\ 2 \\ 1 \end{bmatrix}$.

*Solution:*

We have now $\begin{bmatrix} 1 & 2 & 3 \\ 3 & 2 & 1 \end{bmatrix}$, then

$$\vec{a} \times \vec{b} = \begin{bmatrix} \det\begin{bmatrix} 2 & 3 \\ 2 & 1 \end{bmatrix} \\ -\det\begin{bmatrix} 1 & 3 \\ 3 & 1 \end{bmatrix} \\ \det\begin{bmatrix} 1 & 2 \\ 3 & 2 \end{bmatrix} \end{bmatrix} = \begin{bmatrix} -4 \\ 8 \\ -4 \end{bmatrix}$$

Area $= \left\| \vec{a} \times \vec{b} \right\| = \sqrt{(-4)^2 + (8)^2 + (-4)^2} = \sqrt{16 + 64 + 16} = \sqrt{96}$.

**Theorem 1.10.2:** The volume $V$ of a parallelepiped with vectors $\vec{a}, \vec{b}$, and $\vec{c}$ as adjacent edges is $V = \det\left(\vec{a} \cdot \left(\vec{b} \times \vec{c}\right)\right)$.

### EXAMPLE 1.10.4

Find the volume $V$ of a parallelepiped having $\vec{a} = 2\vec{i} - 3\vec{j} + \vec{k}$, $\vec{b} = 4\vec{j} - 3\vec{k}$, and $\vec{c} = \vec{i} + 2\vec{j} + \vec{k}$ as adjacent edges.

### Solution:

$$V = \det\left(\vec{a} \cdot \left(\vec{b} \times \vec{c}\right)\right)$$

$$V = \det\begin{bmatrix} 2 & -3 & 1 \\ 0 & 4 & -3 \\ 1 & 2 & 1 \end{bmatrix} = 2\det\begin{bmatrix} 4 & -3 \\ 2 & 1 \end{bmatrix} + 3\det\begin{bmatrix} 0 & -3 \\ 1 & 1 \end{bmatrix} + 1\det\begin{bmatrix} 0 & 4 \\ 1 & 2 \end{bmatrix}$$

$$V = 2(10) + 3(3) + 1(-4) = 25.$$

**NOTE** When $\vec{a}, \vec{b}$, and $\vec{c}$ have the same initial point, they lie in the same plane

$$\Leftrightarrow \vec{a} \cdot \left(\vec{b} \times \vec{c}\right) = \det\begin{bmatrix} a_1 & a_2 & a_3 \\ b_1 & b_2 & b_c \\ c_1 & c_2 & c_3 \end{bmatrix} = 0. \text{ That is, the volume of the parallelepiped is}$$

$0 \Leftrightarrow \vec{a}, \vec{b}$, and $\vec{c}$ are coplanar.

Again, we may use the *Maple*$^{TM}$ packages *linalg*, *LinearAlgebra*, or *Student [VectorCalculus]* to perform many of the vector operations. An example follows with *linalg*.

```
> with(linalg) :
> a := vector([-2, 0, 1]);
```

$$a := \begin{bmatrix} -2 & 0 & 1 \end{bmatrix}$$

```
> b := vector([-1, 3, 0]);
```

$$b := \begin{bmatrix} -1 & 3 & 0 \end{bmatrix}$$

```
> crossprod(a, b);
```

$$\begin{bmatrix} -3 & -1 & -6 \end{bmatrix}$$

```
> dotprod(a, b);
```

$$2$$

```
> angle(a, b);
```

$$\arccos\left(\frac{1}{25}\sqrt{5}\sqrt{10}\right)$$

Also, we may use MATLAB package, using the following commands.

\>> a=[-2,0,1];

\>> b=[-1,3,0];

\>> CP=cross(a, b)

CP =

   -3   -1   -6

\>> DP=dot(a, b)

DP =

   2

\>> CosTheta = dot(a,b)/(norm(a)*norm(b));

\>> ThetaInDegrees = acos(CosTheta)*180/pi

ThetaInDegrees =

   73.5701

## EXERCISES 1.10

**1.10.1** Evaluate the $\det \begin{bmatrix} 3 & 2 & -1 \\ -2 & -4 & 1 \\ 5 & 8 & 0 \end{bmatrix}$.

**1.10.2** Evaluate the det $\begin{bmatrix} a & 0 & 0 \\ b & e & 0 \\ c & d & f \end{bmatrix}$.

**1.10.3** Evaluate the det $\begin{bmatrix} a & b & c \\ 0 & e & d \\ 0 & 0 & f \end{bmatrix}$.

**1.10.4** Show that det $\begin{bmatrix} 0 & 0 & c \\ 0 & b & e \\ a & d & f \end{bmatrix} = -cba$.

**1.10.5** Solve for $d$ in the equation Solve for $d$ in the equation
$$\det \begin{bmatrix} 1-d & 1 & 1 \\ 2 & d & -4 \\ 3 & 1 & 0 \end{bmatrix} = 4.$$

**1.10.6** Determine the values of $z$ for which the det $\begin{bmatrix} z-6 & 0 & 0 \\ 0 & z & -1 \\ 0 & 4 & z-4 \end{bmatrix} = 0$.

**1.10.7** Evaluate the det $\begin{bmatrix} t & -3 & 9 \\ 2 & 4 & t+1 \\ 1 & t^2 & 3 \end{bmatrix}$.

**1.10.8** Show that det $\begin{bmatrix} a^2-b^2 & a+b & a \\ a-b & 1 & 1 \\ a-b & 1 & b \end{bmatrix} = 0$.

**1.10.9** Solve the following equation for $c$ in the determinant
$$\det \begin{bmatrix} c-1 & 1 & 0 \\ 2 & -1 & 1 \\ 3c & 0 & 4 \end{bmatrix} = 1.$$

**1.10.10** Prove det $\begin{bmatrix} 1 & 1 & 1 \\ a & b & c \\ a^2 & b^2 & c^2 \end{bmatrix} = (b-c)(c-a)(a-b)$.

**1.10.11** Prove $\det \begin{bmatrix} a & b & c \\ b & c & a \\ c & a & b \end{bmatrix} = 3abc - a^3 - b^3 - c^3.$

**1.10.12** Evaluate the $\det \begin{bmatrix} 1 & 1 & 1 \\ x & y & z \\ x+2 & y+3 & z+4 \end{bmatrix}.$

**1.10.13** Find the values of $\lambda$ that satisfy $\det \begin{bmatrix} 1 & 1 & 0 \\ 0 & \lambda & 1 \\ 0 & 1 & \lambda \end{bmatrix} = 0.$

**1.10.14** Find the values of $\lambda$ that satisfy $\det \begin{bmatrix} 1-\lambda & 0 & -1 \\ 1 & 2-\lambda & 1 \\ 3 & 3 & -\lambda \end{bmatrix} = 0.$

**1.10.15** Find the $\vec{a} \cdot (\vec{b} \times \vec{c})$ of the vectors $\vec{a} = 2\vec{i} - \vec{j} + 3\vec{k}$, $\vec{b} = \vec{i} + 4\vec{j} - \vec{k}$, and $\vec{c} = 2\vec{j} + 3\vec{k}.$

**1.10.16** Find the constant $a$ such that the vectors $\vec{a} = 2\vec{i} - \vec{j} + \vec{k}$, $\vec{b} = \vec{i} + 2\vec{j} - 3\vec{k}$, and $\vec{c} = 3\vec{i} + a\,\vec{j} + 5\vec{k}$ are coplanar.

## 1.11 SOME SOLID GEOMETRY

In this section, we examine some examples and prove some theorems of three-dimensional geometry.

### EXAMPLE 1.11.1

Cube ABCDD′C′B′A ABCDD′C′B′A in Fig. 1.11.1 has side of length $a$. $M$ is the midpoint of edge [BB′] and $N$ is the midpoint of edge [B′C′]. Prove that $\overrightarrow{AD'} \parallel \overrightarrow{MN}$ and find the area of the quadrilateral MND′A .

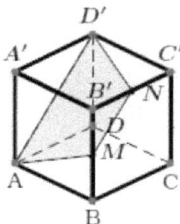

**FIGURE 1.11.1** Example 1.11.1.

### Solution:

By the Pythagorean Theorem, $\left\|\overrightarrow{AD'}\right\| = a\sqrt{2}$. Because they are diagonals that belong to parallel faces of the cube, $\overrightarrow{AD'} \parallel \overrightarrow{BC'}$. Now, M and N are the midpoints of the sides [B'B] and [B'C'] of $\Delta B'C'B$, and hence $\overrightarrow{MN} \parallel \overrightarrow{BC'}$ by example 1.1.6. The aforementioned example also gives $\left\|\overrightarrow{MN}\right\| = \frac{1}{2}\left\|\overrightarrow{AD'}\right\| = \frac{a\sqrt{2}}{2}$. In consequence, $\overrightarrow{AD'} \parallel \overrightarrow{MN}$. This means that the four points A, D', M, N are all on the same plane. Hence MND'A is a trapezoid with bases of length $a\sqrt{2}$ and $\frac{a\sqrt{2}}{2}$, see Fig. 1.11.2.

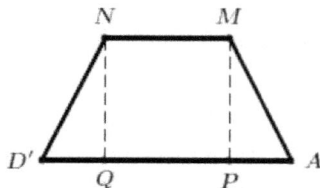

**FIGURE 1.11.2** Example 1.11.1.

From the figure

$$\left\|\overrightarrow{D'Q}\right\| = \left\|\overrightarrow{AP}\right\| = \frac{1}{2}\left(\left\|\overrightarrow{AD'}\right\| - \left\|\overrightarrow{MN}\right\|\right) = \frac{a\sqrt{2}}{4}.$$

Also, by the Pythagorean theorem,

$$\left\|\overrightarrow{D'N}\right\| = \sqrt{\left\|\overrightarrow{D'C'}\right\|^2 + \left\|\overrightarrow{C'N}\right\|^2} = \sqrt{a^2 + \frac{a^2}{4}} = \frac{a\sqrt{5}}{2}.$$

The height of this trapezoid is thus

$$\left\|\overrightarrow{NQ}\right\| = \sqrt{\frac{5a^2}{4} - \frac{a^2}{8}} = \frac{3a}{2\sqrt{2}}.$$

The area of the trapezoid is finally,

$$\frac{3a}{2\sqrt{2}} \cdot \left(\frac{a\sqrt{2} + \frac{a\sqrt{2}}{2}}{2}\right) = \frac{9a^2}{8}.$$

Let us prove a three-dimensional version of Thales' Theorem.

**Theorem 1.11.1 (Thales' theorem):**   If two lines are cut by three parallel planes, their corresponding segments are proportional.

See Fig. 1.11.3.

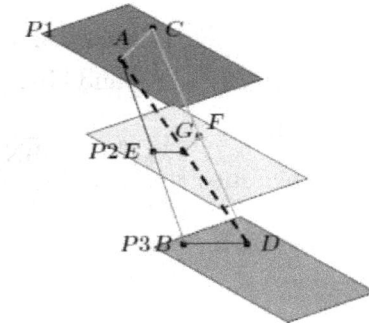

**FIGURE 1.11.3** Thales' theorem in 3D.

**Proof:**

Given the lines $\overleftrightarrow{AB}$ and $\overleftrightarrow{CD}$, we must prove that

$$\frac{\overline{AE}}{\overline{EB}} = \frac{\overline{CF}}{\overline{FD}}.$$

Draw line $\overleftrightarrow{AD}$ cutting plane $P2$ in $G$. The plane containing points A, B, and D intersects plane $P2$ in the line $\overleftrightarrow{EG}$. Similarly, the plane containing points A,

C, and D intersects plane $P2$ in the line $\overleftrightarrow{GF}$. Since $P2$ and $P3$ are parallel planes, $\overleftrightarrow{EG} \| \overleftrightarrow{BD}$, and so by Thales' Theorem on the plane (theorem 1.2.2)

$$\frac{\overline{AE}}{\overline{EB}} = \frac{\overline{AG}}{\overline{GD}}.$$

Similarly, since $P1$ and $P2$ are parallel, $\overleftrightarrow{AC} \| \overleftrightarrow{GF}$ and

$$\frac{\overline{CF}}{\overline{FD}} = \frac{\overline{AG}}{\overline{GD}}.$$

It follows that

$$\frac{\overline{AE}}{\overline{EB}} = \frac{\overline{CF}}{\overline{FD}},$$

as needed to be shown.

### EXAMPLE 1.11.2

In cube $ABCDD'C'B'A'$ of edge of length a, as in Fig. 1.11.4, the points M and N are located on diagonals $[AB']$ and $[BC']$ such that $\overrightarrow{MN}$ is parallel to the face ABCD of the cube. If $\left\| \overrightarrow{MN} \right\| = \frac{\sqrt{5}}{3} \left\| \overrightarrow{AB} \right\|$, find the ratios $\frac{\left\| \overrightarrow{AM} \right\|}{\left\| \overrightarrow{AB'} \right\|}$ and $\frac{\left\| \overrightarrow{BN} \right\|}{\left\| \overrightarrow{BC'} \right\|}$.

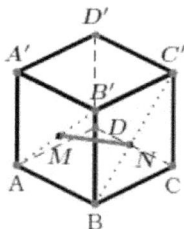

**FIGURE 1.11.4** Example 1.11.2.

### Solution:

There is a unique plane parallel $P$ to face ABCD and containing M. Since $\overrightarrow{MN}$ is parallel to face ABCD, $P$ also contains N. The intersection of $P$ with the cube produces a lamina $A''B''C''D''$, as in Fig. 1.11.5.

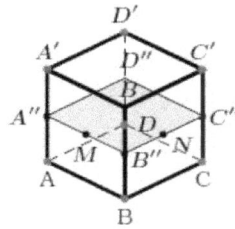

**FIGURE 1.11.5** Example 1.11.2.

First, notice that $\left\|\overrightarrow{AB'}\right\| = \left\|\overrightarrow{BC'}\right\| = a\sqrt{2}$ . Put

$$\frac{\left\|\overrightarrow{AM}\right\|}{\left\|\overrightarrow{AB'}\right\|} = x \Rightarrow \frac{\left\|\overrightarrow{MB'}\right\|}{\left\|\overrightarrow{AB'}\right\|} = \frac{\left\|\overrightarrow{AB'}\right\| - \left\|\overrightarrow{AM}\right\|}{\left\|\overrightarrow{AB'}\right\|} = 1 - x.$$

Now, as $\Delta B'AB \sim \Delta B'MB''$ and $\Delta BC'B' \sim \Delta BNB''$ .

$$\frac{\left\|\overrightarrow{MB'}\right\|}{\left\|\overrightarrow{AB'}\right\|} = \frac{\left\|\overrightarrow{B''B'}\right\|}{\left\|\overrightarrow{BB'}\right\|}, \quad \frac{\left\|\overrightarrow{MB'}\right\|}{\left\|\overrightarrow{AB'}\right\|} = \frac{\left\|\overrightarrow{MB''}\right\|}{\left\|\overrightarrow{AB}\right\|} \Rightarrow \left\|\overrightarrow{MB''}\right\| = (1-x)a,$$

$$\frac{\left\|\overrightarrow{BB''}\right\|}{\left\|\overrightarrow{BB'}\right\|} = \frac{\left\|\overrightarrow{AM}\right\|}{\left\|\overrightarrow{AB'}\right\|}, \quad \frac{\left\|\overrightarrow{B''N}\right\|}{\left\|\overrightarrow{B'C'}\right\|} = \frac{\left\|\overrightarrow{BB''}\right\|}{\left\|\overrightarrow{BB'}\right\|} \Rightarrow \left\|\overrightarrow{B''N}\right\| = xa.$$

Since $\left\|\overrightarrow{MN}\right\| = \frac{\sqrt{5}}{3}a$ , by the Pythagorean theorem,

$$\left\|\overrightarrow{MN}\right\|^2 = \left\|\overrightarrow{MB''}\right\|^2 + \left\|\overrightarrow{B''N}\right\|^2 \Rightarrow \frac{5}{9}a^2 = (1-x)^2 a^2 + x^2 a^2 \Rightarrow x \in \left\{\frac{1}{3}, \frac{2}{3}\right\}.$$

There are two possible positions for the segment, giving the solutions

$$\frac{\left\|\overrightarrow{AM}\right\|}{\left\|\overrightarrow{AB'}\right\|} = \frac{\left\|\overrightarrow{BN}\right\|}{\left\|\overrightarrow{BC'}\right\|} = \frac{1}{3}, \quad \frac{\left\|\overrightarrow{AM}\right\|}{\left\|\overrightarrow{AB'}\right\|} = \frac{\left\|\overrightarrow{BN}\right\|}{\left\|\overrightarrow{BC'}\right\|} = \frac{2}{3}.$$

**EXAMPLE 1.11.3**

Find the direction ratios and direction cosines of a point $\begin{pmatrix} 1 \\ \\ \\ \end{pmatrix}$ (4, 5, -2) in 3D

geometry

## EXERCISES 1.11

**1.11.1**  In a regular tetrahedron with vertices A, B, C, D and with $\left\|\overrightarrow{AB}\right\| = a$, points M and N are the midpoints of the edges [AB] and [CD], respectively.

    **1.**  Find the length of the segment [MN].

    **2.**  Find the angle between the lines [MN] and [BC].

**1.11.2**  In cube ABCDD'C'B'A' of edge of length $a$, find the distance between the lines that contain the diagonals [A'B] and [AC].

**1.11.3**  Prove that the sum of the squares on the diagonals of a parallel-epiped is equal to four times the sum of the squares on the three coterminous edges.

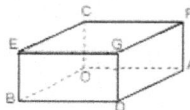

**FIGURE 1.11.6** Exercise 1.11.2.

**1.11.4**  In a quadrilateral $ABCD$, prove that

$$AB^2 + BC^2 + CD^2 + DA^2 = AC^2 + BD^2 + 4PQ^2$$

where, $P$ and $Q$ are the middle points of the diagonals $AC$ and $BD$.

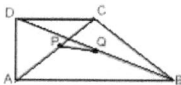

**1.11.5**  Let $S$ be the solid in three-dimensional space consisting of all points $(x, y, z)$ satisfying the following system of six conditions:

$$x \geq 0, y \geq 0, z \geq 0,$$
$$x + y + z \leq 11,$$
$$2x + 4y + 3z \leq 36,$$
$$2x + 3z \leq 24.$$

Determine the number of vertices and the number of edges of $S$.

## 1.12 CAVALIERI AND THE PAPPUS–GULDIN RULES

**Theorem 1.12.1 (Cavalieri's principle):** All planar regions with cross sections of proportional length at the same height have an area in the same proportion. All solids with cross sections of proportional areas at the same height have their volume in the same proportion.

**Proof:**

We only provide the proof for the second statement, as the proof for the first is similar. Cut any two such solids by horizontal planes that produce cross sections of area $A(x)$ and $cA(x)$, where $c > 0$ is the constant of proportionality, at an arbitrary height $x$ above a fixed base. From elementary calculus, we know that $\int_{x_1}^{x_2} A(x)dx$ and $\int_{x_1}^{x_2} cA(x)dx$ give the volume of the portion of each solid cut by all horizontal planes as x runs over some interval $[x_1; x_2]$. As $\int_{x_1}^{x_2} A(x)dx = \int_{x_1}^{x_2} cA(x)dx$, the corresponding volumes must also be proportional.

### EXAMPLE 1.12.1

Use Cavalieri's principle in order to deduce that the area enclosed by the ellipse with equation $\dfrac{x^2}{a^2} + \dfrac{y^2}{b^2} = 1, a > 0, b > 0$, is $\pi ab$.

*Solution:*

Consider the circle with equation $x^2 + y^2 = a^2$, as in Fig. 1.12.1. Then, for $y > 0$, $y = \sqrt{a^2 - x^2}$, $y = \dfrac{b}{a}\sqrt{a^2 - x^2}$.

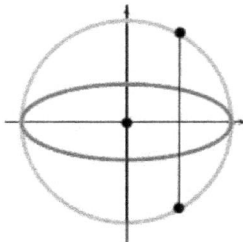

**FIGURE 1.12.1** Ellipse and circle.

The corresponding ordinate for the ellipse and the circle are proportional, and hence, the corresponding chords for the ellipse and the circle will be proportional. By Cavalieri's first principle,

Area of the ellipse $= \dfrac{b}{a}$ ( Area of the circle)

$$= \frac{b}{a}\left(\pi a^2\right)$$

$$= \pi a b .$$

### EXAMPLE 1.12.2

Use Cavalieri's principle in order to deduce that the volume of a sphere with radius $a$ is $\dfrac{4}{3}\pi a^3$ .

### *Solution:*

The following method is due to Archimedes, who was so proud of it that he wanted a sphere inscribed in a cylinder on his tombstone. We need to recall that the volume of a right circular cone with base radius $a$ and height $h$ is $\dfrac{\pi a^2 h}{3}$ .

Consider a hemisphere of radius $a$, as in Fig. 1.12.2.

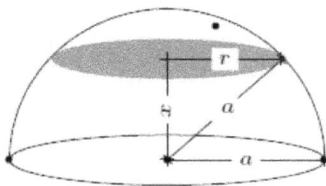

**FIGURE 1.12.2** Hemisphere.

Cut a horizontal slice at height $x$, producing a circle of radius $r$. By the Pythagorean theorem, $x^2 + r^2 = a^2$, and so this circular slab has area $\pi r^2 = \pi \left( a^2 - x^2 \right)$. Now, consider a punctured cylinder of base radius $a$ and height $a$, as in Fig. 1.12.3, with a cone of height $a$ and base radius $a$ cut from it. A horizontal slab at height $x$ is an annular region of area $\pi a^2 - \pi x^2$, which agrees with a horizontal slab for the sphere at the same height.

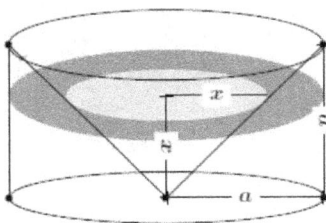

**FIGURE 1.12.3** Punctured cylinder.

By Cavalieri's principle,

Volume of the hemisphere = Volume of the punctured cylinder

$$= \pi a^2 - \frac{\pi a^3}{3}$$

$$= \frac{2\pi a^3}{3}.$$

It follows that the volume of the sphere is $2\left( \dfrac{2\pi a^3}{3} \right) = \dfrac{4\pi a^3}{3}$.

Essentially, the same method of proof as Cavalieri's principle gives the next result.

**Theorem 1.12.2 (Pappus–Guldin Rule):** The area of the lateral surface of a solid of revolution is equal to the product of the length of the generating curve on the side of the axis of revolution and the length of the path described by the center of gravity of the generating curve under a full revolution. The volume of a solid of revolution is equal to the product of the area of the generating plane on one side of the revolution axis and the length of the path described by the center of gravity of the area under a full revolution about the axis.

## EXAMPLE 1.12.3

Since the center of gravity of a circle is at its center, by the Pappus–Guldin rule, the surface area of the torus with the generating circle having radius $r$, and radius of gyration $R$ (as in Fig. 1.12.4) is $(2\pi r)(2\pi R) = 4\pi^2 rR$. Also, the volume of the solid torus is $(\pi r^2)(2\pi R) = 2\pi^2 r^2 R$.

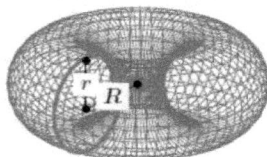

**FIGURE 1.12.4** A torus.

## EXERCISES 1.12

**1.12.1** Use the Pappus–Guldin Rule to find the lateral area and the volume of a right circular cone with base radius $r$ and height $h$.

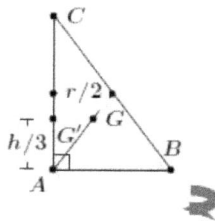

**FIGURE 1.12.5** Generating a cone, Exercise 1.12.1.

**1.12.2** Use the Pappus–Guldin rule to find the lateral area and the volume of a rectangular cylinder with base radius $r$ and height $h$

**1.12.3** Use the Pappus–Guldin rule to find the lateral area and the volume of a semicircle sphere with base radius $r$ and height $h$.

**1.12.4** Find the volume for spherical cap with radius $r$ and height $h$, with base area $A$.

**1.12.5** A large plastic balloon with a thin metal coating used for satellite communications has a diameter of 60 m. Find its area and volume.

**1.12.6** Tell whether the statement true or false.

1. A radius of a small circle of a sphere is a radius of the sphere.

2. A diameter of a great circle of a sphere is the diameter of the sphere.

## 1.13 DIHEDRAL ANGLES AND PLATONIC SOLIDS

**Definition 1.13.1** When two half-planes intersect in space they intersect on a line. The portion of space bounded by the half-planes and the line is called the *dihedral angle*. The intersecting line is called the *edge* of the dihedral angle and each of the two half-planes of the dihedral angle is called a *face*. See Fig. 1.13.1.

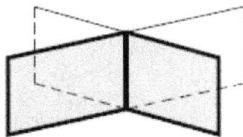

**FIGURE 1.13.1** Dihedral angles.

**Definition 1.13.2** The *rectilinear angle* of a dihedral angle is the angle whose sides are perpendicular to the edge of the dihedral angle at the same point, on each of the faces. See Fig. 1.13.2.

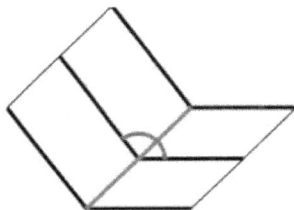

**FIGURE 1.13.2** Rectilinear of a dihedral angle.

All the rectilinear angles of a dihedral angle measure the same. Hence the measure of a dihedral angle is the measure of any one of its rectilinear angles.

In analogy to dihedral angles, we now define polyhedral angles.

**Definition 1.13.3**    The opening of three or more planes that meet at a common point is called a *polyhedral angle* or *solid angle*. In the particular case of three planes, we use the term *trihedral angle*. The common point is called the *vertex* of the polyhedral angle. Each of the intersecting lines of two consecutive planes is called an *edge* of the polyhedral angle. The portion of the planes lying between consecutive edges are called the *faces* of the polyhedral angle. The angles formed by adjacent edges are called *face angles*. A polyhedral angle is said to be *convex* if the section made by a plane cutting all its edges forms a convex polygon.

In the trihedral angle of Fig. 1.13.3, $V$ is the vertex, $\Delta VAB$, $\Delta VBC$, $\Delta VCA$ are faces. Also, notice that in any polyhedral angle, any two adjacent faces form a dihedral angle.

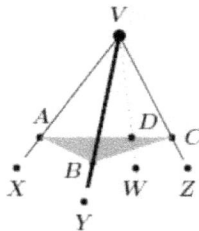

**FIGURE 1.13.3** Trihedral angle.

**Theorem 1.13.1:**    The sum of any two face angles of a trihedral angle is greater than the third face angle.

**Proof:**    Consider Fig. 1.13.3. If $\angle ZVX$ is smaller or equal to in size than either $\angle XVY$ or $YVZ$, then we are done, so assume that, say, $\angle ZVX > XVY$. We must demonstrate that

$$\angle XVY + \angle YVZ > \angle ZVX.$$

Since we are assuming that $\angle ZVX > XVY$, we may draw, in $\angle XVY$ the line segment $[VW]$ such that $\angle XVW = \angle XVY$.

Through any point $D$ of the segment $[VW]$, draw $\triangle ADC$ on the plane $P$ containing the points $V$, $X$, $Z$. Take the point $B \in [VY]$ so that $VD = VB$. Consider now the plane containing the line segment $[AC]$ and the point $B$. Observe that $\triangle AVD \cong AVB$. Hence $AD = AB$. Now, by the triangle inequality in $\triangle ABC$, $AB + BC > CA$. This implies that $\angle BVC > \angle DVC$. Hence

$$\angle AVB + \angle BVC = \angle AVD + \angle BVC$$

$$> \angle AVD + \angle DVC$$

$$= \angle AVC,$$

which proves that $\angle XVY + \angle YVZ > \angle ZVX$, as wanted.

**Theorem 1.13.2**   The sum of the face angles of any convex polyhedral angle is less than $2\pi$ radians.

**Proof:**   Let the polyhedral angle have $n$ faces and vertex $V$. Let the faces be cut by a plane, intersecting the edges at the points $A_1, A_2, \ldots A_n$, say. An illustration can be seen in Fig. 1.13.4, where for convenience, we have depicted only five edges.

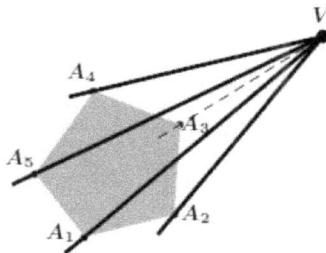

*FIGURE 1.13.4* Polyhedral angle.

Observe that the polygon $A_1 A_2 \ldots A_n$ is convex and that the sum of its interior angles is $\pi(n-2)$.

We would like to prove that

$$\angle A_1 V A_2 + \angle A_2 V A_3 + \angle A_3 V A_4 + \cdots + \angle A_{n-1} V A_n + \angle A_n V A_1 < 2\pi .$$

Now, let $A_{k-1}$, $A_k$, $A_{k+1}$, be three consecutive vertices of the polygon $A_1 A_2 \ldots A_n$. This notation means that $A_{k-1} A_k A_{k+1}$, represents any of the $n$ triplets $A_1 A_2 A_3, A_2 A_3 A_4, A_3 A_4 A_5, \ldots, A_{n-2} A_{n-1} A_n, A_{n-1} A_n A_1, A_n A_1 A_2$, that is, we let $A_0 = A_n$, $A_{n+1} = A_1$, $A_{n+2} = A_2$, etc. Consider the trihedral angle with vertex $A_k$ and whose face angles at $A_k$ are $\angle A_{k-1} A_k A_{k+1}$, $\angle V A_k A_{k-1}$, and $\angle V A_k A_{k+1}$, as in Fig. 1.13.5.

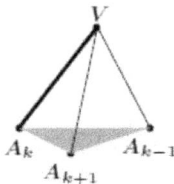

**FIGURE 1.13.5** $A, A_k, A_{k+1}$ are three consecutive vertices.

Observe that as $k$ ranges from 1 through $n$, the sum

$$\sum_{1 \le k \le n} \angle A_{k-1} A_k A_{k+1} = \pi (n-2),$$

be of the interior angles of the polygon $A_1 A_2 \cdots A_n$. By Theorem 1.13.1,

$$\angle V A_k A_{k-1} + \angle V A_k A_{k+1} > \angle A_{k-1} A_k A_{k+1}.$$

Thus

$$\sum_{1 \le k \le n} V A_k A_{k-1} + \angle V A_k A_{k+1} > \sum_{1 \le k \le n} \angle A_{k-1} A_k A_{k+1} = \pi (n-2).$$

Also,

$$\sum_{1 \le k \le n} V A_k A_{k+1} + \angle V A_{k+1} A_k + \angle A_k V A_{k+1} = \pi n,$$

since this is summing the sum of the angles of the n triangles of the faces. But clearly

$$\sum_{1 \le k \le n} V A_k A_{k+1} = \sum_{1 \le k \le n} \angle V A_{k+1} A_k,$$

since one sum adds the angles in one direction and the other in the opposite direction. For the same reason,

$$\sum_{1 \le k \le n} VA_k A_{k-1} = \sum_{1 \le k \le n} \angle VA_k A_{k+1} \; .$$

Hence

$$\sum_{1 \le k \le n} \angle A_k VA_{k+1} = \pi n - \sum_{1 \le k \le n} \left( VA_k A_{k+1} + \angle VA_{k+1} A_k \right)$$

$$= \pi n - \sum_{1 \le k \le n} \left( VA_k A_{k+1} + \angle VA_k A_{k-1} \right)$$

$$= \pi n - \pi \left( n - 2 \right)$$

$$= 2\pi ,$$

as we needed to show.

**Definition 1.13.4**   A *Platonic solid* is a polyhedron having congruent regular polygon as faces and having the same number of edges meeting at each corner.

Suppose a regular polygon with $n \ge 3$ sides is a face of a platonic solid with $m \ge 3$ faces meeting at a corner. Since each interior angle of this polygon measures $\dfrac{\pi (n-2)}{n}$, we must have in view of Theorem 1.13.2,

$$m\left( \frac{\pi (n-2)}{n} \right) < 2\pi \Rightarrow m(n-2) < 2n \Rightarrow (m-2)(n-2) < 4.$$

Since $n \ge 3$ and $m \ge 3$, the preceding inequality only holds for five pairs $(n,m)$. Appealing to Euler's formula for polyhedrons, which states that $V + F = E + 2$, where $V$ is the number of vertices, $F$ is the number of faces, and $E$ is the number of edges of a polyhedron, we obtain the values in the following table.

| *m* | *n* | *S* | *E* | *F* | Name of regular polyhedron |
|---|---|---|---|---|---|
| 3 | 3 | 4 | 6 | 4 | Tetrahedron or regular Pyramid. |
| 4 | 3 | 8 | 12 | 6 | Hexahedron or cube. |
| 3 | 4 | 6 | 12 | 8 | Octahedron. |
| 5 | 3 | 20 | 30 | 12 | Dodecahedron. |
| 3 | 5 | 12 | 30 | 20 | Icosahedron. |

Thus, there are at most five Platonic solids. That there are exactly five can be seen by explicit construction. Figs. 1.13.6 through 1.13.10 depict the Platonic solids.

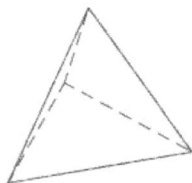

**FIGURE 1.13.6** Tetrahedron.

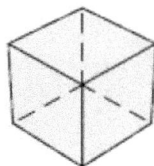

**FIGURE 1.13.7** Cube or hexahedron.

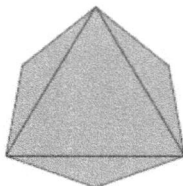

**FIGURE 1.13.8** Octahedron.

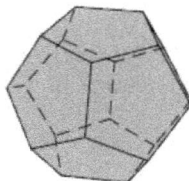

**FIGURE 1.13.9** Dodecahedron.

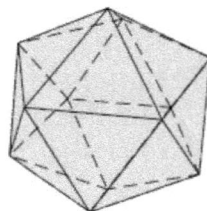

**FIGURE 1.13.10** Icosahedron.

## EXERCISES 1.13

**1.13.1** What are the five regular polyhedrons known as Platonic solids?

**1.13.2** Each Platonic solid by pairs $(m,n)$, where $m$ is the number of edges of each face (or the number of vertices of each face), and $n$ is the number of faces meeting at each vertex (or the number of edges meeting at each vertex). Write the formula for the total number of vertices $(V)$, edges $(E)$, and faces $(F)$, in terms of $m$ and $n$.

**1.13.3** Each Platonic solid by pairs $(m,n)$, where $m$ is the number of edges of each face (or the number of vertices of each face), and $n$ is the number of faces meeting at each vertex (or the number of edges meeting at each vertex). Write the formula for the dihedral angle, $\theta$, of the solid $(m,n)$.

**1.13.4** Each Platonic solid by pairs $(m,n)$, where $m$ is the number of edges of each face (or the number of vertices of each face), and $n$ is the number of faces meeting at each vertex (or the number of edges meeting at each vertex). Write the formula for the surface area $A$, and the volume $V$, of the solid $(m,n)$ with edge length $g$ and in radius $r$.

## 1.14 SPHERICAL TRIGONOMETRY

Consider a point $B(x, y, z)$ in Cartesian coordinates. From $O(0, 0, 0)$, we draw a straight line to $B(x, y, z)$, and let its distance be $\rho$. We measure its inclination from the positive $z$-axis. Let us say it is an angle of $\varphi$, $\varphi \in [0;\pi]$ radians, as in Fig. 1.14.1.

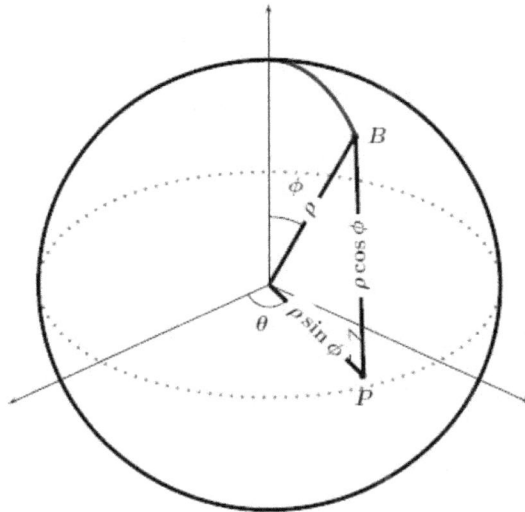

**FIGURE 1.14.1** Spherical coordinates.

Observe that $z = \rho \ cos \ \varphi$. We now project the line segment $[OB]$ onto the $xy$- plane in order to find the polar coordinates of $x$ and $y$. Let $\theta$ be angle that this projection makes with the positive $x$-axis.

Since $OP = \rho \ sin \ \varphi$, we find $x = \rho \ cos \ \theta \ sin \ \varphi$, $y = \rho \ sin \ \theta \ sin \ \varphi$.

**Definition 1.14.1** Given a point $(x, y, z)$ in Cartesian coordinates, its *spherical coordinates* are given by

$$x = \rho \cos \theta \sin \varphi, \; y = \rho \sin \theta \sin \varphi, \; z = \rho \cos \varphi.$$

Here $\varphi$ is the *polar angle*, measured from the positive $z$-axis, and $\theta$ is the *azimuthal angle*, measured from the positive $x$-axis. By convention, $0 \leq \theta \leq 2\pi$ and $0 \leq \varphi \leq \pi$.

Spherical coordinates are extremely useful when considering regions, which are symmetric about a point.

**Definition 1.14.2** If a plane intersects with a sphere, the intersection will be a circle. If this circle contains the center of the sphere, we call it a *great circle*. Otherwise, we talk of a *small circle*. The *axis* of any circle on a sphere is the diameter of the sphere, which is normal to the plane containing the circle. The endpoints of such a diameter are called the *poles* of the circle.

NOTE *The radius of a great circle is the radius of the sphere. The poles of a great circle are equally distant from the plane of the circle, but this is not the case in a small circle. By the pole of a small circle, we mean the closest pole to the plane containing the circle. A pole of a circle is equidistant from every point of the circumference of the circle.*

**Definition 1.14.3** Given the center of the sphere, and any two points on the surface of the sphere, a plane can be drawn. This plane will be unique if and only if the points are not diametrically opposite. In the case where the two points are not diametrically opposite, the great circle formed is split into a larger and a smaller arc by the two points. We call the smaller arc the *geodesic* joining the two points. If the two points are diametrically opposite then every plane containing the line forms with the sphere a great circle, and the arcs formed are then of equal length. In this case, we take any such arc as a geodesic.

**Definition 1.14.4** A *spherical triangle* is a triangle on the surface of a sphere all whose vertices are connected by geodesics. The three arcs of great circles, which form a spherical triangle, are called the sides of the spherical triangle; the angles formed by the arcs at the points where they meet are called the *angles* of the spherical triangle.

If $A, B, C$ are the vertices of a *spherical triangle*, it is customary to label the opposite arcs with the same letter name, but in lowercase.

**NOTE** *A spherical triangle has then six angles: three vertex angles $\angle A, \angle B, \angle C$, and three arc angles, $\angle a, \angle b, \angle c$. Observe that if O is the center of the sphere then*

$$\angle a = \angle\left(\overrightarrow{OB}, \overrightarrow{OC}\right), \quad \angle b = \angle\left(\overrightarrow{OC}, \overrightarrow{OA}\right), \quad \angle c = \angle\left(\overrightarrow{OA}, \overrightarrow{OB}\right),$$

and

$$\angle A = \angle\left(\overrightarrow{OA} \times \overrightarrow{OB}, \overrightarrow{OA} \times \overrightarrow{OC}\right), \quad \angle B = \angle\left(\overrightarrow{OB} \times \overrightarrow{OC}, \overrightarrow{OB} \times \overrightarrow{OA}\right), \quad \angle C = \angle\left(\overrightarrow{OC} \times \overrightarrow{OA}, \overrightarrow{OC} \times \overrightarrow{OB}\right).$$

**Theorem 1.14.1:** Let $\triangle ABC$ be a spherical triangle. Then

$$cosa\ cosb + sina\ sinb\ cosC = cosc.$$

**Proof:** Consider a spherical triangle $ABC$ with $A\left(x_1, y_1, z_1\right)$, $B\left(x_2, y_2, z_2\right)$, and let $O$ be the center and $\rho$ be the radius of the sphere. In spherical coordinates, this is, say,

$$z_1 = \rho\ cos\theta_1, \qquad\qquad x_1 = \rho\ sin\theta_1\ cos\varphi_1, \qquad\qquad y_1 = \rho\ sin\theta_1\ sin\varphi_1$$

$$z_2 = \rho\ cos\theta_2, \qquad\qquad x_2 = \rho\ sin\theta_2\ cos\varphi_2, \qquad\qquad y_2 = \rho\ sin\theta_2\ sin\varphi_2;$$

By a rotation, we may assume that the $z$-axis passes through $C$. Then the following quantities give the square of the distance of the line segment $[AB]$:

$$\left(x_1 - x_2\right)^2 + \left(y_1 - y_2\right)^2 + \left(z_1 - z_2\right)^2, \quad \rho^2 + \rho^2 - 2\rho^2 cos\angle\left(AOB\right).$$

Since $x_1^2 + y_1^2 + z_1^2 = \rho^2$, $x_2^2 + y_2^2 + z_2^2 = \rho^2$, we gather that $x_1 x_2 + y_1 y_2 + z_1 z_2 = \rho^2 cos\angle\left(AOB\right)$. Therefore we obtain

$$cos\theta_2\ cos\theta_1 + sin\theta_2\ sin\theta_1\ cos\left(\varphi_1 - \varphi_2\right) = cos\angle\left(AOB\right),$$

that is,

$$cosa\ cosb + sina\ sinb\ cosC = cos\ c.$$

**Theorem 1.14.2:** Let $I$ be the dihedral angle of two adjacent faces of a regular polyhedron. Then

$$\sin\frac{I}{2} = \frac{\cos\dfrac{\pi}{n}}{\sin\dfrac{\pi}{m}}.$$

**Proof:** Let $AB$ be the edge common to the two adjacent faces, $C$ and $D$ the centers of the faces; bisect $AB$ at $E$, and join $CE$ and $DE$; $CE$ and $DE$ will be perpendicular to $AB$, and the angle $CED$ is the angle of inclination of the two adjacent faces; we shall denote it by $I$. In the plane containing $CE$ and $DE$ draw $CO$ and $DO$ at right angles to $CE$ and $DE$, respectively, and meeting at $O$; about $O$ as center describe a sphere meeting $OA$, $OC$, $OE$ at $a$, $c$, $e$ respectively, so that $cae$ forms a spherical triangle. Since $AB$ is perpendicular to $CE$ and $DE$, it is perpendicular to the plane $CED$, therefore the plane $AOB$ which contains $AB$ is perpendicular to the plane $CED$; hence the angle $cea$ of the spherical triangle is a right angle. Let m be the number of ides in each face of the polyhedron, $n$ the number of the plane angles which form each solid angle. Then the angle $ace = ACE = \dfrac{2\pi}{2m} = \dfrac{\pi}{m}$; and the angle $cae$ is half one of the $n$ equal angles formed on the sphere round $a$, that is, $cae = \dfrac{2\pi}{2n} = \dfrac{\pi}{n}$. From the right-angled triangle $cae$

$$cos\ cae = cos\ cOe\ sin\ ace,$$

that is

$$\cos\frac{\pi}{2} = \cos\left(\frac{\pi}{2} - \frac{I}{2}\right)\sin\frac{\pi}{m};$$

therefore,

$$\sin\frac{I}{2} = \frac{\cos\dfrac{\pi}{n}}{\sin\dfrac{\pi}{m}}.$$

**Theorem 1.14.3:** Let $r$ and $R$ be, respectively, the radii of the inscribed and circumscribed spheres of a regular polyhedron. Then

$$r = \frac{a}{2}\cot\frac{\pi}{m}\tan\frac{I}{2}, \quad R = \frac{a}{2}\tan\frac{I}{2}\tan\frac{\pi}{n}.$$

Here $a$ is the length of any edge of the polyhedron, and $I$ is the dihedral angle of any two faces.

**Proof:** Let the edge $AB = a$, let $OC = r$ and $OA = R$, so that $r$ is the radius of the inscribed sphere, and $R$ is the radius of the circumscribed sphere. Then

$$CE = AE\cot ACE = \frac{a}{2}\cot\frac{\pi}{m},$$

$$r = CE\tan CEO = CE\tan\frac{I}{2} = \frac{a}{2}\cot\tan\frac{\pi}{m}\tan\frac{I}{2};$$

also

$$r = R\cos aOc = R\cot eca\cot eac = R\cot\frac{\pi}{m}\cot\frac{\pi}{n};$$

therefore

$$R = r\tan\frac{\pi}{m}\tan\frac{\pi}{n} = \frac{a}{2}\tan\frac{I}{2}\tan\frac{\pi}{n}.$$

From the previous formula, we now easily find that the volume of the pyramid, which has one face of the polyhedron for base and $O$ for vertex is $\frac{r}{3}\cdot\frac{ma^2}{4}\cot\frac{\pi}{m}$, and therefore the volume of the polyhedron is $\frac{mFra^2}{12}\cot\frac{\pi}{m}$.

Furthermore, the area of one face of the polyhedron is $\frac{ma^2}{4}\cot\frac{\pi}{m}$, and therefore the surface area of the polyhedron is $\frac{mFa^2}{4}\cot\frac{\pi}{m}$.

## EXERCISES 1.14

**1.14.1**  The four vertices of a regular tetrahedron are

$$V_1 = \begin{pmatrix} 1 \\ 0 \\ 0 \end{pmatrix}, \ V_2 = \begin{pmatrix} -1/2 \\ \sqrt{3}/2 \\ 0 \end{pmatrix}, \ V_3 = \begin{pmatrix} -1/2 \\ -\sqrt{3}/2 \\ 0 \end{pmatrix}, \ V_4 = \begin{pmatrix} 0 \\ 0 \\ \sqrt{2} \end{pmatrix}.$$

What is the cosine of the dihedral angle between any pair of faces of the tetrahedron?

**1.14.2**  Consider a tetrahedron whose edge measures $a$. Show that its volume is $\dfrac{a^3\sqrt{2}}{12}$, its surface area is $a^2\sqrt{3}$, and that the radius of the inscribed sphere is $\dfrac{a\sqrt{6}}{12}$.

**1.14.3**  Consider a cube whose edge measures $a$. Show that its volume is $a^3$, its surface area is $6a^2$, and that the radius of the inscribed sphere is $\dfrac{a}{2}$.

**1.14.4**  Consider an octahedron whose edge measures $a$. Show that its volume is $\dfrac{a^3\sqrt{2}}{3}$, its surface area is $2a^2\sqrt{3}$, and that the radius of the inscribed sphere is $\dfrac{a\sqrt{6}}{6}$.

**1.14.5**  Consider a dodecahedron whose edge measures $a$. Show that its volume is $\dfrac{a^3}{4}\left(15+7\sqrt{5}\right)$, its surface area is $3a^2\sqrt{25+10\sqrt{5}}$, and that the radius of the inscribed sphere is $\dfrac{a}{4}\sqrt{10+22\sqrt{\dfrac{1}{5}}}$.

**1.14.6**  Consider an icosahedron whose edge measures $a$. Show that its volume is $\dfrac{5a^3}{12}\left(5+\sqrt{5}\right)$, its surface area is $5a^2\sqrt{3}$, and that the radius of the inscribed sphere is $\dfrac{a}{12}\left(5\sqrt{3}+\sqrt{15}\right)$.

## 1.15 CANONICAL SURFACES

In this section, we consider various surfaces that we shall periodically encounter in subsequent sections. Just like in one-variable Calculus it is important to identify the equation and the shape of a line, a parabola, a circle, etc., it will become important for us to be able to identify certain families of often-occurring surfaces. We shall explore both their Cartesian and their parametric form. We remark that in order to parameterize curves ("one-dimensional entities"), we needed one parameter and that in order to parameterize surfaces we shall need two parameters.

Let us start with the plane. Recall that if $a$, $b$, $c$ are real numbers, not all zero, then the Cartesian equation of a plane with normal vector $\begin{bmatrix} a \\ b \\ c \end{bmatrix}$ and passing through the point $(x_0, y_0, z_0)$ is

$$a(x - x_0) + b(y - y_0) + c(z - z_0) = 0.$$

If we know that the vectors $\vec{u}$ and $\vec{v}$ are on the plane (parallel to the plane) then with the parameters $p$, $q$, the equation of the plane is

$$x - x_0 = pu_1 + qv_1,$$
$$y - y_0 = pu_2 + qv_2,$$
$$z - z_0 = pu_3 + qv_3.$$

**Definition 1.15.1** A surface $S$ consisting of all lines parallel to a given line $\Delta$ and passing through a given curve $\Gamma$ is called a *cylinder*. The line $\Delta$ is called the *directrix* of the cylinder.

NOTE    *To recognize whether a given surface is a cylinder we look at its Cartesian equation. If it is of the form $f(A, B) = 0$, where A, B are secant planes, then the curve is a cylinder. Under these conditions, the lines generating S will be parallel to the line of equation $A = 0$, $B = 0$.. In practice, if one of the variables x, y, or z is missing, then the surface is a cylinder, whose directrix will be the axis of the missing coordinate.*

## EXAMPLE 1.15.1

Fig. 1.15.1 shows the cylinder with Cartesian equation $x^2 + y^2 = 1$. One starts with the circle $x^2 + y^2 = 1$ on the $xy$-plane and moves it up and down the $z$-axis. A parameterization for this cylinder is the following:

$$x = \cos v, \quad y = \sin v, \quad z = u, \quad u \in \mathbb{R}, v \in [0; 2\pi].$$

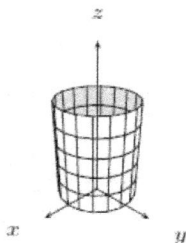

**FIGURE 1.15.1** Circular cylinder $x^2 + y^2 = 1$.

The *Maple*$^{TM}$ commands to graph this surface are:

```
> with(plots) :
> implicitplot3d(x^2 + y^2 = 1, x = -1 ..1, y = -1 ..1, z = -10 ..10);
```

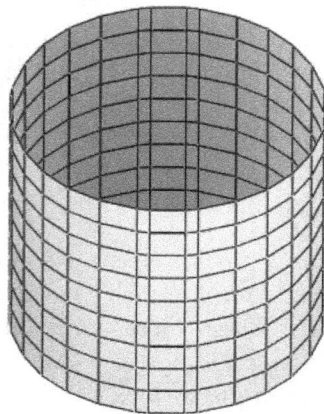

> $plot3d( [\cos(s), \sin(s), t], s = -10 ..10, t = -10 ..10, numpoints = 5001 )$

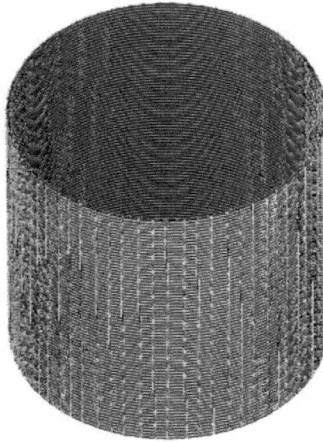

The method of parameterization previously utilized for the cylinder is quite useful when doing parameterizations in space. We refer to it as the method of cylindrical coordinates. In general, we first find the polar coordinates of $x, y$ in the $xy$-plane, and then lift $(x, y, 0)$ parallel to the z-axis to $(x, y, z)$:

$$x = r\cos\theta, \quad y = r\sin\theta, \quad z = z.$$

See Fig. 1.15.2.

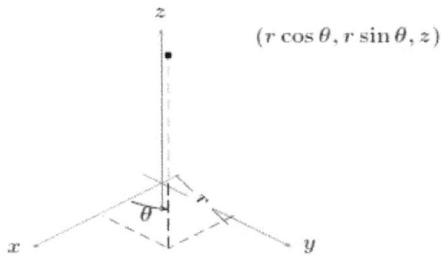

**FIGURE 1.15.2** Cylindrical coordinates.

## EXAMPLE 1.15.2

Fig. 1.15.3 shows the parabolic cylinder with Cartesian equation $z = y^2$. One starts with the parabola $z = y^2$ on the $yz$-plane and moves it up and down the $x$-axis. A parameterization for this parabolic cylinder is the following:

$$x = u, \quad y = v, \quad z = v^2, \quad u \in \mathbb{R}, v \in \mathbb{R}.$$

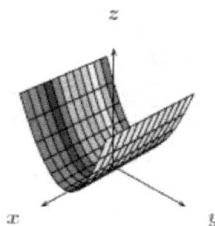

**FIGURE 1.15.3** The parabolic cylinder $z = y^2$.

The *Maple*™ commands to graph this surface are:

```
>  with(plots) :
>
   implicitplot3d(z = y^2, x = -10..10, y = -10..10, z = -10..10, numpoints
      = 5001);
```

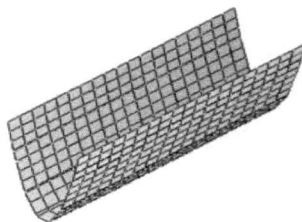

```
>
   plot3d([t, s, s^2], s = -10..10, t = -10..10, numpoints = 5001, axes
      = boxed);
```

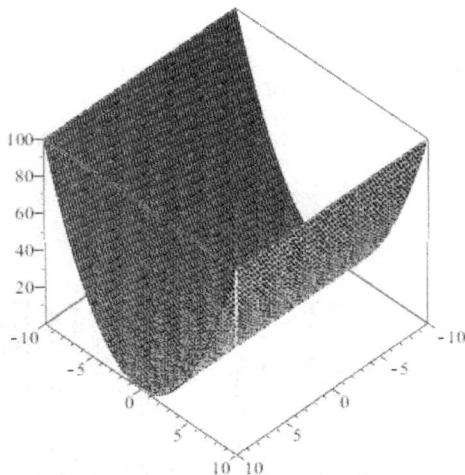

## EXAMPLE 1.15.3

Fig. 1.15.4 shows the hyperbolic cylinder with Cartesian equation $x^2 - y^2 = 1$. One starts with the hyperbola $x^2 - y^2 = 1$ on the $xy$-plane and moves it up and down the $z$-axis. A parameterization for this parabolic cylinder is the following:

$$x = \pm \cosh v, \quad y = \sinh v, \quad z = v, \quad u \in \mathbb{R}, v \in \mathbb{R}.$$

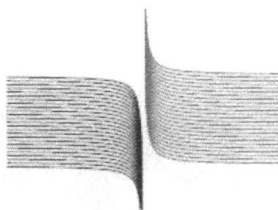

**FIGURE 1.15.4** The hyperbolic cylinder $x^2 - y^2 = 1$.

We need a choice of sign for each of the portions. We have used the fact that $cosh^2 v - sinh^2 v = 1$. The $Maple^{TM}$ commands to graph this surface are:

> *with( plots ) :*

> *implicitplot3d( x^2 − y^2 = 1, x = −10 ..10, y = −10 ..10, z = −10 ..10,*
> *numpoints = 5001 );*

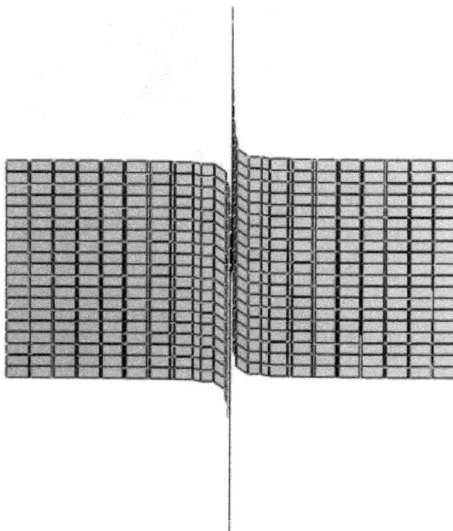

> *plot3d( {[ −cosh(s), sinh(s), t], [cosh(s), sinh(s), t]}, s = −2 ..2, t = −10*
> *..10, numpoints = 5001, axes = boxed );*

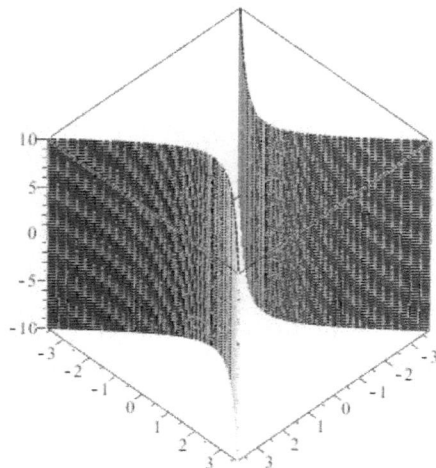

**Definition 1.15.2**  Given a point $\Omega \in \mathbb{R}^3$ (called the apex) and a curve (called the generating curve), the surface $s$ obtained by drawing rays from $\Omega$ and passing through $\Gamma$ is called a *cone*.

**NOTE**  *In practice, if the Cartesian equation of a surface can be put into the form* $f\left(\dfrac{A}{C}, \dfrac{B}{C}\right) = 0$, *where A, B, C, are planes secant at exactly one point, then the surface is a cone, and its apex is given by* $A = 0$, $B = 0$, $C = 0$.

### EXAMPLE 1.15.4

The surface in $\mathbb{R}^3$ implicitly given by $z^2 = x^2 + y^2$ is a cone, as its equation can be put in the form $\left(\dfrac{x}{z}\right)^2 + \left(\dfrac{y}{z}\right)^2 - 1 = 0$. Considering the planes $x = 0$, $y = 0$, $z = 0$, the apex is located at $(0,0,0)$. The graph is shown in Fig. 1.15.5

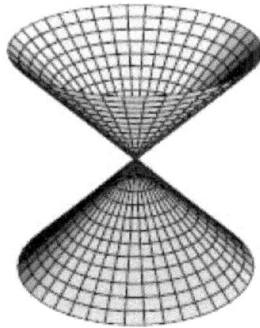

FIGURE 1.15.5  Cone $\dfrac{x^2}{a^2} + \dfrac{y^2}{b^2} = \dfrac{z^2}{c^2}$.

**Definition 1.15.3**  A surface $S$ obtained by making a curve $\Gamma$ turn around a line $\Delta$ is called a *surface of revolution*. We then say that $\Delta$ is the axis of revolution. The intersection of $S$ with a half-plane bounded by $\Delta$ is called a *meridian*.

**NOTE**  *If the Cartesian equation of S can be put in the form* $f(A, \Sigma) = 0$, *where A is a plane and $\Sigma$ is a sphere, then the surface is of revolution. The axis of S is the line passing through the center of $\Sigma$ and perpendicular to the plane A.*

**EXAMPLE 1.15.5**

Find the equation of the surface of revolution generated by revolving the hyperbola

$$x^2 - 4z^2 = 1,$$

about the $z$-axis.

*Solution:*

Let $(x, y, z)$ be a point on $S$. If this point were on the $xz$ plane, it would be on the hyperbola, and its distance to the axis of rotation would be $|x| = \sqrt{1 + 4z^2}$. Anywhere else, the distance of $(x, y, z)$ to the axis of rotation is the same as the distance of $(x, y, z)$ to $(0, 0, z)$, that is $\sqrt{x^2 + y^2}$. We must have

$$\sqrt{x^2 + y^2} = \sqrt{1 + 4z^2},$$

which is to say

$$x^2 + y^2 - 4z^2 = 1.$$

This surface is called a hyperboloid of one sheet. See Fig. 1.15.6.

**FIGURE 1.15.6** One-sheet hyperboloid $\dfrac{z^2}{c^2} = \dfrac{x^2}{a^2} + \dfrac{y^2}{b^2} - 1$.

Observe that when $z = 0$, $x^2 + y^2 = 1$ is a circle on the $xy$-plane. When $x = 0, y^2 - 4z^2 = 1$, is a hyperbola on the $yz$-plane. When $y = 0, x^2 - 4z^2 = 1$ is a hyperbola o the $xz$-plane.

A parameterization for this hyperboloid is

$$x = \sqrt{1 + 4u^2} \cos v, \quad y = \sqrt{1 + 4u^2} \sin v, \quad z = u, \quad u \in \mathbb{R}, v \in [0; 2\pi].$$

**EXAMPLE 1.15.6**

The circle $(y-a)^2 + z^2 = r^2$, on the $yz$-plane ($a$, $r$ are positive real numbers) is revolved around the $z$-axis, forming a torus $T$. Find the equation of this torus.

**Solution:**

Let $(x, y, z)$ be a point on $T$. If this point were on the $yz$- plane, it would be on the circle, and of the distance to the axis of rotation would be $y = a + \mathrm{sgn}(y-a)\sqrt{r^2 - z^2}$, where $\mathrm{sgn}(t)$ (with $\mathrm{sgn}(t) = -1$ if $t < 0, \mathrm{sgn}(t) = 1$ if $t > 0$, and $\mathrm{sgn}(0) = 0$) is the sign of $t$. Anywhere else, the distance from $(x, y, z)$ to the $z$-axis is the distance of this point to the point $(x, y, z)$: $\sqrt{x^2 + y^2}$.

We must have

$$x^2 + y^2 = \left(a + \mathrm{sgn}(y-a)\sqrt{r^2 - z^2}\right)^2 = a^2 + 2a\,\mathrm{sgn}(y-a)\sqrt{r^2 - z^2} + r^2 - z^2.$$

Rearranging

$$x^2 + y^2 + z^2 - a^2 - r^2 = 2a\,\mathrm{sgn}(y-a)\sqrt{r^2 - z^2},$$

or

$$\left(x^2 + y^2 + z^2 - \left(a^2 + r^2\right)\right)^2 = 4a^2 r^2 - 4a^2 z^2.$$

Since $\left(\mathrm{sgn}(y-a)\right)^2 = 1$, (it could not be 0, why?). Rearranging again,

$$\left(x^2 + y^2 + z^2\right)^2 - 2\left(a^2 + r^2\right)\left(x^2 + y^2\right) + 2\left(a^2 - r^2\right)z^2 + \left(a^2 - r^2\right)^2 = 0.$$

The equation of the torus thus, is of the fourth degree, and its graph appears in Fig. 1.15.7.

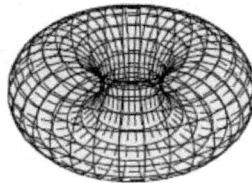

**FIGURE 1.15.7** The torus.

A parameterization for the torus generated by revolving the circle $(y-a)^2 + z^2 = r^2$ around the $z$-axis is

$$x = a\cos\theta + r\cos\theta\cos a, \quad y = a\sin\theta + r\sin\theta\cos a, \quad z = r\sin a, \quad \text{with}$$

$$(\theta, a) \in [-\pi; \pi]^2.$$

## EXAMPLE 1.15.7

The surface $z = x^2 + y^2$ is called an *elliptic paraboloid*. The equation clearly requires that $z \geq 0$. For fixed $z = c$, $c > 0$, $x^2 + y^2 = c$ is a *circle*. When $y = 0$, $z = x^2$ is a parabola on the $xz$-plane. When $x = 0$, $z = y^2$ is a parabola on the $yz$-plane. See Fig. 1.15.8. The following is a parameterization of this paraboloid:

$$x = \sqrt{u}\cos v, \quad y = \sqrt{u}\sin v, \quad z = u, \quad u \in [0; +\infty[, v \in [0; 2\pi].$$

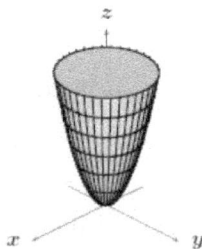

**FIGURE 1.15.8** Paraboloid $z = \dfrac{x^2}{a^2} + \dfrac{y^2}{b^2}$.

## EXAMPLE 1.15.8

The surface $z = x^2 - y^2$ is called a *hyperbolic paraboloid* or *saddle*. If $z = 0$, $x^2 - y^2 = 0$ is a pair of lines in the $xy$- plane. When $y = 0$, $z = x^2$ is a parabola on the $xz$- plane. When $x = 0$, $z = -y^2$ is a parabola on the $yz$-plane. See Fig. 1.15.9. The following is a parameterization of this hyperbolic paraboloid:

$$x = u, \quad y = v, \quad z = u^2 - v^2, \quad u \in \mathbb{R}, v \in \mathbb{R}.$$

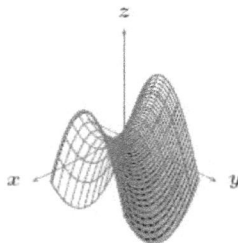

**FIGURE 1.15.9** Hyperbolic paraboloid $z = \dfrac{x^2}{a^2} - \dfrac{y^2}{b^2}$ .

## EXAMPLE 1.15.9

The surface $z^2 = x^2 + y^2 + 1$ is called a *hyperboloid of two sheets*. For $x^2 + y^2 < 1$, $z^2 < 0$, is impossible, and hence, there is no graph when $x^2 + y^2 < 1$. When $y = 0, z^2 - x^2 = 1$, is a hyperbola on the $xz$-plane. When $x = 0$, $z^2 - y^2 = -1$ is a hyperbola on the $yz$-plane. When $z = c$ is a constant, then the $x^2 + y^2 = c^2 + 1$, are circles. See Fig. 1.15.10. The following is a parameterization for the top sheet of this hyperboloid of two sheets

$$x = u \cos v, \quad y = u \sin v, \quad z = u^2 + 1, \quad u \in \mathbb{R}, v \in [0; 2\pi],$$

and the following parameterizes the bottom sheet,

$$x = u \cos v, \quad y = u \sin v, \quad z = -u^2 - 1, \quad u \in \mathbb{R}, v \in [0; 2\pi].$$

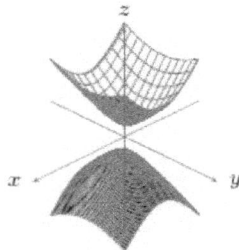

**FIGURE 1.15.10** Two-sheet hyperboloid $\dfrac{z^2}{c^2} = \dfrac{x^2}{a^2} + \dfrac{y^2}{b^2} + 1$ .

### EXAMPLE 1.15.10

The surface $z^2 = x^2 + y^2 - 1$ is called a *hyperboloid of one sheet*. For $x^2 + y^2 < 1$, $z^2 < 0$ is impossible, and hence there is no graph when $x^2 + y^2 < 1$. When $y = 0$, $z^2 - x^2 = -1$ is a hyperbola on the $xz$-plane. When $x = 0$, $z^2 - y^2 = -1$ is a hyperbola on the $yz$-plane. When $z = c$ is a constant, then the $x^2 + y^2 = c^2 + 1$ are circles. See Fig. 1.15.11. The following is a parameterization for this hyperboloid of one sheet

$$x = \sqrt{u^2 + 1}\cos v, \quad y = \sqrt{u^2 + 1}\sin v, \quad z = u, \quad u \in \mathbb{R}, v \in [0; 2\pi].$$

**FIGURE 1.15.11** One-sheet hyperboloid $\dfrac{z^2}{c^2} = \dfrac{x^2}{a^2} + \dfrac{y^2}{b^2} - 1$.

### EXAMPLE 1.15.11

Let $a, b, c$ be strictly positive real numbers. The surface $\dfrac{x^2}{a^2} + \dfrac{y^2}{b^2} + \dfrac{z^2}{c^2} = 1$ is called an *ellipsoid*. For $z = 0$, $\dfrac{x^2}{a^2} + \dfrac{y^2}{b^2} = 1$ is an ellipse on the $xy$ plane. When $y = 0$, $\dfrac{x^2}{a^2} + \dfrac{z^2}{c^2} = 1$ is an ellipse on the $xz$-plane. When $x = 0$, $\dfrac{y^2}{b^2} + \dfrac{z^2}{c^2} = 1$ is an ellipse on the $yz$-plane. See Fig. 1.15.12. We may parameterize the ellipsoid using spherical coordinates:

$$x = a\cos\theta\sin\phi, \quad y = b\sin\theta\sin\phi, \quad z = c\cos\phi, \quad \theta \in [0; 2\pi], \phi \in [0; \pi].$$

**FIGURE 1.15.12** The ellipsoid $\frac{x^2}{a^2}+\frac{y^2}{b^2}+\frac{z^2}{c^2}=1$.

## EXERCISES 1.15

**1.15.1** Find the equation of the surface of revolution $S$ generated by revolving the ellipse $4x^2+z^2=1$ about the $z$-axis.

**1.15.2** Find the equation of the surface of revolution generated by revolving the line $3x+4y=1$ about the $y$-axis.

**1.15.3** Describe the surface parameterized by
$\phi(u,\,v)\mapsto(v\cos u,\,v\sin u,\,au),\,(u,v)\in(0,2\pi)\times(0,\,1),\,a>0.$

**1.15.4** Describe the surface parameterized by
$\phi(u,\,v)=\left(au\cos v,\,bu\sin v,\,u^2\right),\,(u,v)\in(1,+\infty)\times(0,2\pi),\,a,b>0.$

**1.15.5** Use Maple to show the cylinder with Cartesian equation $3x^2+5y^2=1$. One starts with the circle $3x^2+5y^2=1$ on the

$xy$-plane and moves it up and down the z-axis, where $-1\le x\le1$, $-1\le y\le1$, and $-10\le z\le10$.

**1.15.6** Demonstrate that the surface in $\mathbb{R}^3$ $S:e^{x2+y2+z2}-(x+z)e^{-2xz}=0,$ implicitly defined is a cylinder.

**1.15.7** Show that the surface in $\mathbb{R}^3$ implicitly defined by
$x^4+y^4+z^4-4xyz(x+y+z)=1$ is a surface of revolution, and find its axis of revolution.

**1.15.8** Show that the surface $S$ in $\mathbb{R}^3$ given implicitly by the equation $\frac{1}{x-y}+\frac{1}{y-z}+\frac{1}{z-x}=1$ is a cylinder and find the direction of its directrix.

**1.15.9** Show that the surface $S$ in $\mathbb{R}^3$ implicitly defined as
$xy+yz+zx+x+y+z+1=0$ is of revolution and find its axis.

**1.15.10** Demonstrate that the face in $\mathbb{R}^3$ given implicitly by $z^2 - xy = 2z - 1$ is a cone.

**1.15.11** (**Putnam Exam 1970**): Determine, with proof, the radius of the largest circle, which can lie on the ellipsoid $\dfrac{x^2}{a^2} + \dfrac{y^2}{b^2} + \dfrac{z^2}{c^2} = 1$,

$a > b > c > 0$.

**1.15.12** The hyperboloid of one sheet in Fig. 1.15.13 has the property that if it is cut by planes at $z = \pm 2$, its projection on the $xy$ plane produces the ellipse $x^2 + \dfrac{y^2}{4} = 1$, and if it is cut by a at z = 0, its projection on the $xy$-plane produces the ellipse $4x^2 + y^2 = 1$. Find its equation.

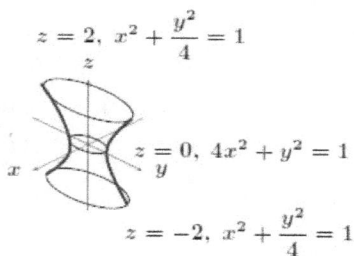

*FIGURE 1.15.13* Exercise 1.15.12.

## 1.16 PARAMETRIC CURVES IN SPACE

In analogy to curves on the plane, we now define curves in space.

**Definition 1.16.1** Let $[a;b] \subseteq \mathbb{R}$. A *parametric curve* representation r of a curve $\Gamma$ is a function $r: [a;b] \to \mathbb{R}^3$, with $r(t) = \begin{pmatrix} x(t) \\ y(t) \\ z(t) \end{pmatrix}$, and such that $r([a;b]) = \Gamma$. $r(a)$ is the *initial point* of the curve and $r(b)$ its *terminal point*.

A curve is *closed* if its initial point and its final point coincide. The *trace* of the curve r is the set of all images of r, that is, $\Gamma$. The length of the curve is $\int_\Gamma \| d\, \vec{r} \|$.

**EXAMPLE 1.16.1**

The trace of $r(t) = \vec{i} \cos t + \vec{j} \sin t + \vec{k} t$ is known as a *cylindrical helix*. See Fig. 1.16.1.

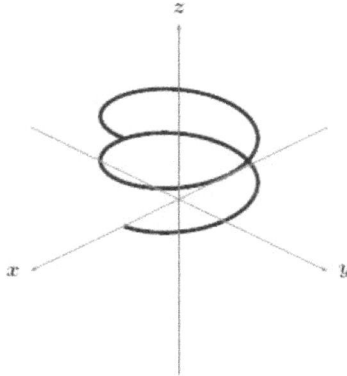

**FIGURE 1.16.1** Helix.

To find the length of the helix as $t$ traverses the interval $[0; 2\pi]$, first observe that

$$\left\| d\,\vec{x} \right\| = \left\| (\sin t)^2 + (-\cos t)^2 + 1 \right\| dt = \sqrt{2}\ dt\,,$$

and thus the length is

$$\int_0^{2\pi} \sqrt{2}\ dt = 2\pi \sqrt{2}\,.$$

The *Maple*$^{TM}$ commands to graph this curve and to find its length are:

```
> with(plots) :
> with(Student[VectorCalculus]) :
> spacecurve([cos(t), sin(t), t], t = 0 ..2 * Pi, axes = normal);
```

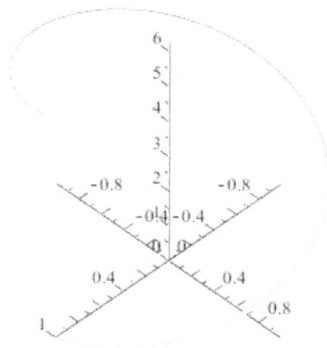

> *PathInt( 1, [x, y, z] = Path( ⟨cos(t), sin(t), t⟩, 0 ..2 \* Pi ));*

$$2\sqrt{2}\,\pi$$

## EXAMPLE 1.16.2

Find a parametric representation for the curve resulting by the intersection of the plane $3x + y + z = 1$ and the cylinder $x^2 + 2y^2 = 1$ in $\mathbb{R}^3$.

*Solution:*

The projection of the intersection of the plane $3x + y + z = 1$ and the cylinder is the ellipse $x^2 + 2y^2 = 1$ on the $xy$-plane. This ellipse can be parameterized as

$$x = \cos t, \quad y = \frac{\sqrt{2}}{2}\sin t, \quad 0 \le t \le 2\pi.$$

From the equation of the plane,

$$z = 1 - 3x - y = 1 - 3\cos t - \frac{\sqrt{2}}{2}\sin t.$$

Thus we may take the parameterization

$$\mathbf{r}(t) = \begin{bmatrix} x(t) \\ y(t) \\ z(t) \end{bmatrix} = \begin{bmatrix} \cos t \\ \dfrac{\sqrt{2}}{2}\sin t \\ 1 - 3\cos t - \dfrac{\sqrt{2}}{2}\sin t \end{bmatrix}.$$

**EXAMPLE 1.16.3**

Let $a$, $b$, $c$ be strictly positive real numbers. Consider the region $\mathfrak{R} = \left\{ (x,y,z) \in \mathbb{R}^3 : |x| \le a, \ |y| \le b, \ z = c \right\}$. A point $P$ moves along the ellipse

$$\frac{x^2}{a^2} + \frac{y^2}{b^2} = 1, \ z = c + 1,$$

once around, and acts as a source light projecting a shadow of $\mathfrak{R}$ onto the $xy$-plane. Find the area of this shadow.

***Solution:***

First consider the same problem as $P$ moves around the circle $x^2 + y^2 = 1$, $z = c + 1$, and the region $\mathfrak{R}' = \left\{ (x,y,z) \in \mathbb{R}^3 : |x| \le 1, \ |y| \le 1, \ z = c \right\}$. See Fig. 1.16.2.

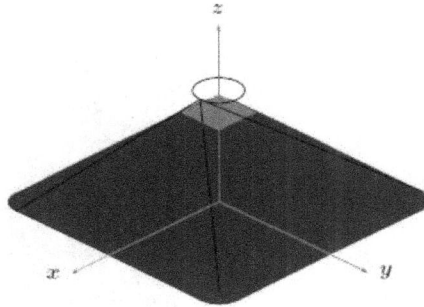

**FIGURE 1.16.2** Example 1.16.3.

For fixed $P(u,v,c+1)$ on the circle, the image of $\mathfrak{R}'$ ( a $2 \times 2$ square) on the $xy$- plane is $a(2c+2) \times (2c+2)$ square with center at the point $Q(-cu,-cv,0)$ (see Fig. 1.16.3).

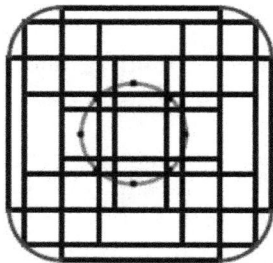

**FIGURE 1.16.3** Example 1.16.3.

As $P$ moves along the circle, $Q$ moves along the circle with equation $x^2 + y^2 = c^2$ on the $x^2 + y^2 = c^2$ on the $xy$-plane (Fig. 1.16.3), being the center of $a(2c+2) \times (2c+2)$ square. This creates a region as in Fig. 1.16.4, where each quarter circle has radius $c$, and the central square has side $2c+2$, of area

$$\pi c^2 + 4(c+1)^2 + 8c(c+1).$$

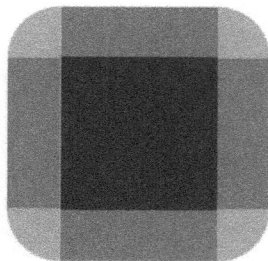

**FIGURE 1.16.4** Example 1.16.3.

Resizing to a region

$$\Re = (x, y, z) \in \mathbb{R}^3 : |x| \le a, |y| \le b, z = c,$$

and an ellipse

$$\frac{x^2}{a^2} + \frac{y^2}{b^2} = 1, \quad z = c+1,$$

we use instead of $c+1$, $a(c+1)$ (parallel to the $x$-axis) and $b(c+1)$ (parallel to the $y$-axis), so that the area shadowed is

$$\pi ab(c+1)^2 + 4ab(c+1)^2 + 4abc(c+1) = c^2 ab(\pi + 12) + 16abc + 4ab.$$

## EXERCISES 1.16

**1.16.1** Let $C$ be the curve in $\mathbb{R}^3$ defined by $x = t^2$, $y = 4t^{3/2}$, $z = 9t$, $t \in [0; +\infty]$.

Calculate the distance along $C$ from $(1, 4, 9)$ to $(16, 32, 36)$.

**1.16.2** Consider the surfaces in $\mathbb{R}^3$ implicitly defined by

$$z - x^2 - y^2 - 1 = 0, \quad z + x^2 + y^2 - 3 = 0.$$

Describe as vividly as possible these surfaces and their intersection, if they at all intersect. Find a parametric equation for the curve on which they intersect, if they at all intersect.

**1.16.3** Consider the space curve $\vec{r} : t \mapsto \begin{bmatrix} \dfrac{t^4}{1+t^2} \\ \dfrac{t^3}{1+t^2} \\ \dfrac{t^2}{1+t^2} \end{bmatrix}$. Let $t_k$, $1 \le k \le 4$ ncn-

zero real numbers.

Prove that $\vec{r}(t_1)$, $\vec{r}(t_2)$, $\vec{r}(t_3)$, and $\vec{r}(t_4)$ are coplanar if and only if

$$\frac{1}{t_1} + \frac{1}{t_2} + \frac{1}{t_3} + \frac{1}{t_4} = 0.$$

**1.16.4** Give a parameterization for the part of the ellipsoid $x^2 + \dfrac{y^2}{9} + \dfrac{z^2}{4} = 1$, which lies on top of the plane $x + y + z = 0$.

**1.16.5** Find the parametric equations that represent the curve, which is the intersection of the surfaces

$$y^2 + z^2 = 16 \text{ and } x = 8 - y^2 - z.$$

**1.16.6**  Let $a$ be a real number parameter, and consider the planes

$$P_1 : ax + y + z = -a,$$
$$P_2 : x - ay + az = -1.$$

Let $l$ be their intersection line.

1.  Find a direction vector for $l$.

2.  As $a$ varies through $\mathbb{R}$, $l$ describes a surface $S$ in $\mathbb{R}^3$. Let $(x, y, z)$ be the point of intersection of this surface and the plane $z = c$. Find an equation relating $x$ and $y$.

3.  Find the volume bounded by the two planes, $x = 0$, and $x = 1$, and the surface $S$ as $c$ varies.

## 1.17 MULTIDIMENSIONAL VECTORS

We briefly describe space in $n$-dimensions. The ideas expounded earlier about the plane and space carry almost without change.

**Definition 1.17.1**  $\mathbb{R}^n$ is the $n$-dimensional space, the collection

$$\mathbb{R}^n = \left\{ \begin{pmatrix} x_1 \\ x_2 \\ \vdots \\ x_n \end{pmatrix} : x_k \in \mathbb{R} \right\}.$$

**Definition 1.17.2**  If $\vec{a}$ and $\vec{b}$ are two vector in $\mathbb{R}^n$ their vector sum $\vec{a} + \vec{b}$ is defined by the coordinatewise addition

$$\vec{a} + \vec{b} = \begin{bmatrix} a_1 + b_1 \\ a_2 + b_2 \\ \vdots \\ a_n + b_n \end{bmatrix}.$$

**Definition 1.17.3** A real number $a \in \mathbb{R}$ will be called a *scalar*. If $a \in \mathbb{R}$ and $\vec{\mathbf{a}} \in \mathbb{R}^n$, we define *scalar multiplication* of a vector and a scalar by the coordinatewise

$$a \ \vec{\mathbf{a}} = \begin{bmatrix} a \, a_1 \\ a \, a_2 \\ \vdots \\ a \, a_n \end{bmatrix}.$$

**Definition 1.17.4** The *standard ordered basis* for $\mathbb{R}^n$ is the collection of vectors $\{\vec{\mathbf{e}}_1, \vec{\mathbf{e}}_2, \dots \vec{\mathbf{e}}_n\}$ with $\vec{\mathbf{e}}_k = \begin{bmatrix} 0 \\ \vdots \\ 1 \\ \vdots \\ 0 \end{bmatrix}$.

(a 1 in the $k$ slot and 0's everywhere else). Observe that

$$\sum_{k=1}^{n} a_k \vec{\mathbf{e}}_k = \begin{bmatrix} a_1 \\ a_2 \\ \vdots \\ a_n \end{bmatrix}.$$

**Definition 1.17.5** Given vectors $\vec{\mathbf{a}}$, $\vec{\mathbf{b}}$ of $\mathbb{R}^n$ their *dot product* is

$$\vec{\mathbf{a}} \cdot \vec{\mathbf{b}} = \begin{bmatrix} a_1 \\ a_2 \\ \vdots \\ a_n \end{bmatrix} \cdot \begin{bmatrix} b_1 \\ b_2 \\ \vdots \\ b_n \end{bmatrix} = a_1 b_1 + a_2 b_2 + \cdots + a_n b_n = \sum_{k=1}^{n} a_k b_k.$$

**NOTE** *The norm of a vector in $\mathbb{R}^n$ is given by* $\|\vec{x}\| = \sqrt{\vec{x} \cdot \vec{x}}$ .

We now establish one of the most useful inequalities in analysis.

**Theorem 1.17.1 (Cauchy–Bunyakovsky–Schwartz Inequality):**  Let $\vec{x}$ and $\vec{y}$ be any two vectors in $\mathbb{R}^n$. Then we have

$$\left|\vec{x}\bullet\vec{y}\right| \le \left\|\vec{x}\right\| \left\|\vec{y}\right\|.$$

**Proof:** Since the norm of any vector is non-negative, we have

$$\left\|\vec{x}+t\vec{y}\right\| \ge 0 \Leftrightarrow \left(\vec{x}+t\vec{y}\right)\bullet\left(\vec{x}+t\vec{y}\right) \ge 0$$

$$\Leftrightarrow \vec{x}\bullet\vec{x} + 2t\vec{x}\bullet\vec{y} + t^2\vec{y}\bullet\vec{y} \ge 0$$

$$\Leftrightarrow \left\|\vec{x}\right\|^2 + 2t\vec{x}\bullet\vec{y} + t^2\left\|\vec{y}\right\|^2 \ge 0.$$

This last expression is a quadratic polynomial in t, which is always non-negative. As such its discriminant must be non-positive, that is,

$$\left(2\vec{x}\bullet\vec{y}\right)^2 - 4\left(\left\|\vec{x}\right\|^2\right)\left(\left\|\vec{y}\right\|^2\right) \le 0 \Leftrightarrow \left|\vec{x}\bullet\vec{y}\right| \le \left\|\vec{x}\right\|\left\|\vec{y}\right\|,$$

giving the theorem.

**NOTE**  *The preceding proof works for any vector space (cf. below) that has an inner product.*

The form of the Cauchy-Bunyakovsky-Schwarz most useful to us will be

$$\left|\sum_{k=1}^{n} x_k y_k\right| \le \left(\sum_{k=1}^{n} x_k^2\right)^{1/2} \left(\sum_{k=1}^{n} y_k^2\right)^{1/2}, \tag{1.22}$$

or real numbers $x_k$, $y_k$.

**Corollary 1.17.1 (Triangle Inequality):**  Let $\vec{a}$ and $\vec{b}$ be any two vectors in $\mathbb{R}^n$. Then we have

$$\left\|\vec{a}+\vec{b}\right\| \le \left\|\vec{a}\right\| + \left\|\vec{b}\right\|.$$

**Proof:**

$$\left\|\vec{a}+\vec{b}\right\|^2 = \left(\vec{a}+\vec{b}\right)\bullet\left(\vec{a}+\vec{b}\right)$$
$$= \vec{a}\bullet\vec{a}+2\vec{a}\bullet\vec{b}+\vec{b}\bullet\vec{b}$$
$$\leq \left\|\vec{a}\right\|^2 + 2\left\|\vec{a}\right\|\bullet\left\|\vec{b}\right\| + \left\|\vec{b}\right\|^2$$
$$= \left(\left\|\vec{a}\right\|+\left\|\vec{b}\right\|\right)^2,$$

from where the desired result follows.

Again, the preceding proof is valid in any vector space that has a norm. We now consider a generalization of the Euclidean norm. Given $p>1$ and $\vec{x}\in\mathbb{R}^n$, we put

$$\left\|\vec{x}\right\|_p = \left(\sum_{k=1}^{n}\left|x_k\right|^p\right)^{\frac{1}{p}} \tag{1.23}$$

Clearly,

$$\left\|\vec{x}\right\|_p \geq 0 \tag{1.24}$$

$$\left\|\vec{x}\right\|_p = 0 \Leftrightarrow \vec{x}=\vec{0} \tag{1.25}$$

$$\left\|a\,\vec{x}\right\|_p = \left|a\right|\,\left\|\vec{x}\right\|_p,\ a\in\mathbb{R} \tag{1.26}$$

We now prove analogues of the Cauchy–Bunyakovsky–Schwarz and the Triangle Inequality for $\left\|\cdot\right\|_p$. For this, we need the following lemma.

**1.17.1 Lemma (Young's Inequality):** Let $p>1$ and put $\dfrac{1}{p}+\dfrac{1}{q}=1$. Then for $(a,b)\in\left([0;+\infty[\right)^2$, we have $ab\leq\dfrac{(a)^p}{p}+\dfrac{(b)^q}{q}$.

**Proof:**

Let $0<k<1$, and consider the function

$$f:\begin{matrix}[0;+\infty[\ \to\ \mathbb{R}\\ x\to x^k-k(x-1)\end{matrix}$$

Th $0 = f'(x) = kx^{k-1} - k \Leftrightarrow x = 1$.     Since     $f''(x) = k(k-1)x^{k-2} < 0$     for

$0 < k < 1, x \geq 0, x = 1$ is a maximum point. Hence $f(x) \leq f(1)$ for $x \geq 0$, that

is $x^k \leq 1 + k(x-1)$. Letting $k = \dfrac{1}{p}$ and $x = \dfrac{(a)^p}{(b)^q}$, we deduce

$\dfrac{a}{(b)^{\frac{q}{p}}} \leq 1 + \dfrac{1}{p}\left(\dfrac{(a)^p}{(b)^q} - 1\right)$. Rearranging gives,

$$ab \leq (b)^{1+\frac{p}{q}} + \dfrac{(a)^p (b)^{1+\frac{p}{q}-p}}{p} - \dfrac{(b)^{1+\frac{p}{q}}}{p}$$

From where we obtain the inequality.

**Definition 1.17.6**  Let $\vec{x}$ and $\vec{y}$ be two non-zero vectors in a vector space over the real numbers. Then the angle $\widehat{(\vec{x},\vec{y})}$ between the is given by the relation

$$\cos\widehat{(\vec{x},\vec{y})} = \dfrac{\vec{x}\bullet\vec{y}}{\|\vec{x}\|\|\vec{y}\|}.$$

This expression agrees with the geometry in the case of the dot product for $\mathbb{R}^2$ and $\mathbb{R}^3$.

## EXAMPLE 1.17.1

Assume that $a_k, b_k, c_k, \ k = 1, \ldots, n$, are positive real numbers. Show that

$$\left(\sum_{k=1}^{n} a_k b_k c_k\right)^4 \leq \left(\sum_{k=1}^{n} a_k^4\right)\left(\sum_{k=1}^{n} b_k^4\right)\left(\sum_{k=1}^{n} c_k^2\right)^2.$$

***Solution:***

Using CBS on $\displaystyle\sum_{k=1}^{n} (a_k b_k) c_k$, once we obtain

$$\sum_{k=1}^{n} a_k b_k c_k \leq \left(\sum_{k=1}^{n} a_k^2 b_k^2\right)^{1/2} \left(\sum_{k=1}^{n} c_k^2\right)^{1/2}.$$

Using CBS again on $\left( \sum\limits_{k=1}^{n} a_k^2 b_k^2 \right)^{1/2}$, we obtain

$$\sum_{k=1}^{n} a_k b_k c_k \leq \left( \sum_{k=1}^{n} a_k^2 b_k^2 \right)^{1/2} \left( \sum_{k=1}^{n} c_k^2 \right)^{1/2}$$

$$\leq \left( \sum_{k=1}^{n} a_k^4 \right)^{1/4} \left( \sum_{k=1}^{n} b_k^4 \right)^{1/4} \left( \sum_{k=1}^{n} c_k^2 \right)^{1/2},$$

which gives the required inequality.

We now use the CBS inequality to establish another important inequality.

**Lemma 1.17.2**   Let $a_k > 0$, $q_k > 0$, with $\sum\limits_{k=1}^{n} q_k = 1$. Then

$$\lim_{x \to 0} \log \left( \sum_{k=1}^{n} q_k a_k^x \right)^{1/x} = \sum_{k=1}^{n} q_k \log a_k.$$

**Proof:** Recall that $\log(1+x) \sim x$ as $x \to 0$. Thus

$$\lim_{x \to 0} \log \left( \sum_{k=1}^{n} q_k a_k^x \right)^{1/x} = \lim_{x \to 0} \frac{\log \left( \sum\limits_{k=1}^{n} q_k a_k^x \right)}{x}$$

$$= \lim_{x \to 0} \frac{\sum\limits_{k=1}^{n} q_k \left( a_k^x - 1 \right)}{x}$$

$$= \lim_{x \to 0} \sum_{k=1}^{n} q_k \frac{\left( a_k^x - 1 \right)}{x}$$

$$= \sum_{k=1}^{n} q_k \log a_k.$$

**Theorem 1.17.2 (Arithmetic Mean–Geometric Mean Inequality):**   Let $a_k \geq 0$. Then

$$\sqrt[n]{a_1 a_2 \cdots a_n} \leq \frac{a_1 + a_2 + \cdots + a_n}{n}.$$

**Proof:** If $b_k \geq 0$, then by CBS

$$\frac{1}{n}\sum_{k=1}^{n} b_k \geq \left(\frac{1}{n}\sum_{k=1}^{n}\sqrt{b_k}\right)^2. \tag{1.27}$$

Successive applications of Equation (1.23) yield the monotone decreasing sequence

$$\frac{1}{n}\sum_{k=1}^{n} a_k \geq \left(\frac{1}{n}\sum_{k=1}^{n}\sqrt{a_k}\right)^2 \geq \left(\frac{1}{n}\sum_{k=1}^{n}\sqrt[4]{a_k}\right)^4 \geq \cdots,$$

which by Lemma 1.17.1 has a limit

$$\exp\left(\frac{1}{n}\sum_{k=1}^{n}\log a_k\right) = \sqrt[n]{a_1 a_2 \cdots a_n} \; ,$$

giving

$$\sqrt[n]{a_1 a_2 \cdots a_n} \leq \frac{a_1 + a_2 + \cdots + a_n}{n} \; ,$$

as wanted.

## EXAMPLE 1.17.2

For any positive integer $n > 1$, we have $1 \cdot 3 \cdot 5 \cdots (2n - 1) < n^n$. For, by AMGM,

$$1 \cdot 3 \cdot 5 \cdots (2n - 1) < \left(\frac{1 + 3 + 5 + \cdots + (2n - 1)}{n}\right)^n = \left(\frac{n^2}{n}\right)^n = n^n.$$

Notice that since the factors are unequal we have strict inequality.

**Definition 1.17.7** Let $a_1 > 0$, $a_2 > 0$, $\ldots$, $a_n > 0$. Their *harmonic mean* is given by

$$\frac{n}{\dfrac{1}{a_1} + \dfrac{1}{a_2} + \cdots + \dfrac{1}{a_n}}.$$

As a corollary to AMGM, we obtain

**Corollary 1.17.2 (Harmonic Mean–Geometric Mean Inequality)**   Let $b_1 > 0$, $b_2 > 0$, ..., $b_n > 0$. Then

$$\frac{n}{\dfrac{1}{b_1} + \dfrac{1}{b_2} + \cdots + \dfrac{1}{b_n}} \leq \left(b_1 b_2 \cdots b_n\right)^{1/n}.$$

**Proof:** This follows by putting $a_k = \dfrac{1}{b_k}$ in Theorem 1.17.2. Then

$$\left(\frac{1}{b_1}\frac{1}{b_2}\cdots\frac{1}{b_n}\right)^{1/n} \leq \frac{\dfrac{1}{b_1} + \dfrac{1}{b_2} + \cdots + \dfrac{1}{b_n}}{n}.$$

Combining Theorem 1.17.2 and Corollary 1.17.2, we deduce

**Corollary 1.17.3 (Harmonic Mean–Arithmetic Mean Inequality)**   Let $b_1 > 0$, $b_2 > 0$, ..., $b_n > 0$. Then

$$\frac{n}{\dfrac{1}{b_1} + \dfrac{1}{b_2} + \cdots + \dfrac{1}{b_n}} \leq \frac{b_1 + b_2 + \cdots + b_n}{n}.$$

**EXAMPLE 1.17.3**

Let $a_k > 0$, and $s = a_1 + a_2 + \cdots + a_n$. Prove that

$$\sum_{k=1}^{n} \frac{s}{s - a_k} \geq \frac{n^2}{n-1}.$$

and

$$\sum_{k=1}^{n} \frac{a_k}{s - a_k} \geq \frac{n}{n-1}.$$

### Solution:

Put $b_k = \dfrac{s}{s - a_k}$. Then

$$\sum_{k=1}^{n} \frac{1}{b_k} = \sum_{k=1}^{n} \frac{s - a_k}{n} = n - 1,$$

and from Corollary 1.17.3,

$$\frac{n}{n-1} \leq \frac{\displaystyle\sum_{k=1}^{n} \frac{s}{s - a_k}}{n},$$

from where the first inequality is proved. Since $\dfrac{s}{s - a_k} - 1 = \dfrac{a_k}{s - a_k}$, we have

$$\sum_{k=1}^{n} \frac{a_k}{s - a_k} = \sum_{k=1}^{n} \left( \frac{s}{s - a_k} - 1 \right)$$

$$= \sum_{k=1}^{n} \left( \frac{s}{s - a_k} \right) - n$$

$$\geq \frac{n^2}{n-1} - n$$

$$= \frac{n}{n-1}.$$

## EXERCISES 1.17

**1.17.1**   The *Arithmetic Mean Geometric Mean Inequality* says that if $a_k \geq 0$, then

$$\left( a_1 a_2 \cdots a_n \right)^{1/n} \leq \frac{a_1 + a_2 + \cdots + a_n}{n}.$$

Equality occurs if and only if $a_1 = a_2 = \cdots = a_n$. In this exercise, you will follow the steps of a proof by George Polya.

**1.**   Prove that $\forall x \in \mathbb{R}, x \leq e^{x-1}$.

**2.** Put $A_k = \dfrac{na_k}{a_1 + a_2 + \cdots + a_n}$, and $G_n = a_1 a_2 \cdots a_n$. Prove that

$$A_1 A_2 \cdots A_n = \frac{n^n G_n}{\left(a_1 + a_2 + \cdots + a_n\right)^n}, \text{ and that}$$

$$A_1 + A_2 + \cdots + A_n = n.$$

**3.** Deduce that $G_n \leq \left(\dfrac{a_1 + a_2 + \cdots + a_n}{n}\right)^n$.

**4.** Prove the AMGM inequality by assembling the preceding results.

**1.17.2** Demonstrate that if $x_1, x_2, \ldots, x_n$ are strictly positive real numbers then

$$\left(x_1 + x_2 + \cdots + x_n\right)\left(\frac{1}{x_1} + \frac{1}{x_2} + \cdots + \frac{1}{x_n}\right) \geq n^2.$$

**1.17.3** (**USAMO 1978**): Let $a, b, c, d, e$ be real numbers such that $a + b + c + d + e = 8$, $a^2 + b^2 + c^2 + d^2 + e^2 = 16$. Maximize the value of $e$.

**1.17.4** Find all the positive real numbers $a_1 \leq a_2 \leq \ldots \leq a_n$ such that

$$\sum_{k=1}^{n} a_k = 96, \quad \sum_{k=1}^{n} a_k^2 = 144, \quad \sum_{k=1}^{n} a_k^3 = 1216.$$

**1.17.5** Demonstrate that for integer $n > 1$, we have

$$n! < \left(\frac{n+1}{2}\right)^n.$$

**1.17.6** Prove the sequence $x_n = \left(1 + \dfrac{1}{n}\right)^n$, $n = 1, 2, \ldots$ is strictly increasing.

**1.17.7** Let $a_k > 0$. Use the /Cauchy-Bunyakovsky-Schwarz Inequality to show that $\left(\displaystyle\sum_{k=}^{n} a_k^2\right)\left(\displaystyle\sum_{k=1}^{n} \frac{1}{a_k^2}\right) \geq n^2.$

**1.17.8** Let $\vec{a} \in \mathbb{R}^n$ be a fixed vector. Demonstrate that $\chi = \{\vec{x} \in \mathbb{R}^n : \vec{a} \bullet \vec{x} = 0\}$ is a subspace of $\mathbb{R}^n$.

**1.17.9** Let $\vec{a}_i \in \mathbb{R}^n, 1 \leq i \leq k,\ k \leq n$, be $k$ non-zero vectors such that $\vec{a}_i \bullet \vec{a}_j = 0$ for $i \neq j$. Prove that these $k$ vectors are linearly independent.

**1.17.10** Let $a_k \geq 0,\ 1 \leq k \leq n$ be arbitrary. Prove that

$$\left(\sum_{k=1}^{n} a_k\right)^2 \leq \frac{n(n+1)(2n+1)}{6} \sum_{k=1}^{n} \frac{a_k^2}{k^2}.$$

# DIFFERENTIATION

Based on the understanding of the concepts of vectors and parametric curves from the previous chapter, in this chapter, we focus on the differentiation of functions of several variables. We mainly discuss some topology, multivariable functions, limits and continuity, the definition of the derivative, the Jacobi matrix, gradients and directional derivatives, Levi-Civitta and Einstein, exterma, and Lagrange multipliers.

## 2.1 SOME TOPOLOGY

**Definition 2.1.1** Let $a \in \mathbb{R}^n$ and let $\varepsilon > 0$. An *open ball* centered at a of radius $\varepsilon$ is the set

$$B_\varepsilon(a) = \left\{ x \in \mathbb{R}^n : \|x - a\| < \varepsilon \right\}.$$

An *open box* is a Cartesian product of open intervals

$$]a_1; b_1[ \times ]a_2; b_2[ \times \cdots \times ]a_{n-1}; b_{n-1}[ \times ]a_n; b_n[,$$

where $a_k, b_k$ are real numbers.

### EXAMPLE 2.1.1

An open ball in $\mathbb{R}$ is an open interval, an open ball in $\mathbb{R}^2$ is an open disk (see Figure 2.1.1) and an open ball in $\mathbb{R}^3$ is an open sphere (see Figure 2.1.2). An

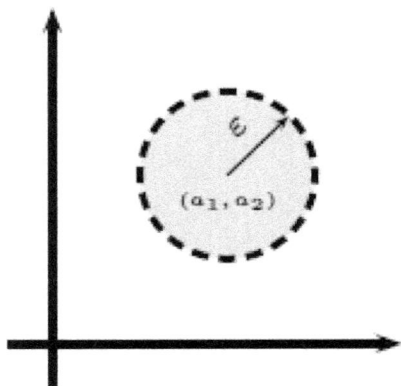

**FIGURE 2.1.1** Open ball in $\mathbb{R}^2$.

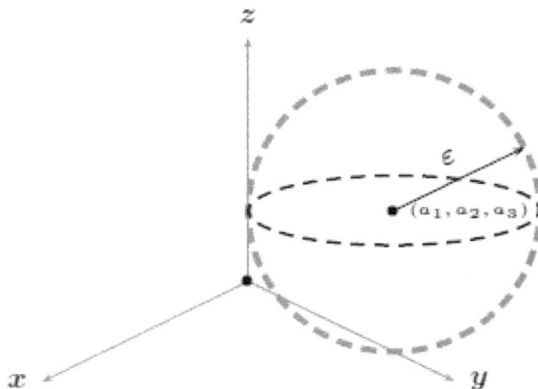

**FIGURE 2.1.2** Open ball in $\mathbb{R}^3$.

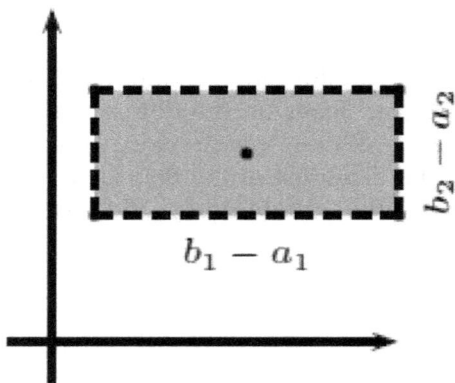

**FIGURE 2.1.3** Open rectangle in $\mathbb{R}^2$.

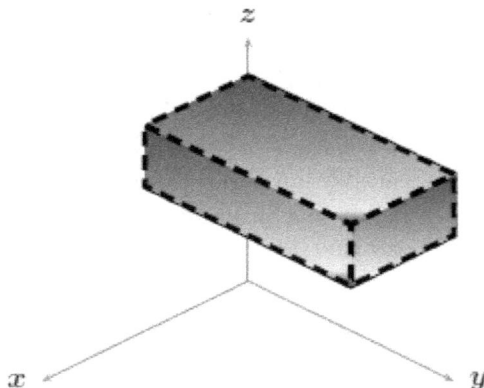

**FIGURE 2.1.4** Open rectangle in $\mathbb{R}^3$.

open box in $\mathbb{R}$ is an open interval, an open box in $\mathbb{R}^2$ is a rectangle without its boundary (see Figure 2.1.3) and an open box in $\mathbb{R}^3$ is a box without its boundary (see Figure 2.1.4).

**Definition 2.1.2** A set $S \subseteq \mathbb{R}^n$ is said *open* if for very point belonging to it we can surround the point by a sufficiently small open ball so that at this ball lay completely within the set. That is, $\forall a \in S, \exists \varepsilon > 0$ such that $B_\varepsilon(a) \subseteq S$.

**EXAMPLE 2.1.2**

The region $]-1;1[$ is open in $\mathbb{R}$. The interval $]-1;1]$ is not open; however, as no interval centered at 1 is totally contained in $]-1;1]$.

**EXAMPLE 2.1.3**

The region $]-1;1[\times]0;+\infty[$ is open in $\mathbb{R}^2$.

**EXAMPLE 2.1.4**

The ellipsoidal region $\{(x,y)\in\mathbb{R}^2 : x^2+4y^2<4\}$ open in $\mathbb{R}^2$.

You will recognize that open boxes, open ellipsoids and their unions, and finite intersections are open sets in $\mathbb{R}^n$.

**Definition 2.1.3**  A set $S\subseteq\mathbb{R}^n$ is said *closed* in $\mathbb{R}$, as its complement, $\mathbb{R}\setminus S$ is open.

**EXAMPLE 2.1.5**

The closed interval $[-1;\ 1]$ is closed in $\mathbb{R}$, as its complement, $\mathbb{R}\setminus[-1;1]=]-\infty;-1[\cup]1;\infty[$ is open in $\mathbb{R}$. However, the interval $]-1;1]$ is neither open nor closed in $\mathbb{R}$.

**EXAMPLE 2.1.6**

The region $[-1;\ 1]\times[0;\ +1[\times[0;\ 2]$ is closed in $\mathbb{R}^3$.

**Definition 2.1.3**  A point $P$ in a set $S\subset\mathbb{R}^n$ is called an *interior point* of $S$ if and only if there is some open ball with center $P$, which contains only points of $S$.

NOTE   *A set is called an open set if and only if all its points are interior points.*

## EXERCISES 2.1

**2.1.1** Determine whether the following sub-sets of $\mathbb{R}^2$ are open, closed, or neither, in $\mathbb{R}^2$.

1. $A = \left\{ (x,y) \in \mathbb{R}^2 : 2 \leq x \leq 4, 1 \leq y \leq 5 \right\}$

2. $B = \left\{ (x,y) \in \mathbb{R}^2 : 0 \leq x^2 + y^2 \leq 4 \right\}$

3. $C = \left\{ (x,y) \in \mathbb{R}^2 : x > 0, y < \sin\left(\dfrac{1}{x}\right) \right\}$

4. $D = \left\{ (x,y) \in \mathbb{R}^2 : x > 0, y > 0 \right\}$

5. $E = \left\{ (x,y) \in \mathbb{R}^2 : x^2 + y^2 < r^2 \right\}$, for $r > 0$

6. $F = \left\{ (x,y) \in \mathbb{R}^2 : (x-a)^2 + (y-b)^2 < r^2 \right\}$, for $r > 0$ and $(a,b) \in \mathbb{R}^2$

7. $G = \left\{ (x,y) \in \mathbb{R}^2 : a < x < b, c < y < d \right\}$ for a rectangle $= \left[ (a,b),(c,d) \right]$, where $(a,b) \in \mathbb{R}^2$ and $(c,d) \in \mathbb{R}^2$

8. $H = \left\{ (x,y) \in \mathbb{R}^2 : 2 \geq y^2 - 4x > 1 \right\}$

9. $I = \left\{ (x,y) \in \mathbb{R}^2 : 2y + x > -1 \right\}$

10. $J = \left\{ (x,y) \in \mathbb{R}^2 : x^2 + y^2 < 5, x^2 - y^2 > 1 \right\}$

11. $K = \left\{ (x,y) \in \mathbb{R}^2 : x^2 + 4y^2 > 4 \right\}$

12. $L = \left\{ (x,y) \in \mathbb{R}^2 : x^2 + 4y^2 \geq 4, x^2 + 4y^2 \leq 16 \right\}$

13. $M = \left\{ (x,y) \in \mathbb{R}^2 : x \geq 2y, y \geq 2x \right\}$

14. $N = \left\{ (x,y) \in \mathbb{R}^2 : x^2 + y^2 = 1 \right\}$

15. $L = \left\{ (x,y) \in \mathbb{R}^2 : \dfrac{x^2}{2} + \dfrac{y^2}{3} < 1 \right\}$

16. $T = \left\{ (x,y) \in \mathbb{R}^2 : |x| < 1, |y| < 1 \right\}$

17. $V = \left\{ (x,y) \in \mathbb{R}^2 : |x| \leq 1, |y| \leq 1 \right\}$

**18.** $V = \left\{ (x,y) \in \mathbb{R}^2 : x^2 + y \geq 0 \right\}$

**19.** $V = \left\{ (x,y) \in \mathbb{R}^2 : x = y \right\}$

**20.** $V = \left\{ (x,y) \in \mathbb{R}^2 : xy > 0 \right\}$

**21.** $V = \left\{ (x,y) \in \mathbb{R}^2 : y = |x-1| + 2 - x \right\}$

**22.** $V = \left\{ (x,y) \in \mathbb{R}^2 : x^2 > y \right\}$

**2.1.2**   Show that the set $A = \left\{ (x,y) \in \mathbb{R}^2 : x^2 + y^2 < r^2 \right\}$, for $r > 0$ is open in $\mathbb{R}^2$.

**HINT**   *If $a > 0$ and $b > 0$, let $\delta > 0$ and $\varepsilon$ be the smaller of $a$ and $b$, and consider $B_\delta\,(a,b)$.*

**2.1.3**   Determine whether the following sub-sets of $\mathbb{R}^3$ are open, closed, or neither, in $\mathbb{R}^3$.

**1.** $A = \left\{ (x,y,z) \in \mathbb{R}^3 : 3 < \left( x^2 + y^2 + z^2 \right)^{\frac{1}{2}} < 6 \right\}$

**2.** $B = \left\{ (x,y,z) \in \mathbb{R}^3 : x^2 + 2y^2 + 3z^2 < 6 \right\}$

**3.** $C = \left\{ (x,y,z) \in \mathbb{R}^3 : x^2 + y^2 < 7, z \equiv 0 \right\}$

**4.** $D = \left\{ (x,y,z) \in \mathbb{R}^3 : (x-a)^2 + (y-b)^2 + (z-c)^2 < \varepsilon^2 \right\}$ with center $(a,b,c)$ and radius $\varepsilon$

**5.** $E = \left\{ (x,y,z) \in \mathbb{R}^3 : (x-a)^2 + (y-b)^2 + (z-c)^2 \leq \varepsilon \right\}$ with center $(a,b,c)$ and radius $\varepsilon$

**6.** $F = \left\{ (x,y,z) \in \mathbb{R}^3 : (x-1)^2 + (y-2)^2 + (z+3)^2 \leq 5 \right\}$

**7.** $B = \left\{ (x,y,z) \in \mathbb{R}^3 : x^2 + y^2 + z^2 \leq 5 \right\}$

**8.** $L = \left\{ (x,y,z) \in \mathbb{R}^3 : \dfrac{x^2}{a^2} + \dfrac{y^2}{b^2} + \dfrac{z^2}{c^2} \leq 1 \right\}$

**9.** $V = \left\{ (x,y,z) \in \mathbb{R}^3 : x^2 + y^2 - z^2 = 0 \right\}$

**10.** $M = \left\{ (x,y,z) \in \mathbb{R}^3 : x^2 + y^2 + z^2 - 4z = 0 \right\}$

**11.** $T = \{(x, y, z) \in \mathbb{R}^3 : x \geq 0, y \geq 0, z \geq 0\}$

**12.** $G = \{(x, y, z) \in \mathbb{R}^3 : \frac{xy}{z}\}$

**13.** $C = \{(x, y, z) \in \mathbb{R}^3 : xy > z\}$

**14.** $B = \{(x, y, z) \in \mathbb{R}^3 : (x - y)^2 = z^2\}$

**15.** $T = \{(x, y, z) \in \mathbb{R}^3 : 1 < x^2 + y^2 + z^2 < 2\}$

**2.1.4**   Determine whether the following statements are true or false.

**1.** The set of $\{(x) \in \mathbb{R} : 0 < x < 1\}$ is open in $\mathbb{R}$.

**2.** If $a$ and $b > 0$, the open rectangle $\{(x, y) \in \mathbb{R}^2 : 0 < x < a, 0 < y < b\}$ is an open set in $\mathbb{R}^2$.

**3.** The interaction of two open sets is an open set.

**4.** The union of two closed sets is a closed set.

**5.** The interaction of two closed sets is a closed set.

**6.** An open ball is a convex set.

**7.** A set that contains its boundary is called a *closed set.*

**2.1.5**   Prove that the union of two open sets is an open set.

**HINT**   *A set is called an* open set *if and only if all its points are interior points.*

**2.1.6**   Let $p(x, y)$ be a polynomial with real coefficients in the real variables $x$ and $y$, defined over the entire plane $\mathbb{R}^2$. What are the possibilities for the image (range) of $p(x, y)$?

**2.1.7**   A set of points of complex number $(z = x + iy)$ is given. Determine whether the set is open, closed, or neither.

**1.** $V$ is the set of all $z$ satisfying $|z - 2| \leq |z + 5i|$.

**2.** The set $M$ consists of all $z$ with $\text{Im}(z) < 9$.

**3.** $D$ is the set of all z such that $1 < \text{Re}(z) \leq 6$.

**4.** The set $C$ consists of all $z$ such that $\text{Re}(z) > (\text{Im}(z))^2$.

5.  $B$ is the set of all numbers $x + iy$ with $x$ and $y$ any rational numbers

6.  $A$ is the set of all $z$ satisfying $|z - i| < 7$.

7.  $F$ is the set of all $z$ satisfying $|z - 2i| < 9$.

8.  The set $T$ consists of all $z = x + iy$ with $x > 0$.

9.  The set $K$ consists of all $z = x + iy$ with $x \geq 0$.

10. The set $E$ consists of all $z = x + iy$ with $-1 \leq x \leq 1$ and $y > 0$.

## 2.2  MULTIVARIABLE FUNCTIONS

Let $A \subseteq \mathbb{R}^n$. For most of this course, our concern will be functions of the form

$$f : A \rightarrow \mathbb{R}^m.$$

If $m = 1$, we say that $f$ is a *scalar field*. If $m \geq 2$, we say that $f$ is a *vector field*.

We would like to develop a calculus analogous to the situation in $\mathbb{R}$. In particular, we would like to examine limits, continuity, differentiability, and integrability of multivariable functions. Needless to say, the introduction of more variables greatly complicates the analysis. For example, recall that the graph of a function $f : A \rightarrow \mathbb{R}^m$, $A \subseteq \mathbb{R}^n$ is the set

$$\{(\mathrm{x}, f(\mathrm{x})) : x \in A\} \subseteq \mathbb{R}^{n+m}.$$

If $m + n > 3$, we have an object of more than three-dimensions. In this case, $n = 2$, $m = 1$, we have a tri-dimensional surface. We will now briefly examine this case.

**Definition 2.2.1**  Let $A \subseteq \mathbb{R}^2$ and let $f : A \rightarrow \mathbb{R}$ be a function. Given $c \in \mathbb{R}$, the *level curve* (or *contour*) at $z = c$ is the curve resulting from the intersection of the surface $z = f(x, y)$ and the plane $z = c$, if there is such a curve.

### EXAMPLE 2.2.1

The level curves of the surface $f(x, y) = x^2 + y^2$ (an elliptic paraboloid) are the concentric circles $x^2 + y^2 = c$, $c > 0$.

## EXAMPLE 2.2.2

Sketch the level curve for $f(x, y) = \frac{1}{3}\sqrt{12 - 6x^2 - 2y^2}$ .

### Solution:

The *Maple*™ commands to graph this function.

> $with(plots)$ :

>

$$contourplot\left(\frac{1}{3}\left(12 - 6x^2 - 2y^2\right)^{\frac{1}{2}}, x = -2..2, y = -2..2, color = blue\right)$$

The graph is shown in Figure 2.2.1(a).

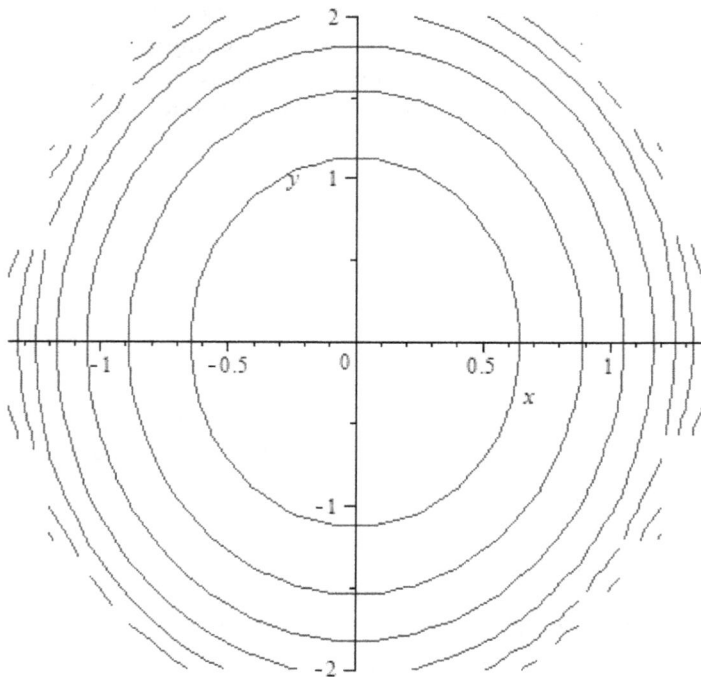

**FIGURE 2.2.1 (a)** Level curve of $f(x, y) = \frac{1}{3}\sqrt{12 - 6x^2 - 2y^2}$ in 2D with Maple.

>

$$plot3d\left(\frac{1}{3}(12 - 6x^2 - 2y^2)^{\frac{1}{2}}, x = -2..2, y = -2..2, axes = frame, style \right.$$

$$\left. = contour, color = blue \right)$$

The three-dimensional plot is presented in Figure 2.2.1(b).

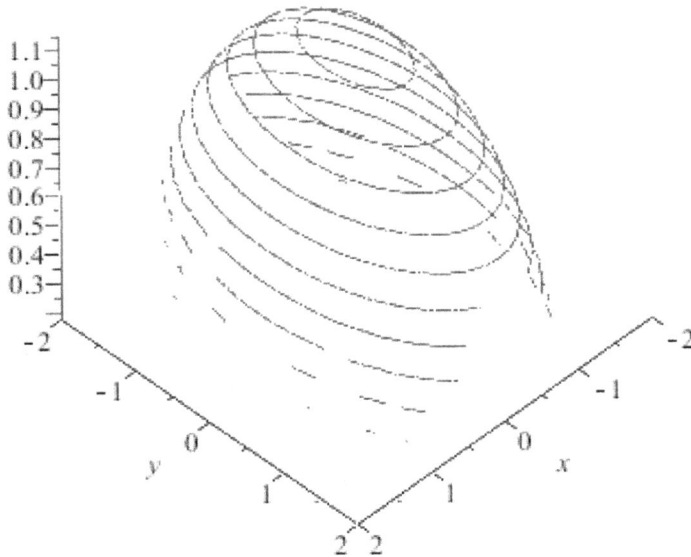

**FIGURE 2.2.1  (b)** Level curve of $f(x, y) = \dfrac{1}{3}\sqrt{12 - 6x^2 - 2y^2}$ in 3D with Maple.

The *MATLAB* commands to graph this function.

```
>> syms x y
>> f =(1/3*sqrt(12-(6*x^2)-2*y^2))

f =

1/3*(12-6*x^2-2*y^2)^(1/2)

>> ezcontour(f, [-2, 2, -2, 2])
```

The graph is shown in Figure 2.2.2(a).

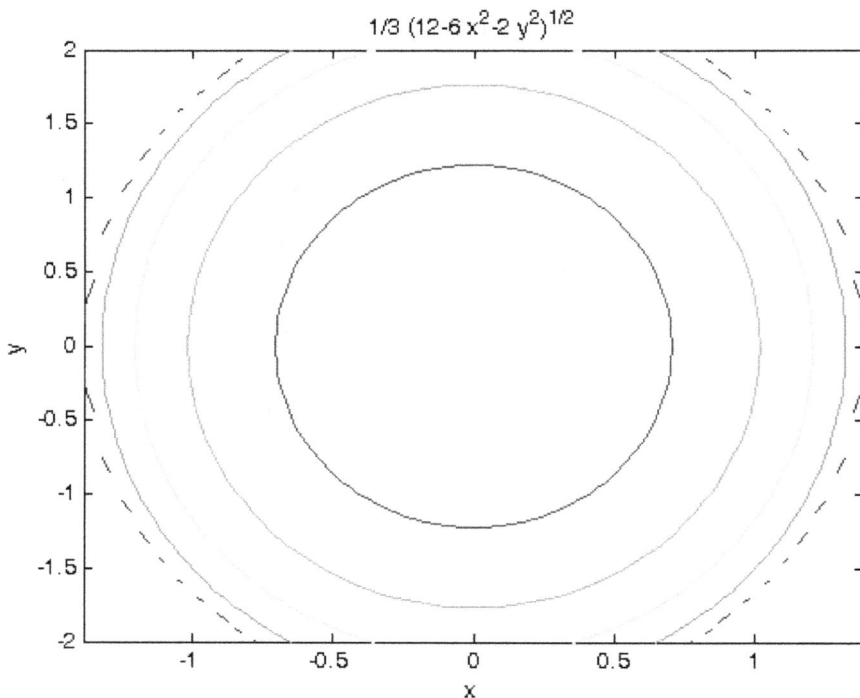

**FIGURE 2.2.2** **(a)** Level curve of $f(x, y) = \dfrac{1}{3}\sqrt{12 - 6x^2 - 2y^2}$ in 2D with MATLAB.

```
>> syms x y
>> f =(1/3*sqrt(12-(6*x^2)-2*y^2))

f =

1/3*(12-6*x^2-2*y^2)^(1/2)

ezsurf(f, [-2, 2, -2, 2])
```

The surface plot is presented in Figure 2.2.2(b).

$$\tfrac{1}{3}\,(12\text{-}6\,x^2\text{-}2\,y^2)^{1/2}$$

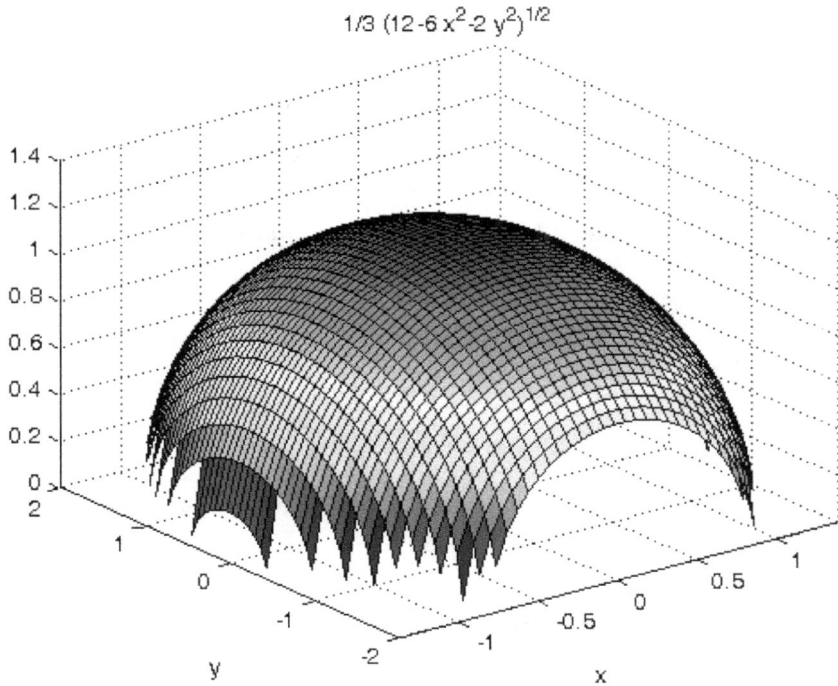

**FIGURE 2.2.2 (b)** Level curve of $f(x,y) = \dfrac{1}{3}\sqrt{12 - 6x^2 - 2y^2}$ in surface form with MATLAB.

**Definition 2.2.2**  Let $A \subseteq \mathbb{R}^3$ and let $f : A \to \mathbb{R}$ be a function. Given $c \in \mathbb{R}$, the *level surface* at $w = c$ is the surface resulting from the intersection of the surface $w = f(x,y,z)$ and the plane $w = c$, if there is such a surface.

In other words, the *level surface* of a function $w = f(x,y,z)$ is the surface in a rectangular three variables coordinate system $(xyz)$ defined by $f(x,y,z) = c$, where $c$ is any constant.

**EXAMPLE 2.2.3**

Sketch the level surface for $f(x,y,z) = x^2 + 4y^2 + 16z^2$.

***Solution:***

The *Maple*$^{\text{Tm}}$ commands to graph this function.

> $f := (x, y, z) \rightarrow x^2 + 4y^2 + 16z^2;$

$$f := (x, y, z) \rightarrow x^2 + 4y^2 + 16z^2$$

> $s1 := implicitplot3d(f(x, y, z) = 2, x = -2..2, y = -2..2, z = -2..2, style = wireframe, color = blue) :$

> $display([s1], axes = framed, scaling = constrained);$

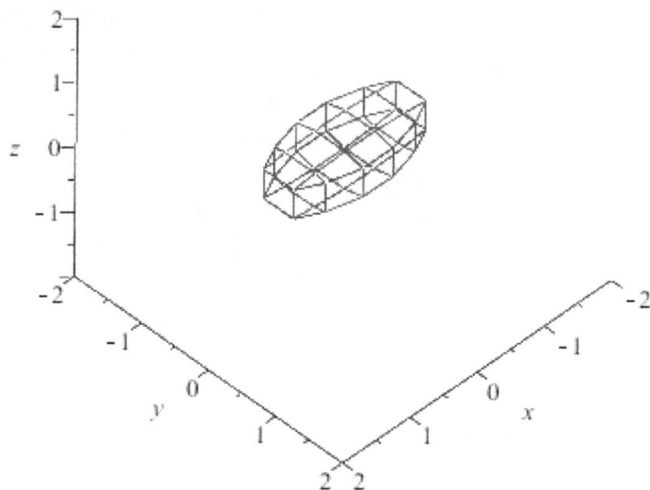

**FIGURE 2.2.3** Level surface $f(x, y, z) = x^2 + 4y^2 + 16z^2$.

## EXERCISES 2.2

**2.2.1** Sketch the level curves in 2D and 3D views for the following maps.

1. $(x, y) \mapsto x + y$
2. $(x, y) \mapsto xy$
3. $(x, y) \mapsto x^3 - y$
4. $(x, y) \mapsto x^2 + 4y^2$
5. $(x, y) \mapsto y^2 - x^2$
6. $(x, y) \mapsto \min(|x|, |y|)$

7. $(x,y) \mapsto \sin(x^2 + y^2)$

8. $(x,y) \mapsto \cos(x^2 - y^2)$

9. $(x,y) \mapsto 5 - x^2 - y^2$

10. $(x,y) \mapsto ye^x$

11. $(x,y) \mapsto \sin x \sin y$

12. $(x,y) \mapsto \ln(x^2 + y^2 - 1)$

13. $(x,y) \mapsto \tan^{-1}\left(\dfrac{y}{x+1}\right)$

14. $(x,y) \mapsto x^{2/3} + y^{2/3}$

15. $(x,y) \mapsto (x+1)^2 + y^2$

**2.2.2**   Sketch the level surfaces for the following maps.

1. $(x,y,z) \mapsto x + y + z$

2. $(x,y,z) \mapsto xyz$

3. $(x,y,z) \mapsto \min(|x|,|y|,|z|)$

4. $(x,y,z) \mapsto x^2 + y^2$

5. $(x,y,z) \mapsto x^2 + 4y^2$

6. $(x,y,z) \mapsto \sin(z - x^2 - y^2)$

7. $(x,y,z) \mapsto x^2 + y^2 + z^2$

8. $(x,y,z) \mapsto \cos^{-1}\sqrt{\dfrac{y-z}{y+z}}$

9. $(x,y,z) \mapsto x^2 + y^2 - z$

10. $(x,y,z) \mapsto \sin\left(\dfrac{x+z}{1-y}\right)$

11. $(x,y,z) \mapsto \ln(x - 2y - 3z + 4)$

12. $(x,y,z) \mapsto \tan^{-1}\left(\dfrac{x+z}{y}\right)$

**2.2.3**    Describe geometrically how a surface $z = g(x,y)$ would have to be transformed in order to obtain each of the following surfaces $z = f(x,y)$, where is

1.  $f(x,y) = g(x,y) + 2$
2.  $f(x,y) = 2g(x,y)$
3.  $f(x,y) = -g(x,y)$
4.  $f(x,y) = 2 - g(x,y)$
5.  $f(x,y) = g(-x,y)$
6.  $f(x,y) = g(2x,y)$
7.  $f(x,y) = -g(-x,-y)$

**2.2.4**    Let $v(t)$ be a strictly increasing function of $t$, and let $f(x,y) = v\big(g(x,y)\big)$. How are the level curves of $g(x,y)$ and $f(x,y)$ related?

## 2.3   LIMITS AND CONTINUITY

We start this section with the notion of *limit*.

**Definition 2.3.1**    A function $f : \mathbb{R}^n \to \mathbb{R}^m$ is said to have a *limit* $L \in \mathbb{R}^m$ at $a \in \mathbb{R}^n$ if $\forall \varepsilon > 0 \exists \delta > 0$ such that

$$0 < \|x - a\| < \delta \Rightarrow \|f(x) - L\| < \varepsilon.$$

In such a case, we write

$$\lim_{x \to a} f(x) = L.$$

The notions of infinite limits, limits at infinity, and continuity at a point are analogously defined. Limits in more than one dimension are perhaps trickier to find, as one must approach the test point from infinitely many directions.

**EXAMPLE 2.3.1**

Find $\displaystyle\lim_{(x,y) \to (0,0)} \frac{x^2 y}{x^2 + y^2}$.

### Solution:

We use the *sandwich theorem*. Observe that, $0 \le x^2 \le x^2 + y^2$, and so $0 \le \dfrac{x^2}{x^2 + y^2} \le 1$. Thus

$$\lim_{(x,y)\to(0,0)} 0 \le \lim_{(x,y)\to(0,0)} \left| \frac{x^2 y}{x^2 + y^2} \right| \le \lim_{(x,y)\to(0,0)} |y|,$$

And hence

$$\lim_{(x,y)\to(0,0)} \frac{x^2 y}{x^2 + y^2} = 0.$$

The *Maple*™ commands to graph this surface and find this limit appear in the following. Notice that *Maple* is unable to find the limit and so the limit unevaluated.

> *with( plots )* :

> *plot3d*$\left( x^2 \cdot y / \left( x^2 + y^2 \right), x = -10 ..10, y = -10 ..10, axes = boxed, color = \jmath$
        $^2 + y^2, style = surface \right);$

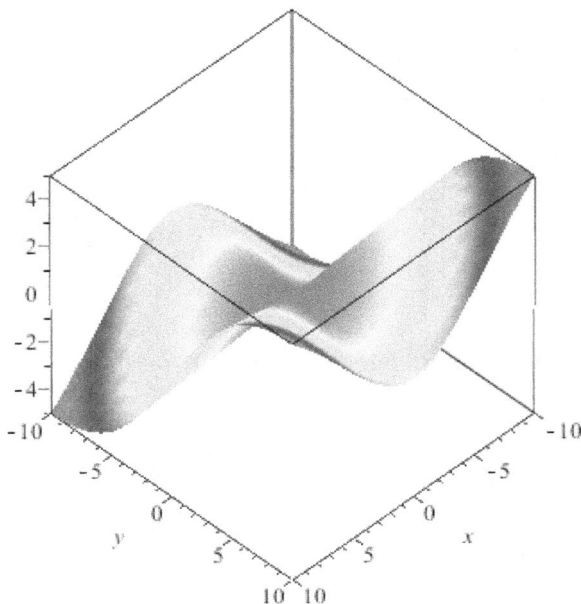

**FIGURE 2.3.1** Example 2.3.1 for 3D plot of $(x, y) \mapsto \dfrac{x^2 y}{x^2 + y^2}$.

> $limit(x^2 \cdot y / (x^2 + y^2), \{x = 0, y = 0\});$

$$limit\left( \frac{x^2 y}{x^2 + y^2}, \{x = 0, y = 0\} \right)$$

### EXAMPLE 2.3.2

Find $\lim\limits_{(x,y) \to (0,0)} \dfrac{x^5 y^3}{x^6 + y^4}.$

**Solution:**

Either $|x| \le |y|$ or $|x| \ge |y|$. Observe that if $|x| \le |y|$, then $\left| \dfrac{x^5 y^3}{x^6 + y^4} \right| \le \dfrac{y^8}{y^4} = y^4.$

If $|y| \le |x|$, then $\left| \dfrac{x^5 y^3}{x^6 + y^4} \right| \le \dfrac{x^8}{x^6} = x^2.$

Thus

$$\left| \frac{x^5 y^3}{x^6 + y^4} \right| \le \max\left( y^4, x^2 \right) \le y^4 + x^2 \to 0,$$

As $(x,y) \to (0,0)$.

Alternative:

Let $X = x^3$, $Y = y^2$.

$$\left| \frac{x^5 y^3}{x^6 + y^4} \right| \le \frac{X^{5/3} Y^{3/2}}{X^2 + Y^2}.$$

Passing to polar coordinate $X = \rho \cos\theta, Y = \rho \sin\theta$, we obtain

$$\left| \frac{x^5 y^3}{x^6 + y^4} \right| = \frac{X^{5/3} Y^{3/2}}{X^2 Y^2} = \rho^{5/3 + 3/2 - 2} |\cos\theta|^{5/3} |\sin\theta|^{3/2} \le \rho^{7/6} \to 0,$$

as $(x,y) \to (0,0)$.

The *Maple*™ commands to graph this surface appear as follows.

> *with*( *plots* ) :

>
$$plot3d\left(x^5 \cdot y^3 / \left(x^6 + y^4\right), x = -10 ..10, y = -10 ..10, axes = boxed, color = x^2 + y^2, style = surface\right);$$

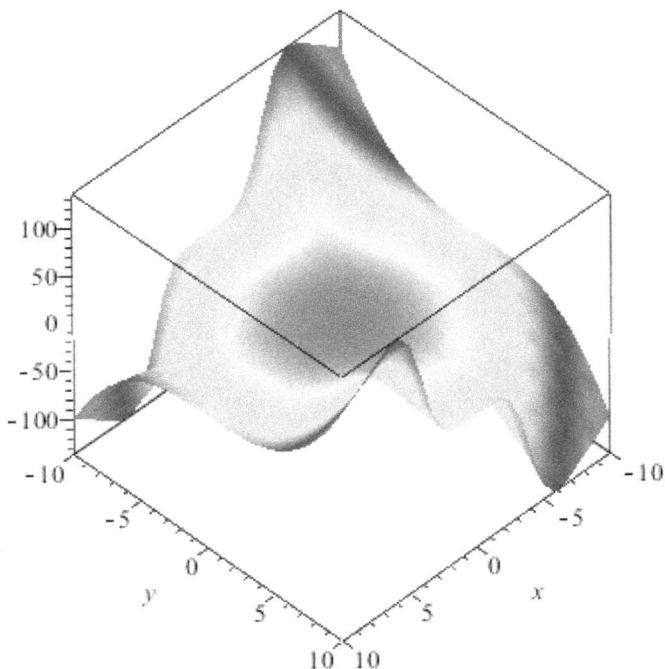

**FIGURE 2.3.2** Example 2.3.2 for 3D plot of $(x, y) \mapsto \dfrac{x^5 y^3}{x^6 + y^4}$

## EXAMPLE 2.3.3

Find $\lim\limits_{(x,y)\to(0,0)} \dfrac{1 + x + y}{x^2 - y^2}$.

### *Solution:*

When $y = 0$,

$$\frac{1 + x}{x^2} \to +\infty,$$

As $x \to 0$.

When $x = 0$,

$$\frac{1+y}{-y^2} \to -\infty,$$

As $y \to 0$.

The limit does not exist.

The *Maple*™ commands to graph this surface are as follows.

```
>  with(plots) :
>  plot3d( (1 + x + y)/(x² - y²), x = -10 ..10, y = -10 ..10, axes = boxed, style
          = surface )
```

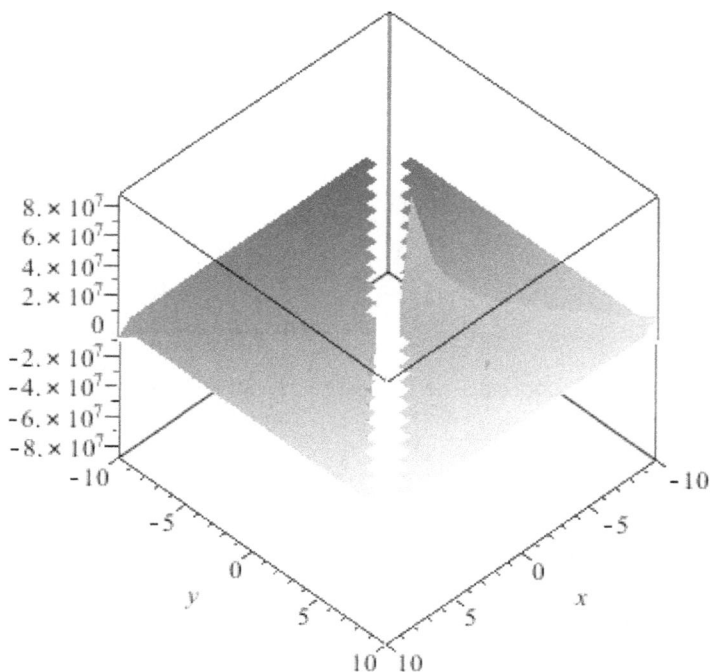

**FIGURE 2.3.3** Example 2.3.3 for 3D plot of $(x, y) \mapsto \dfrac{1+x+y}{x^2 - y^2}$.

## EXAMPLE 2.3.4

Find $\displaystyle\lim_{(x,y)\to(0,0)}\frac{xy^6}{x^6+y^8}$.

### Solution:

Putting $x=t^4, y=t^3$, we find

$$\frac{xy^6}{x^6+y^8}=\frac{1}{2t^2}\to+\infty,$$

as $t\to 0$. But when $y=0$, the function is 0. Thus the limit does not exist.

The $Maple^{\text{TM}}$ commands to graph this surface appear below.

> $with(plots)$ :

>

$plot3d\left(\dfrac{x\cdot y^6}{x^6+y^8}, x=-10\,..10, y=-10\,..10, axes=boxed, style\right.$

$\left.\qquad = surface\right)$

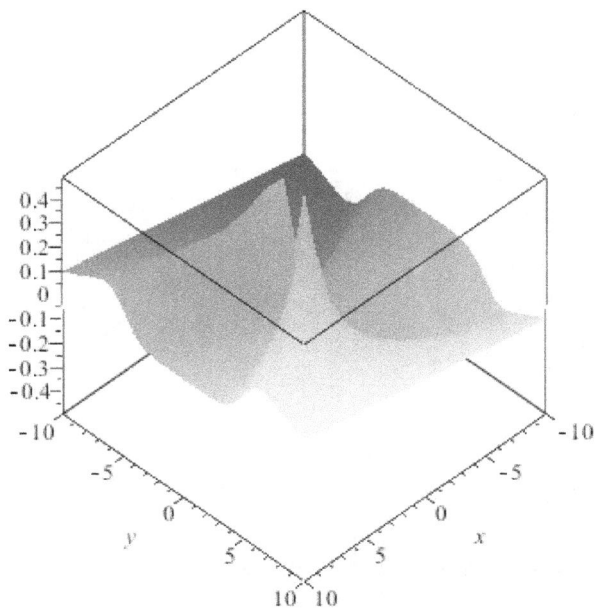

**FIGURE 2.3.4** Example 2.3.4 for 3D plot of $(x,y)\mapsto\dfrac{xy^6}{x^6+y^8}$.

## EXAMPLE 2.3.5

Find $\lim\limits_{(x,y)\to(0,0)} \dfrac{\left((x-1)^2+y^2\right)\log_e\left((x-1)^2+y^2\right)}{|x|+|y|}$.

### *Solution:*

When $y = 0$, have

$$\frac{2(x-1)^2\ln\left(|1-x|\right)}{|x|} \sim -\frac{2x}{x},$$

And so the function does not have a limit at $(0,0)$.

The *Maple*™ commands to graph this surface appear as follows.

> *with*(*plots*) :
>

$$plot3d\left(\frac{\left((x-1)^2+y^2\right)\ln\left((x-1)^2+y^2\right)}{|x|+|y|}, x=-10..10, y=-10..10,\right.$$

$$\left.axes = boxed, style = surface\right)$$

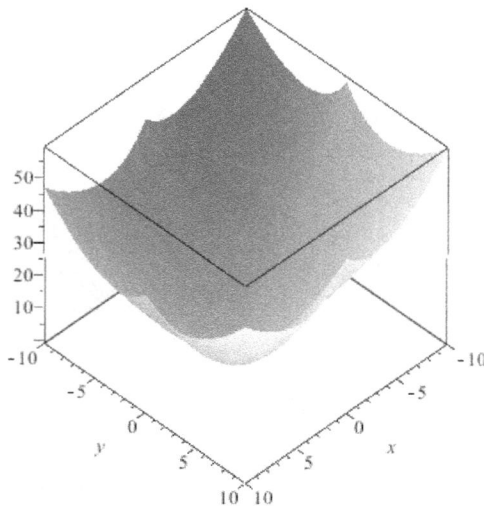

**FIGURE 2.3.5** Example 2.3.5 for 3D plot of $(x,y) \mapsto \dfrac{\left((x-1)^2+y^2\right)\log_e\left((x-1)^2+y^2\right)}{|x|+|y|}$.

## EXAMPLE 2.3.6

Find $\displaystyle\lim_{(x,y)\to(0,0)} \frac{\sin\left(x^4\right)+\sin\left(y^4\right)}{\sqrt{x^4+y^4}}$.

### *Solution:*

$$\sin\left(x^4\right)+\sin\left(y^4\right)\le x^4+y^4,$$

and so

$$\left|\frac{\sin(x^4)+\sin(y^4)}{\sqrt{x^4+y^4}}\right|\le\sqrt{x^4+y^4}\to 0,$$

as $(x,y)\to(0,0)$.

The *Maple*™ commands to graph this surface appear as follows.

> $with(plots)$ :
>

$plot3d\left(\dfrac{\sin\left(x^4\right)+\sin\left(y^4\right)}{\sqrt{x^4+y^4}}, x=-10..10, y=-10..10, axes = boxed,\right.$

$\left.style = surface\right)$

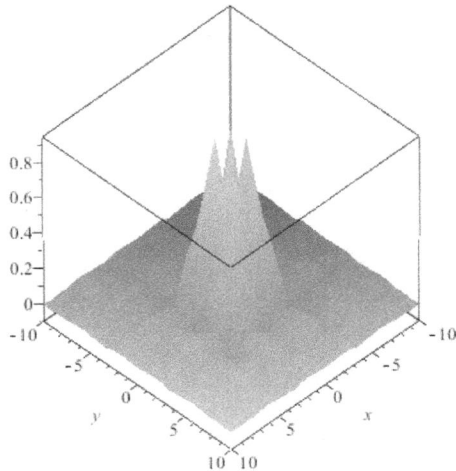

**FIGURE 2.3.6** Example 2.3.6 for 3D plot of $(x,y)\mapsto\dfrac{\sin\left(x^4\right)+\sin\left(y^4\right)}{\sqrt{x^4+y^4}}$.

### EXAMPLE 2.3.7

Find $\displaystyle\lim_{(x,y)\to(0,0)} \frac{\sin(x)-y}{x-\sin(y)}$.

### *Solution:*

When $y = 0$, we obtain

$$\frac{\sin x}{x} \to 1,$$

As $x \to 0$. When $y = x$, the function is identically -1. Thus the limit does not exist.

The *Maple*™ commands to graph this surface appear as follows.

```
>  with(plots) :
>
   plot3d( sin(x) - y / x - sin(y), x = -10 ..10, y = -10 ..10, axes = boxed, style
      = surface )
```

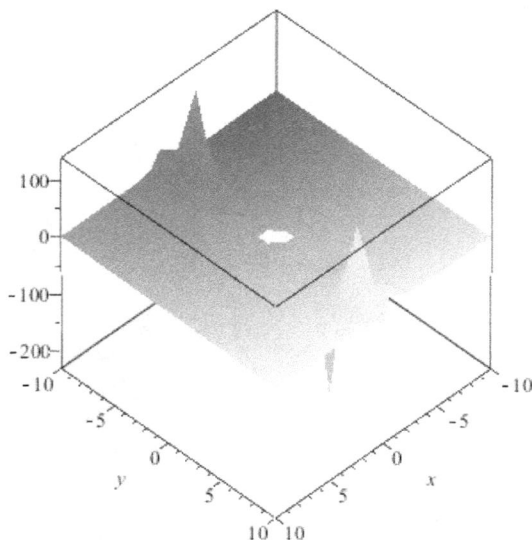

**FIGURE 2.3.7** Example 2.3.7 for 3D plot of $(x,y) \mapsto \dfrac{\sin(x)-y}{x-\sin(y)}$.

If $f : \mathbb{R}^2 \to \mathbb{R}$, it may be that the limits

$$\lim_{y \to y_0}\left(\lim_{x \to x_0} f(x,y)\right), \quad \lim_{x \to x_0}\left(\lim_{y \to y_0} f(x,y)\right),$$

Both exist. These are called the *iterated limits* of $f$ as $(x,y) \to (x_0,y_0)$. The following possibilities might occur.

1.  If $\lim_{(x,y) \to (x_0,y_0)} f(x,y)$ exists, then each of the iterated limits $\lim_{y \to y_0}\left(\lim_{x \to x_0} f(x,y)\right)$ and $\lim_{x \to x_0}\left(\lim_{y \to y_0} f(x,y)\right)$ exists.

2.  If the iterated limits exist and $\lim_{y \to y_0}\left(\lim_{x \to x_0} f(x,y)\right) \neq \lim_{x \to x_0}\left(\lim_{y \to y_0} f(x,y)\right)$, then $\lim_{(x,y) \to (x_0,y_0)} f(x,y)$ does not exist.

3.  It may occur that $\lim_{y \to y_0}\left(\lim_{x \to x_0} f(x,y)\right) = \lim_{x \to x_0}\left(\lim_{y \to y_0} f(x,y)\right)$, but that $\lim_{(x,y) \to (x_0,y_0)} f(x,y)$ does not exist.

4.  It may occur that $\lim_{(x,y) \to (x_0,y_0)} f(x,y)$ exists, but one of the iterated limits does not exist.

**NOTE**   *If you get two or more different values for $\lim_{(x,y) \to (x_0,y_0)} f(x,y)$ as approach $(x_0,y_0)$ along different paths, then $\lim_{(x,y) \to (x_0,y_0)} f(x,y)$ does not exist.*

## EXAMPLE 2.3.8

Show that $\lim_{(x,y) \to (0,0)} \dfrac{x-y}{x+y}$ does not exist.

### Solution:

We need to show that $\lim_{y \to y_0}\left(\lim_{x \to x_0} f(x,y)\right) \neq \lim_{x \to x_0}\left(\lim_{y \to y_0} f(x,y)\right)$.

So, $\lim_{x \to 0}\left(\lim_{y \to 0} \dfrac{x-y}{x+y}\right) = \lim_{x \to 0}(1) = 1$ and $\lim_{y \to 0}\left(\lim_{x \to 0} \dfrac{x-y}{x+y}\right) = \lim_{y \to 0}(-1) = -1$.

Thus the iterated limits are not equal and therefore, $\lim_{(x,y) \to (0,0)} \dfrac{x-y}{x+y}$, does not exist.

**NOTE**   *A function $f$ of two variables $x$ and $y$ is a rule that assigns to each ordered pair of real numbers $(x,y)$ in a set of ordered pairs of real numbers (D) a unique real number $f(x,y)$. The set D is the domain of $f$, and the corresponding set of values for $f(x,y)$ is the range of $f$, that is, $\{f(x,y):(x,y) \in D\}$.*

### EXAMPLE 2.3.9

Describe the domain of the function

1. $f(x,y) = x^2 + y^2$
2. $f(x,y) = \ln xy$
3. $f(x,y) = \dfrac{\sqrt{x^2 + y^2 - 16}}{x}$
4. $f(x,y) = \dfrac{x}{\sqrt{16 - x^2 - y^2 - z^2}}$

*Solution:*

1. The entire $xy$-plane.

2. The set of all points $(x,y)$ in the plane for which $xy > 0$. This consists of all points in the first and third quadrants.

3. The set of all points $(x,y)$ laying on or outside the circle $x^2 + y^2 = 16$ except when $x$ equal to zero, that is $D = \{(x,y): x^2 + y^2 - 16 \geq 0, x \neq 0\}$.

4. The set of all points $(x,y,z)$ laying inside a sphere of radius 4 that is centered at the origin.

**Definition 2.3.2**   A function $f(x,y)$ is *continuous at a point* $(x_0, y_0)$ in an open region $R$ if all the following conditions hold:

(**a**) $f(x_0, y_0)$ exists (i.e., $(x_0, y_0)$ is in the domain of $f$.

(**b**) $\displaystyle\lim_{(x,y) \to (x_0, y_0)} f(x,y)$ exists.

(**c**) $\displaystyle\lim_{(x,y) \to (x_0, y_0)} f(x,y) = f(x_0, y_0)$.

If one or more of these three conditions fail to hold, then $f$ is said to be *discontinuous at* $(x_0, y_0)$.

The function $f(x,y)$ is *continuous in the open region* $R$ if it is continuous at every point $(x,y)$ in $R$.

**Definition 2.3.3**    A function $f(x,y)$ defined in a domain D is *continuous in D* if it is continuous at each point of D.

**Theorem 2.3.1**    If $c$ is a real number and $f(x,y)$ and $g(x,y)$ are continuous at $(x_0, y_0)$, then the following functions are also continuous at $(x_0, y_0)$:

1. $cf(x_0, y_0)$
2. $f(x_0, y_0)g(x_0, y_0)$
3. $f(x_0, y_0) \pm g(x_0, y_0)$
4. $\dfrac{f(x_0, y_0)}{g(x_0, y_0)}$, if $(x_0, y_0) \neq 0$

**NOTE**    *The polynomial and rational functions are continuous at every point in their domain D.*

### EXAMPLE 2.3.10

The function $f(x,y) = \dfrac{x - 3y}{x^2 + y^2}$, is continuous at every point in its domain, which means that $f(x,y)$, is continuous at each point in the $xy$-plane except at the point $(0,0)$.

## EXERCISES 2.3

**2.3.1**    Sketch the domain of definition of $(x,y) \mapsto \sqrt{4 - x^2 - y^2}$.

**2.3.2**    Sketch the domain of definition of $(x,y) \mapsto \log(x + y)$.

**2.3.3**    Sketch the domain of definition of $(x,y) \mapsto \dfrac{1}{x^2 + y^2}$.

**2.3.4**    Describe the domain and the range of the function.

1. $f(x,y) = \sqrt{4 - x^2 - y^2}$

2. $f(x,y) = \ln(4 - x - y)$

3. $f(x,y) = \dfrac{5}{\sqrt{x^2 - y}}$

4. $f(x,y) = \tan^{-1}\left(\dfrac{x}{y}\right)$

5. $f(x,y) = \cos^{-1}(x - y)$

6. $f(x,y,z) = \dfrac{xy}{z}$

**2.3.5** Find $\displaystyle\lim_{(x,y)\to(0,0)} \left(x^2 + y^2\right)\sin\left(\dfrac{1}{xy}\right)$.

**2.3.6** Find $\displaystyle\lim_{(x,y)\to(0,2)} \dfrac{\sin xy}{x}$.

**2.3.7** Find $\displaystyle\lim_{(x,y)\to(0,0)} \sqrt{x^2 + y^2}\, \sin\dfrac{1}{x^2 + y^2}$.

**2.3.8** Find $\displaystyle\lim_{(x,y)\to(-2,1)} \left(xy^3 - xy + 3y^2\right)$.

**2.3.9** Find $\displaystyle\lim_{(x,y)\to(0,0)} \dfrac{\sin\left(x^2 + y^2\right)}{3x^2 + 3y^2}$.

**2.3.10** Find $\displaystyle\lim_{(x,y)\to(0,0)} \dfrac{e^{x^2+y^2} - 1}{x^2 + y^2}$.

**2.3.11** Find $\displaystyle\lim_{(x,y,z)\to(0,0,0)} \dfrac{x + y + z}{x^2 + y^2 + z^2}$.

**2.3.12** Demonstrate that

$$\lim_{(x,y,z)\to(0,0,0)} \dfrac{x^2 y^2 z^2}{x^2 + y^2 + z^2} = 0.$$

**2.3.13** Prove that $\displaystyle\lim_{(x,y)\to(0,0)} \dfrac{x^3 y}{x^2 + y^4} = 0$.

**2.3.14** Prove that $\displaystyle\lim_{(x,y)\to(0,0)} \dfrac{2xy^2}{x^2 + y^2} = 0$.

**2.3.15** Show that $\displaystyle\lim_{(x,y)\to(0,0)} \frac{xy}{x^2+y^2}$ does not exist by considering the line $y = x$ as one path and the $x$-axis to the origin another.

**2.3.16** Let $f : \mathbb{R}^2 \to \mathbb{R}$ be such that $f(x,y) = \dfrac{x}{y}$ for $y \neq 0$. Show that $f(x,y)$ is discontinuous at any point $(x_0, 0) \in \mathbb{R}^2$.

**2.3.17** Show that if $g(x,y)$ is continuous or discontinuous where

$$g(x,y) = \begin{cases} \dfrac{x^2 - y^2}{x^2 + y^2} & (x,y) \neq (0,0) \\ 0 & (x,y) = (0,0) \end{cases}.$$

**2.3.18** Describe the largest set $L$ on which the $f$ is continuous.

1. $f(x,y) = \sqrt{x - y + 3}$

2. $f(x,y) = \dfrac{y^2 + 5xy + x^2}{x - y^2}$

**2.3.19** For what $c$ will the function

$$f(x,y) \begin{cases} \sqrt{1 - x^2 - 4y^2}, & \text{if } x^2 + 4y^2 \leq 1, \\ c, & \text{if } x^2 + 4y^2 > 1 \end{cases}$$

be continuous everywhere on the $xy$-plane?

## 2.4   DEFINITION OF THE DERIVATIVE

Before we begin, let us introduce some necessary notation. Let $f : \mathbb{R} \to \mathbb{R}$ be a function. We write $f(h) = o(h)$ if $f(h)$ goes faster to 0 than $h$, that is, $\displaystyle\lim_{h\to 0} \frac{f(h)}{h} = 0$. For example, $h^3 + 2h^2 = o(h)$, since

$$\lim_{h\to 0} \frac{h^3 + 2h^2}{h} = \lim_{h\to 0} h^2 + 2h = 0.$$

We now define the derivative in the multidimensional space $\mathbb{R}^n$. Recall that in one variable, a function $g : \mathbb{R} \to \mathbb{R}$ is said to be differentiable at $x = a$ if the limit

$$\lim_{x \to a} \frac{g(x) - g(a)}{x - a} = g'(a)$$

exists. The limit condition above is equivalent to saying that

$$\lim_{x \to a} \frac{g(x) - g(a) - g'(a)(x - a)}{x - a} = 0,$$

or equivalently,

$$\lim_{h \to 0} \frac{g(a + h) - g(a) - g'(a)(h)}{h} = 0.$$

We may write this as

$$g(a + h) - g(a) = g'(a)(h) + o(h).$$

The preceding analysis provides an analogue definition for the higher-dimensional case. Observe that since we may not divide by vectors, the corresponding definition in higher dimensions involves quotients of norms.

**Definition 2.4.1**    Let $A \subseteq \mathbb{R}^n$. A function $f : A \to \mathbb{R}^m$ is said to be *differentiable* at a $\in A$ if there is a linear transformation, called the *derivative of $f$* at a, $D_a(f) : \mathbb{R}^n \to \mathbb{R}^m$ such that

$$\lim_{x \to a} \frac{\|f(x) - f(a) - D_a(f)(x - a)\|}{\|x - a\|} = 0.$$

Equivalently, $f$ is differentiable at a if there is a linear transformation $D_a(f)$ such that

$$f(a + h) - f(a) = D_a(f)(h) + o(\|h\|),$$

as $h \to 0$.

**NOTE**    *The condition for differentiability at* a *is equivalent to*

$$f(x) - f(a) = D_a(f)(x - a) + o(\|x - a\|),$$

as $x \to 0$.

**Theorem 2.4.1**  If $A$ is an open set in definition 2.4.1, $D_a(f)$ is uniquely determined.

**Proof:**

Let $L : \mathbb{R}^n \to \mathbb{R}^m$ be another linear transformation satisfying the Definition 2.4.1. We must prove that $\forall v \in \mathbb{R}^n$, $L(v) = D_a(f)(v)$. Since $A$ is open, $a + h \in A$ for sufficiently small $\|h\|$. By definition, as $h \to 0$, we have

$$f(a+h) - f(a) = D_a(f)(h) + o(\|h\|).$$

and

$$f(a+h) - f(a) = L(h) + o(\|h\|).$$

Now, observe that,

$$D_a(f)(v) - L(v) = D_a(f)(h) - f(a+h) + f(a) + f(a+h) - f(a) - L(h).$$

By the triangle inequality,

$$\|D_a(f)(v) - L(v)\| \le \|D_a(f)(h) - f(a+h) + f(a)\| + \|f(a+h) - f(a) - L(h)\|$$
$$= o(\|h\|) + o(\|h\|)$$
$$= o(\|h\|),$$

as $h \to 0$. This means

$$\|L(v) - D_a(f)(v)\| \to 0,$$

i.e., $L(v) = D_a(f)(v)$, completing the proof.

**NOTE**  *If $A = \{a\}$, a singleton, then $D_a(f)$ is not uniquely determined. For $\|x - a\| < \delta$ holds only for $x = a$, and so $f(x) = f(a)$. Any linear transformation $T$ will satisfy the definition, as $T(x - a) = T(0) = 0$, and $\| f(x) - f(a) - T(x - a) \| = \|0\| = 0$, identically.*

**EXAMPLE 2.4.1**

If $L : \mathbb{R}^n \to \mathbb{R}^m$ is a linear transformation, then $D_a(L) = L$, for any $a \in \mathbb{R}^n$.

***Solution:***

Since $\mathbb{R}^n$ is an open set, we know that $D_a(L)$ uniquely determined. Thus if $L$ satisfies Definition 2.4.1, then the claim is established. But by linearity $\|L(x) - L(a) - L(x-a)\| = \|L(x) - L(a) - L(x) + L(a)\| = \|0\| = 0$, hence the claim that follows.

**EXAMPLE 2.4.2**

Let

$$f : \begin{array}{ccc} \mathbb{R}^3 \times \mathbb{R}^3 \to & \mathbb{R} \\ (\vec{x}, \vec{y}) \mapsto & \vec{x} \bullet \vec{y} \end{array}$$

be the usual dot product in $\mathbb{R}^3$. Show that $f$ is differentiable and that

$$D_{(\vec{x}, \vec{y})} f\left(\vec{h}, \vec{k}\right) = \vec{x} \bullet \vec{y} + \vec{h} \bullet \vec{k}.$$

***Solution:***

We have

$$f\left(\vec{x} + \vec{h}, \vec{y} + \vec{k}\right) - f(\vec{x}, \vec{y}) = \left(\vec{x} + \vec{h}\right) \bullet \left(\vec{y} + \vec{k}\right) - \vec{x} \bullet \vec{y}$$
$$= \vec{x} \bullet \vec{y} + \vec{x} \bullet \vec{k} + \vec{h} \bullet \vec{y} + \vec{h} \bullet \vec{k} - \vec{x} \bullet \vec{y}$$
$$= \vec{x} \bullet \vec{k} + \vec{h} \bullet \vec{y} + \vec{h} \bullet \vec{k}.$$

As $\left(\vec{h}, \vec{k}\right) \to \left(\vec{0}, \vec{0}\right)$, we have by the Cauchy–Buniakovskii–Schwarz inequality, $\left|\vec{h}, \vec{k}\right| \leq \|\vec{h}\| \|\vec{k}\| = o\left(\|\vec{h}\|\right)$, which proves the assertion.

It is worth knowing that, just like in the one variable case, differentiability at a point implies continuity at that point.

**Theorem 2.4.2**   Suppose $A \subseteq \mathbb{R}^n$ is open and $f : A \to \mathbb{R}^n$ is differentiable on $A$. Then $f$ is continuous on $A$.

**Proof:**

Given $a \in A$, we must show that

$$\lim_{x \to a} f(x) = f(a).$$

Since $f$ is differentiable at $a$, we have

$$f(x) - f(a) = D_a(f)(x-a) + o(\|x-a\|),$$

and so

$$f(x) - f(a) \to 0,$$

as $x \to a$, proving the theorem.

## EXERCISES 2.4

**2.4.1**   Let $L : \mathbb{R}^3 \to \mathbb{R}^3$ be a linear transformation and

$$F : \begin{array}{c} \mathbb{R}^3 \to \mathbb{R}^3 \\ \vec{x} \mapsto \vec{x} \times L(\vec{x}) \end{array}.$$

Show that $F$ is differentiable and that

$$D_x(F)(\vec{h}) = \vec{x} \times L(\vec{h}) + \vec{h} \times L(\vec{x}).$$

**2.4.2**   Let $f : \mathbb{R}^n \to \mathbb{R}$, $n \geq 1$, $f(\vec{x}) = \|\vec{x}\|$ be the usual norm in $\mathbb{R}^n$, with $\|\vec{x}\| = \vec{x} \cdot \vec{x}$. Prove that

$$D_x(f)(\vec{v}) = \frac{\vec{x} \cdot \vec{v}}{\|\vec{x}\|},$$

for $\vec{x} \neq \vec{0}$, but that $f$ is not differential at $\vec{0}$.

## 2.5   THE JACOBI MATRIX

We now establish a way that simplifies the process of finding the derivative of a function at a given point.

**Definition 2.5.1**   Let $A \subseteq \mathbb{R}^n$, $f : A \to \mathbb{R}^m$, and put

$$f(x) = \begin{bmatrix} f_1(x_1, x_2, \ldots, x_n) \\ f_2(x_1, x_2, \ldots, x_n) \\ \vdots \\ f_m(x_1, x_2, \ldots, x_n) \end{bmatrix}.$$

Here $f_i : \mathbb{R}^n \to \mathbb{R}$. The *partial derivative* $\dfrac{\partial f_i}{\partial x_j}(x)$ is defined

$$\frac{\partial f_i}{\partial x_j}(\mathbf{x}) = \lim_{h \to 0} \frac{f_i\left(x_1, x_2, \ldots, x_j + h, \ldots, x_n\right) - f_i\left(x_1, x_2, \ldots, x_j, \ldots, x_n\right)}{h},$$

whenever this limit exists.

To find partial derivatives with respect to the $j$th variable, we simply keep the other variables fixed and differentiate with respect to the $j$th variable.

### EXAMPLE 2.5.1

If $f : \mathbb{R}^3 \to \mathbb{R}$, and $f(x, y, z) = x + y^2 + z^3 + 3xy^2z^3$ then

$$\frac{\partial f}{\partial x}(x, y, z) = 1 + 3y^2z^3,$$

$$\frac{\partial f}{\partial y}(x, y, z) = 2y + 6xyz^3,$$

and

$$\frac{\partial f}{\partial z}(x, y, z) = 3z^2 + 9xy^2z^2.$$

The *Maple*$^{\text{Tm}}$ commands to find these follow.

> $f := (x, y, z) \to x + y\hat{\ }2 + z\hat{\ }3 + 3*x*y\hat{\ }2*z\hat{\ }3;$

$$f := (x, y, z) \to x + y^2 + z^3 + 3xy^2z^3$$

> $diff(f(x, y, z), x);$

$$1 + 3y^2z^3$$

> $diff(f(x, y, z), y);$

$$2y + 6xyz^3$$

> $diff(f(x, y, z), z);$

$$3z^2 + 9xy^2z^2$$

Since the derivative of a function $f_i : \mathbb{R}^n \to \mathbb{R}^m$ is a linear transformation, it can be represented by aid of matrices. The following theorem will allow us to determine the matrix representation for $D_a(f)$ under the standard bases of $\mathbb{R}^n$ and $\mathbb{R}^m$.

**Theorem 2.5.1**   Let

$$f(\mathbf{x}) = \begin{bmatrix} f_1(x_1, x_2, \ldots, x_n) \\ f_2(x_1, x_2, \ldots, x_n) \\ \vdots \\ f_m(x_1, x_2, \ldots, x_n) \end{bmatrix}.$$

Suppose $A \subseteq \mathbb{R}^n$ is an open set and $f : A \to \mathbb{R}^m$ is differentiable. Then each partial derivative $\dfrac{\partial f_i}{\partial x_j}(\mathbf{x})$ exists, and the matrix representation for $D_\mathbf{x}(f)$ with respect to the standard bases of $\mathbb{R}^n$ and $\mathbb{R}^m$ is the *Jacobi matrix*

$$f'(\mathbf{x}) = \begin{bmatrix} \dfrac{\partial f_1}{\partial x_1}(\mathbf{x}) & \dfrac{\partial f_1}{\partial x_2}(\mathbf{x}) & \cdots & \dfrac{\partial f_1}{\partial x_n}(\mathbf{x}) \\ \dfrac{\partial f_2}{\partial x_1}(\mathbf{x}) & \dfrac{\partial f_2}{\partial x_2}(\mathbf{x}) & \cdots & \dfrac{\partial f_2}{\partial x_n}(\mathbf{x}) \\ \vdots & \vdots & \vdots & \vdots \\ \dfrac{\partial f_n}{\partial x_1}(\mathbf{x}) & \dfrac{\partial f_n}{\partial x_2}(\mathbf{x}) & \cdots & \dfrac{\partial f_n}{\partial x_n}(\mathbf{x}) \end{bmatrix}.$$

**Proof:**

Let $\vec{e}_j$, $1 \le j \le n$ be the standard basis for $\mathbb{R}^n$. To obtain the *Jacobi matrix*, we must compute $D_\mathbf{x}(f)(\vec{e}_j)$, which will give us the $j$th column of the *Jacobi matrix*. Let $f'(\mathbf{x}) = (J_{ij})$, and observe that

$$D_\mathbf{x}(f)(\vec{e}_j) = \begin{bmatrix} J_{1j} \\ J_{2j} \\ \vdots \\ J_{nj} \end{bmatrix}.$$

and put $\mathbf{y} = \mathbf{x} + \varepsilon \vec{e}_j$, $\varepsilon \in \mathbb{R}$. Notice that,

$$\frac{\left\| f(\mathbf{y}) - f(\mathbf{x}) - D_\mathbf{x}(f)(\mathbf{y} - \mathbf{x}) \right\|}{\left\| \mathbf{y} - \mathbf{x} \right\|}$$

$$= \frac{\left\| f(x_1, x_2, \ldots, x_j + h, \ldots, x_n) - f(x_1, x_2, \ldots, x_j, \ldots, x_n) - \varepsilon D_\mathbf{x}(f)(\vec{e}_j) \right\|}{|\varepsilon|}.$$

Since the sinistral side $\to 0$ as $\varepsilon \to 0$, the so does the $i$th component of the numerator, and so,

$$\frac{\left| f_i\left(x_1, x_2, \ldots, x_j + h, \ldots, x_n\right) - f_i\left(x_1, x_2, \ldots, x_j, \ldots, x_n\right) - \varepsilon J_{ij} \right|}{|\varepsilon|} \to 0.$$

This entails that

$$J_{ij} = \lim_{\varepsilon \to 0} \frac{f_i\left(x_1, x_2, \ldots, x_j + \varepsilon, \ldots, x_n\right) - f_i\left(x_1, x_2, \ldots, x_j, \ldots, x_n\right)}{\varepsilon} = \frac{\partial f_i}{\partial x_j}(\mathbf{x}).$$

This finishes the proof.

**NOTE**  *Strictly speaking, the Jacobi matrix is not the derivative of a function at a point. It is a matrix representation of the derivative in the standard basis of $\mathbb{R}^n$. We will, however, refer to $f'$ when we mean the Jacobi matrix of $f$.*

**NOTE**  *We will use the symbol $J$ in the exercises to represent* Jacobi determinant *which is $J = \det[f']$.*

### EXAMPLE 2.5.2

Let $f : \mathbb{R}^3 \to \mathbb{R}^2$ be given by

$$f(x, y) = \left(xy + yz, \log_e xy\right).$$

Compute the Jacobi matrix of $f$.

***Solution:***

The Jacobi matrix is the $2 \times 3$ matrix

$$f'(x,y) = \begin{bmatrix} \dfrac{\partial f_1}{\partial x}(x,y) & \dfrac{\partial f_1}{\partial y}(x,y) & \dfrac{\partial f_1}{\partial z}(x,y) \\ \dfrac{\partial f_2}{\partial x}(x,y) & \dfrac{\partial f_2}{\partial y}(x,y) & \dfrac{\partial f_2}{\partial z}(x,y) \end{bmatrix} = \begin{bmatrix} y & x+z & y \\ \dfrac{1}{x} & \dfrac{1}{y} & 0 \end{bmatrix}.$$

### EXAMPLE 2.5.3

Let $f(\rho, \theta, z) = (\rho \cos\theta, \rho \sin\theta, z)$ be the function, which changes from cylindrical coordinates to Cartesian coordinates. We have

$$f'(\rho,\theta,z) = \begin{bmatrix} \cos\theta & -\rho\sin\theta & 0 \\ \sin\theta & \rho\cos\theta & 0 \\ 0 & 0 & 1 \end{bmatrix}.$$

## EXAMPLE 2.5.4

Let $f(\rho,\phi,\theta) = (\rho\cos\theta\sin\phi, \rho\sin\theta\sin\phi, \rho\cos\phi)$ be the function, which changes from spherical coordinates to Cartesian coordinates. We have

$$f'(\rho,\phi,\theta) = \begin{bmatrix} \cos\theta\sin\phi & \rho\cos\theta\cos\phi & -\rho\sin\phi\sin\theta \\ \sin\theta\sin\phi & \rho\sin\theta\cos\phi & \rho\cos\theta\sin\phi \\ \cos\phi & -\rho\sin\phi & 0 \end{bmatrix}.$$

The Jacobi matrix provides a convenient computational tool to compute the derivative of a function at a point. Thus differentiability at a point implies that the partial derivatives of the function exist at the point. The converse, however, is not true.

## EXAMPLE 2.5.5

Let $f : \mathbb{R}^2 \to \mathbb{R}$ be given by

$$f(x,y) = \begin{cases} y \text{ if } x = 0, \\ x \text{ if } y = 0, \\ 1 \text{ if } xy \neq 0. \end{cases}$$

Observe that f is not continuous at $(0,0)$ ($f(0,0) = 0$ but $f(x,y) = 1$ for values arbitrarily close to $(0,0)$), and hence, it is not differentiable there. We have, however, $\frac{\partial f}{\partial x}(0,0) = \frac{\partial f}{\partial y}(0,0) = 1$. Thus even if both partial derivatives exist at $(0,0)$, there is no guarantee that the function will be differentiable at $(0,0)$.

You should also notice that both partial derivatives are not continuous at $(0,0)$. We have, however, the following.

**Theorem 2.5.2**   Let $A \subseteq \mathbb{R}^n$ be an open set, and let $f : \mathbb{R}^n \to \mathbb{R}^m$. Put

$$f = \begin{bmatrix} f_1 \\ f_2 \\ \vdots \\ f_m \end{bmatrix}.$$ If each of the partial derivatives $D_j f_i$ exists and is continuous on $A$,

then $f$ is differentiable on $A$.

The concept of *repeated partial derivatives* is akin to the concept of repeated differentiation. Similarly with the concept of implicit partial differentiation, the following examples should be self-explanatory.

**EXAMPLE 2.5.6**

Let $f(u,v,w) = e^u v \cos w$. Determine $\dfrac{\partial^2}{\partial u \partial v} f(u,v,w)$ at $\left(1, -1, \dfrac{\pi}{4}\right)$.

***Solution:***

We have

$$\frac{\partial^2}{\partial u \partial v}\left(e^u v \cos w\right) = \frac{\partial}{\partial u}\left(e^u \cos w\right) = e^u \cos w,$$

which is $\dfrac{e\sqrt{2}}{2}$ at the desired point.

**EXAMPLE 2.5.7**

The equation $z^{xy} + (xy)^z + xy^2 z^3 = 3$ defines $z$ as an implicit function of $x$ and $y$. Find $\dfrac{\partial z}{\partial x}$ and $\dfrac{\partial z}{\partial y}$ at $(1,1,1)$.

***Solution:***

We have

$$\frac{\partial}{\partial x} z^{xy} = \frac{\partial}{\partial x} e^{xy \log z}$$

$$= \left( y \log z + \frac{xy}{z} \frac{\partial z}{\partial x} \right) z^{xy},$$

$$\frac{\partial}{\partial x} (xy)^z = \frac{\partial}{\partial x} e^{z \log xy}$$

$$= \left( \frac{\partial z}{\partial x} \log xy + \frac{z}{x} \right) (xy)^z,$$

$$\frac{\partial}{\partial x} xy^2 z^3 = y^2 z^3 + 3xy^2 z^2 \frac{\partial z}{\partial x},$$

Hence, at $(1,1,1)$, we have

$$\frac{\partial z}{\partial x} + 1 + 1 + 3 \frac{\partial z}{\partial x} = 0 \Rightarrow \frac{\partial z}{\partial x} = -\frac{1}{2}.$$

Similarly,

$$\frac{\partial}{\partial y} z^{xy} = \frac{\partial}{\partial y} e^{xy \log z}$$

$$= \left( x \log z + \frac{xy}{z} \frac{\partial z}{\partial y} \right) z^{xy},$$

$$\frac{\partial}{\partial y} (xy)^z = \frac{\partial}{\partial y} e^{z \log xy}$$

$$= \left( \frac{\partial z}{\partial y} \log xy + \frac{z}{y} \right) (xy)^z,$$

$$\frac{\partial}{\partial y} xy^2 z^3 = 2xyz^3 + 3xy^2 z^2 \frac{\partial z}{\partial y},$$

Hence, at $(1,1,1)$, we have

$$\frac{\partial z}{\partial y} + 1 + 2 + 3\frac{\partial z}{\partial y} = 0 \Rightarrow \frac{\partial z}{\partial y} = -\frac{3}{4}.$$

Just like in the one-variable case, we have the following rules of differentiation. Let $A \subseteq \mathbb{R}^n$, $B \subseteq \mathbb{R}^m$ be open sets $f, g: A \to \mathbb{R}^m, a \in \mathbb{R}$, be differentiable on $A$, and $h: B \to \mathbb{R}^l$, be differentiable on $B$, and $f(A) \subseteq B$. Then we have

1. **Addition Rule:** $D_x\left((f + a\,g)\right) = D_x(f) + a\,D_x(g)$

2. **Chain Rule:** $D_x\left((h \circ f)\right) = \left(D_{f(x)}(h)\right) \circ \left(D_x(f)\right)$

Since the composition of linear mappings expressed as matrices are matrix multiplication, the Chain Rule takes the alternative form when applied to the Jacobi matrix.

$$(h \circ f)' = (h' \circ f)(f'). \tag{2.1}$$

**EXAMPLE 2.5.8**

Let

$$f(u,v) = \begin{bmatrix} ue^v \\ u + v \\ uv \end{bmatrix},$$

$$h(x,y) = \begin{bmatrix} x^2 + y \\ y + z \end{bmatrix}.$$

Find $(f \circ h)'(x,y)$.

*Solution:*

We have

$$f'(u,v) = \begin{bmatrix} e^v & ue^v \\ 1 & 1 \\ v & u \end{bmatrix},$$

And

$$h'(x,y) = \begin{bmatrix} 2x & 1 & 0 \\ 0 & 1 & 1 \end{bmatrix}.$$

Observe also that

$$f'(h(x,y)) = \begin{bmatrix} e^{y+z} & (x^2+y)e^{y+z} \\ 1 & 1 \\ y+z & x^2+y \end{bmatrix}.$$

Hence

$$(f \circ h)'(x,y) = f'(h(x,y))h'(x,y)$$

$$= \begin{bmatrix} e^{y+z} & (x^2+y)e^{y+z} \\ 1 & 1 \\ y+z & x^2+y \end{bmatrix} \begin{bmatrix} 2x & 1 & 0 \\ 0 & 1 & 1 \end{bmatrix}$$

$$= \begin{bmatrix} 2xe^{y+z} & (1+x^2+y)e^{y+z} & (x^2+y)e^{y+z} \\ 2x & 2 & 1 \\ 2xy+2xz & x^2+2y+z & x^2+y \end{bmatrix}.$$

## EXAMPLE 2.5.9

Let

$$f:\mathbb{R}^2 \to \mathbb{R}, \quad f(u,v) = u^2 + e^v,$$
$$u,v:\mathbb{R}^3 \to \mathbb{R}, \quad u(x,y) = xz, \quad v(x,y) = y+z.$$

Put $h(x,y) = f\begin{bmatrix} u(x,y,z) \\ v(x,y,z) \end{bmatrix}$. Find the partial derivatives of $h$.

*Solution:*

Put $g:\mathbb{R}^3 \to \mathbb{R}^2$, $g(x,y) = \begin{bmatrix} u(x,y) \\ v(x,y) \end{bmatrix} = \begin{bmatrix} xz \\ y+z \end{bmatrix}$. Observe that $h = f \circ g$. Now,

$$g'(x,y) = \begin{bmatrix} z & 0 & x \\ 0 & 1 & 1 \end{bmatrix},$$

$$f'(u,v) = \begin{bmatrix} 2u & e^v \end{bmatrix},$$

$$f'(h(x,y)) = \begin{bmatrix} 2xz & e^{y+z} \end{bmatrix}.$$

Thus

$$\begin{bmatrix} \dfrac{\partial h}{\partial x}(x,y) & \dfrac{\partial h}{\partial y}(x,y) & \dfrac{\partial h}{\partial z}(x,y) \end{bmatrix} = h'(x,y)$$

$$= \left( f'(g(x,y)) \right) \left( g'(x,y) \right)$$

$$= \begin{bmatrix} 2xz & e^{y+z} \end{bmatrix} \begin{bmatrix} z & 0 & x \\ 0 & 1 & 1 \end{bmatrix}$$

$$= \begin{bmatrix} 2xz^2 & e^{y+z} & 2x^2 z + e^{y+z} \end{bmatrix}.$$

Equating components, we obtain

$$\frac{\partial h}{\partial x}(x,y) = 2xz^2,$$

$$\frac{\partial h}{\partial y}(x,y) = e^{y+z},$$

$$\frac{\partial h}{\partial z}(x,y) = 2x^2 z + e^{y+z}.$$

Under certain conditions, we may differentiate under the integral sign.

**Theorem 2.5.3 (Differentiation under the integral sign):** Let $f:[a,b] \times Y \to \mathbb{R}$ be a function, with $[a,b]$ being a closed interval, and $Y$ being a closed and bounded subset of $\mathbb{R}$. Suppose that both $f(x,y)$ and $\dfrac{\partial}{\partial x} f(x,y)$ are continuous in the variable $x$ and $y$ jointly. Then $\displaystyle\int_Y f(x,y)\,dy$ exists as a continuously differentiable function of $x$ on $[a,b]$, with derivative

$$\frac{d}{dx} \int_Y f(x,y)\,dy = \int_Y \frac{\partial}{\partial x} f(x,y)\,dy.$$

**EXAMPLE 2.5.10**

Prove that

$$F(x) = \int_0^{\pi/2} \log\left(\sin^2\theta + x^2\cos^2\theta\right) d\theta = \pi\log\frac{x+1}{2}.$$

***Solution:***

Differentiating under the integral

$$F'(x) = \int_0^{\pi/2} \frac{\partial}{\partial x}\log\left(\sin^2\theta + x^2\cos^2\theta\right) d\theta = \pi\log\frac{x+1}{2}.$$

$$= 2x\int_0^{\pi/2} \frac{\cos^2\theta}{\sin^2\theta + x^2\cos^2\theta}d\theta.$$

The preceding implies that

$$\frac{\left(x^2-1\right)}{2x}\cdot F'(x) = \int_0^{\pi/2} \frac{\left(x^2-1\right)\cos^2\theta}{\sin^2\theta + x^2\cos^2\theta}d\theta$$

$$= \int_0^{\pi/2} \frac{x^2\cos^2\theta + \sin^2\theta - 1}{\sin^2\theta + x^2\cos^2\theta}d\theta$$

$$= \frac{\pi}{2} - \int_0^{\pi/2} \frac{d\theta}{\sin^2\theta + x^2\cos^2\theta}$$

$$= \frac{\pi}{2} - \int_0^{\pi/2} \frac{\sec^2\theta\, d\theta}{\tan^2\theta + x^2}$$

$$= \frac{\pi}{2} - \frac{1}{x}\arctan\frac{\tan\theta}{x}\bigg|_0^{\pi/2}$$

$$= \frac{\pi}{2} - \frac{\pi}{2x},$$

which in turn implies that for $x > 0, x \neq 1$.

$$F'(x) = \frac{2x}{x^2-1}\left(\frac{\pi}{2} - \frac{\pi}{2x}\right) = \frac{\pi}{x+1}.$$

For $x = 1$, one sees immediately that $F'(x) = 2\int_0^{\pi/2} \cos^2\theta\, d\theta = \dfrac{\pi}{2}$, agreeing with the formula. Now

$$F'(x) = \frac{\pi}{x+1} \Rightarrow F(x) = \pi \log(x+1) + C.$$

Since $F(1) = \int_0^{\pi/2} \log 1\, d\theta = 0$, we gather that $C = -\pi \log 2$.

Finally thus

$$F(x) = \pi \log(x+1) - \pi \log 2 = \pi \log \frac{x+1}{2}.$$

Under certain conditions, the interval of integration in the preceding theorem need not be compact.

**EXAMPLE 2.5.11**

Given that $\int_0^{+\infty} \dfrac{\sin x}{x}\, dx = \dfrac{\pi}{2}$, compute $\int_0^{+\infty} \dfrac{\sin^2 x}{x^2}\, dx$.

**Solution:**

Put $I(a) = \int_0^{+\infty} \dfrac{\sin^2 ax}{x^2}\, dx$, with gather that $a \geq 0$. Differentiating both sides with respect to $a$, and making the substitution $u = 2ax$.

$$I'(a) = \int_0^{+\infty} \frac{2x \sin ax \cos ax}{x^2}\, dx$$

$$= \int_0^{+\infty} \frac{\sin 2ax}{x}\, dx$$

$$= \int_0^{+\infty} \frac{\sin u}{u}\, du$$

$$= \frac{\pi}{2}.$$

Integrating each side gives

$$I(a) = \frac{\pi}{2}a + C.$$

Since $I(0) = 0$, we gather that $C = 0$. The desired integral is $I(1) = \frac{\pi}{2}$.

## EXERCISES 2.5

**2.5.1** Find $f_x(x,y) = \frac{\partial}{\partial x}f(x,y)$, $f_y(x,y) = \frac{\partial}{\partial y}f(x,y)$,

$f_x(x,y,z) = \frac{\partial}{\partial x}f(x,y,z)$, $f_y(x,y,z) = \frac{\partial}{\partial y}f(x,y,z)$, and

$f_z(x,y,z) = \frac{\partial}{\partial z}f(x,y,z)$ for the following functions:

**1.** $f(x,y) = \left(x^3 - y^2\right)^3$

**2.** $f(x,y) = x^3\sin\left(\frac{1}{x}\right) + 5y^2$

**3.** $f(x,y) = \begin{cases} \dfrac{3xy}{\left(x^2 + y^2\right)} & \text{if } (x,y) \neq (0,0) \\ 0 & \text{if } (x,y) = (0,0) \end{cases}$

**4.** $f(x,y,z) = \dfrac{x^3 - z^2}{1 + \sin 3y}$

**2.5.2** If $f : \mathbb{R}^3 \to \mathbb{R}$, and $f(x,y,z) = yx\arctan(zx)$. Find $\dfrac{\partial f}{\partial x}, \dfrac{\partial f}{\partial y}$, and $\dfrac{\partial f}{\partial z}$.

**2.5.3** If $f : \mathbb{R}^3 \to \mathbb{R}$, and $f(x,y,z) = \left(x^2 + z^2\right)\log\left(x^2y^2 + 1\right)$. Find $\dfrac{\partial f}{\partial x}, \dfrac{\partial f}{\partial y}$ and $\dfrac{\partial f}{\partial z}$.

**2.5.4** Let $u(x,y)$ and $v(x,y)$ are defined by the equations $x = u\cos v$ and $y = u\sin v$. Find $\dfrac{\partial u}{\partial x}$ and $\dfrac{\partial v}{\partial x}$.

**2.5.5** Let $f : \mathbb{R}^2 \to \mathbb{R}, \; f(x,y) = \min(x, y^2)$. Find $\dfrac{\partial f(x,y)}{\partial x}$ and $\dfrac{\partial f(x,y)}{\partial y}$.

**2.5.6** Prove that if an equation $F(x,y,z) = 0$ defines an implicit differentiable function $f$ of two variables $x$ and $y$ such that $z = f(x,y)$ for all $(x,y)$ in the domain of $f$, D, then

$$\frac{\partial z}{\partial x} = -\frac{F_x(x,y,z)}{F_z(x,y,z)}, \quad \frac{\partial z}{\partial y} = -\frac{F_y(x,y,z)}{F_z(x,y,z)}.$$

**2.5.7** Let $w = u^3 + u^2 v - 3v, u = \sin xy, v = y \ln x$. Find $\dfrac{\partial w}{\partial x}$ and $\dfrac{\partial w}{\partial y}$.

**2.5.8** Let $f : \mathbb{R}^2 \to \mathbb{R}^2$ and $g : \mathbb{R}^3 \to \mathbb{R}^2$ be given by

$$f(x,y) = \begin{bmatrix} xy^2 \\ x^2 y \end{bmatrix}, g(x,y) \begin{bmatrix} x - y + 2z \\ xy \end{bmatrix}.$$

Compute $(f \circ g)'(1,0,1)$, if at all defined. If undefined, explain.

Compute $(g \circ f)'(1,0)$, if at all defined. If undefined, explain.

**2.5.9** Let $f(x,y) = \begin{bmatrix} xy \\ x + y \end{bmatrix}$ and $g(x,y) = \begin{bmatrix} x - y \\ x^2 y^2 \\ x + y \end{bmatrix}$. Find $(g \circ f)'(0,1)$.

**2.5.10** Let $z$ be an implicitly defined function of $x$ and $y$ through the equation $(x + z)^2 + (y + z)^2 = 8$. Find $\dfrac{\partial z}{\partial x}$ at $(1,1,1)$.

**2.5.11** Let $x = r \cos\theta$ and $y = r \sin\theta$. Find the Jacobi matrix $f'(r,\theta)$ and the Jacobi determinant $J(r,\theta)$.

**2.5.12** Let $x = e^u \sin v$ and $y = e^u \cos v$. Find the Jacobi matrix $f'(u,v)$ and the Jacobi determinant $J(u,v)$.

**2.5.13** Let $x = \dfrac{u}{u^2 + v^2}$ and $y = \dfrac{v}{u^2 + v^2}$. Find the Jacobi matrix $f'(u,v)$ and the Jacobi determinant $J(u,v)$.

**2.5.14**   Let $x = u^2 + vw$, $y = 2v + u^2w$, and $z = uvw$. Find the Jacobi matrix $f'(u,v,w)$ and the Jacobi determinant $J(u,v,w)$.

**2.5.15**   Let the transformation of coordinates $x = f(u,v)$ and $y = g(u,v)$ is one-to-one. Find the determinant $\left| \dfrac{\partial(x,y)}{\partial(u,v)} \dfrac{\partial(u,v)}{\partial(x,y)} \right|$.

**2.5.16**   Let $f : \mathbb{R}^3 \to \mathbb{R}^3$ be given by $f(r,\theta,\phi) = (r \sin\phi \cos\theta, r \sin\phi \sin\theta, r \cos\phi)$. Find the Jacobi matrix and Jacobi determinant.

**2.5.17**   Compute $\dfrac{d}{dt} \displaystyle\int_0^1 \left(2x + t^3\right)^2 dx$.

**2.5.18**   Suppose $g : \mathbb{R} \to \mathbb{R}$ is continuous and $a \in \mathbb{R}$ is a constant. Find the partial derivatives with respect to $x$ and $y$ of $f : \mathbb{R}^2 \to \mathbb{R}$, $f(x,y) = \displaystyle\int_a^{x^2 y} g(t)\, dt$.

**2.5.19**   Given that $\displaystyle\int_0^b \dfrac{dx}{x^2 + a^2} = \dfrac{1}{a} \arctan \dfrac{b}{a}$, evaluate $\displaystyle\int_0^b \dfrac{dx}{\left(x^2 + a^2\right)^2}$.

**2.5.20**   Evaluate $\displaystyle\int_0^\infty \dfrac{e^{-ax} \sin x}{x}\, dx$ using differentiation under integral sign.

## 2.6 GRADIENTS AND DIRECTIONAL DERIVATIVES

A function

$$f : \begin{array}{c} \mathbb{R}^n \to \mathbb{R}^m \\ x \mapsto f(x) \end{array}$$

is called a *vector field*. If $m = 1$, it is called a *scalar field*.

**Definition 2.6.1**   Let

$$f : \begin{array}{c} \mathbb{R}^n \to \mathbb{R} \\ x \mapsto f(x) \end{array}$$

be a scalar field. The *gradient* of $f$ is the vector defined and denoted by

$$\nabla f(x) = \begin{bmatrix} \dfrac{\partial f}{\partial x_1}(x) \\[2ex] \dfrac{\partial f}{\partial x_2}(x) \\[1ex] \vdots \\[1ex] \dfrac{\partial f}{\partial x_n}(x) \end{bmatrix}.$$

The *graduation operation* is the operator

$$\nabla = \begin{bmatrix} \dfrac{\partial}{\partial x_1} \\[2ex] \dfrac{\partial}{\partial x_2} \\[1ex] \vdots \\[1ex] \dfrac{\partial}{\partial x_n} \end{bmatrix}.$$

**Theorem 2.6.1**   Let $A \subseteq \mathbb{R}^n$ be open and let $f : A \to \mathbb{R}$ be a scalar field, and assume that $f$ is differentiable in $A$. Let $K \in \mathbb{R}$ be a constant. Then $\nabla f(x)$ is orthogonal to the surface implicitly defined by $f(x) = K$.

**Proof:**

Let

$$c : \begin{array}{c} \mathbb{R} \to \mathbb{R}^n \\ t \mapsto c(t) \end{array}$$

be a curve lying on this surface. Choose $t_0$ so that $c(t_0) = x$. Then

$$(f \circ c)(t_0) = f(c(t)) = K,$$

and using the chain rule

$$f'(c(t_0))c'(t_0) = 0,$$

which translates to

$$(\nabla f(x)) \cdot (c'(t_0)) = 0.$$

Since $c'(t_0)$ is tangent to the surface and its dot product with $\nabla f(x)$ is 0, we conclude that $\nabla f(x)$ is normal to the surface.

**NOTE** *Let $\theta$ be the angle between $\nabla f(x)$ and $c'(t_0)$. Since $\left|(\nabla f(x)) \cdot (c'(t_0))\right| = \left\|\nabla f(x)\right\| \left\|c'(t_0)\right\| \cos\theta$, where $\nabla f(x)$ is the direction in which $f$ is changing the fastest.*

### EXAMPLE 2.6.1

Find a unit vector normal to the surface $x^2y + 3xz^2 = 8$ at the point $(2,0,1)$.

**Solution:**

Given surface is $x^2y + 3xz^2 = 8$, $f(x,y,z) = x^2y + 3xz^2$

$$\nabla f(x,y,z) = \frac{\partial}{\partial x}(x^2y + 3xz^2)\vec{i} + \frac{\partial}{\partial y}(x^2y + 3xz^2)\vec{j} + \frac{\partial}{\partial z}(x^2y + 3xz^2)\vec{k}$$

$$= (2xy + 3z^2)\vec{i} + (x^2)\vec{j} + (6xz)\vec{k}$$

$$\nabla f(2,0,1) = (3)\vec{i} + (4)\vec{j} + (12)\vec{k}$$

Now, the unit vector normal to the surface $x^2y + 3xz^2 = 8$ at the point $(2,0,1)$ is

$$\frac{\nabla f(2,0,1)}{\left|\nabla f(2,0,1)\right|} = \frac{(3)\vec{i} + (4)\vec{j} + (12)\vec{k}}{\sqrt{9 + 16 + 1}} = \frac{3}{\sqrt{26}}\vec{i} + \frac{4}{\sqrt{26}}\vec{j} + \frac{12}{\sqrt{26}}\vec{k}.$$

### EXAMPLE 2.6.2

Find a unit vector normal to the surface $x^3 + y^3 + z = 4$ at the point $(1,1,2)$.

**Solution:**

Here $f(x,y,z) = x^3 + y^3 + z - 4$ has gradient

$$\nabla f(x,y,z) = \begin{bmatrix} 3x^2 \\ 3y^2 \\ 1 \end{bmatrix}$$

which at $(1,1,2)$ is $\begin{bmatrix} 3 \\ 3 \\ 1 \end{bmatrix}$. Normalizing this vector, we obtain

$$\begin{bmatrix} \dfrac{3}{\sqrt{19}} \\[2ex] \dfrac{3}{\sqrt{19}} \\[2ex] \dfrac{1}{\sqrt{19}} \end{bmatrix}.$$

For example, we can determine the gradient and the unit normal function to graph of the function $f(x,y) = \dfrac{3xy}{x^2 + y^2}$ using $Maple^{Tm}$ commands as

> $with(linalg)$ :

> $grad\left( \dfrac{3\,x \cdot y}{(x^2 + y^2)}, [x, y] \right);$

$$\left[ \frac{3y}{x^2 + y^2} - \frac{6x^2 y}{(x^2 + y^2)^2} \quad \frac{3x}{x^2 + y^2} - \frac{6xy^2}{(x^2 + y^2)^2} \right]$$

> $psi := z - \dfrac{3\,x \cdot y}{(x^2 + y^2)};$

$$\psi := z - \frac{3xy}{x^2 + y^2}$$

> $n := grad(psi, [x, y, z]);$

$$n := \left[ -\frac{3y}{x^2 + y^2} + \frac{6x^2 y}{(x^2 + y^2)^2} \quad -\frac{3x}{x^2 + y^2} + \frac{6xy^2}{(x^2 + y^2)^2} \quad 1 \right]$$

> $Nf := normalize(n);$

$$Nf := \left[ \frac{\left( -\dfrac{3y}{x^2 + y^2} + \dfrac{6x\,y}{(x^2 + y^2)^2} \right) |x^4 + 2x^2 y^2 + y^4|}{\sqrt{|x^4 + 2x^2 y^2 + y^4|^2 + 9\,|y|^2\,|x^2 - y^2|^2 + 9\,|x|^2\,|x^2 - y^2|^2}}, \right.$$

$$\frac{\left( -\dfrac{3x}{x^2 + y^2} + \dfrac{6xy^2}{(x^2 + y^2)^2} \right) |x^4 + 2x^2 y^2 + y^4|}{\sqrt{|x^4 + 2x^2 y^2 + y^4|^2 + 9\,|y|^2\,|x^2 - y^2|^2 + 9\,|x|^2\,|x^2 - y^2|^2}},$$

$$\left. \frac{|x^4 + 2x^2 y^2 + y^4|}{\sqrt{|x^4 + 2x^2 y^2 + y^4|^2 + 9\,|y|^2\,|x^2 - y^2|^2 + 9\,|x|^2\,|x^2 - y^2|^2}} \right]$$

## EXAMPLE 2.6.3

Determine the gradient for the function $f(x,y) = \dfrac{\cos(x^2 + y^2)}{x^2 + y^2}$ using $Maple^{Tm}$ and MATLAB commands.

### *Solution:*

$Maple^{Tm}$ commands:

> $with(linalg):$
> $grad\left(\dfrac{\cos(x^2 + y^2)}{(x^2 + y^2)}, [x,y]\right);$

$$\left[ -\frac{2\sin(x^2 + y^2)\,x}{x^2 + y^2} - \frac{2\cos(x^2 + y^2)\,x}{(x^2 + y^2)^2}, \; -\frac{2\sin(x^2 + y^2)\,y}{x^2 + y^2} \right.$$
$$\left. -\frac{2\cos(x^2 + y^2)\,y}{(x^2 + y^2)^2} \right]$$

MATLAB commands:

```
>> syms x y
>> f=(cos(x^2+y^2)/(x^2+y^2))

f =

cos(x^2+y^2)/(x^2+y^2)

>> gradf=jacobian(f,[x,y])

gradf =

[-2*sin(x^2+y^2)*x/(x^2+y^2)-2*cos(x^2+y^2)/(x^2+y^2)^2*x,
-2*sin(x^2+y^2)*y/(x^2+y^2)-2*cos(x^2+y^2)/(x^2+y^2)^2*y]
```

## EXAMPLE 2.6.4

Find the direction of the greatest rate of increase of $f(x,y,z) = xye^z$ at the point $(2,1,2)$.

***Solution:***

The direction is that of the gradient vector. Here

$$\nabla f\left(x,y,z\right) = \begin{bmatrix} ye^z \\ xe^z \\ xye^z \end{bmatrix}$$

which at $\left(2,1,2\right)$ comes $\begin{bmatrix} e^2 \\ 2e^2 \\ 2e^2 \end{bmatrix}$. Normalizing this vector, we obtain

$$\frac{1}{\sqrt{5}} \begin{bmatrix} 1 \\ 2 \\ 2 \end{bmatrix}.$$

### EXAMPLE 2.6.5

Let $f : \mathbb{R}^3 \to \mathbb{R}$ be given by

$$f\left(x,y,z\right) = x + y^2 - z^2.$$

Find the equation of the tangent plane to $f$ at $\left(1,2,3\right)$.

***Solution:***

A vector normal to the plane is $\nabla f\left(1,2,3\right)$. Now

$$\nabla f\left(x,y,z\right) = \begin{bmatrix} 1 \\ 2y \\ -2z \end{bmatrix}$$

which is

$$\begin{bmatrix} 1 \\ 4 \\ -6 \end{bmatrix}$$

at $\left(1,2,3\right)$. The equation of the tangent plane is thus

$$1\left(x-1\right) + 4\left(y-2\right) - 6\left(z-3\right) = 0,$$

Or

$$x + 4y - 6z = -9.$$

**Definition 2.6.2** Let

$$f: \begin{matrix} \mathbb{R}^n \to \mathbb{R}^n \\ x \mapsto f(x) \end{matrix}$$

be a vector field with

$$f(x) = \begin{bmatrix} f_1(x) \\ f_2(x) \\ \vdots \\ f_n(x) \end{bmatrix}.$$

The *divergence of f* is defined and denoted by

$$\operatorname{div} f(x) = \nabla \bullet f(x) = \frac{\partial f_1}{\partial x_1}(x) + \frac{\partial f_2}{\partial x_2}(x) + \ldots + \frac{\partial f_n}{\partial x_n}(x).$$

**EXAMPLE 2.6.6**

If $f(x,y,z) = \left( x^2, y^2, ye^{z^2} \right)$ then

$$\operatorname{div} f(x) = 2x + 2y + 2yze^{z^2}.$$

**Definition 2.6.3** Let $g_k : \mathbb{R}^n \to \mathbb{R}^n, 1 \le k \le n-2$ be vector fields with $g_i = (g_{i1}, g_{i2}, \ldots, g_{in})$. Then the *curl of* $(g_1, g_2, \ldots, g_{n-2})$ is

$$\operatorname{curl}(g_1, g_2, \ldots, g_{n-2})(x) = \det \begin{bmatrix} e_1 & e_2 & \cdots & e_n \\ \dfrac{\partial}{\partial x_1} & \dfrac{\partial}{\partial x_2} & \cdots & \dfrac{\partial}{\partial x_n} \\ g_{11}(x) & g_{12}(x) & \cdots & g_{1n}(x) \\ g_{21}(x) & g_{22}(x) & \cdots & g_{1n}(x) \\ \vdots & \vdots & \vdots & \vdots \\ g_{(n-2)1}(x) & g_{(n-2)2}(x) & \cdots & g_{(n-2)n}(x) \end{bmatrix}.$$

**EXAMPLE 2.6.7**

If $f(x,y,z) = \left( x^2, y^2, ye^{z^2} \right)$, then

$$\operatorname{curl} f(x,y,z) = \nabla \times f(x,y,z) = \left( e^{z^2} \right)\vec{\mathrm{i}}.$$

**EXAMPLE 2.6.8**

If $f(x,y,z,w) = (e^{xyz}, 0, 0, w^2), g(x,y,z,w) = (0,0,z,0)$, then

$$\operatorname{curl}(f,g)(x,y,z,w) = \det \begin{bmatrix} e_1 & e_2 & e_3 & e_4 \\ \dfrac{\partial}{\partial x_1} & \dfrac{\partial}{\partial x_2} & \dfrac{\partial}{\partial x_3} & \dfrac{\partial}{\partial x_4} \\ e^{xyz} & 0 & 0 & w^2 \\ 0 & 0 & z & 0 \end{bmatrix} = \left( xz^2 e^{xyz} \right) e_4.$$

**EXAMPLE 2.6.9**

Find the curl of the vector $\vec{v} = 3xy\vec{i} + 2z\vec{j} + x^3\vec{k}$.

**Solution:**

$$\nabla \times \vec{v} = \det \begin{bmatrix} \vec{i} & \vec{j} & \vec{k} \\ \dfrac{\partial}{\partial x} & \dfrac{\partial}{\partial y} & \dfrac{\partial}{\partial z} \\ 3xy & 2z & x^3 \end{bmatrix} = (0-2)\vec{i} + (3x^2 - 0)\vec{j} + (0-3x)\vec{k} = -2\vec{i} + 3x^2\vec{j} - 3x\vec{k}.$$

**Definition 2.6.4**   Let $A \subseteq \mathbb{R}^n$ be open and let $f : A \to \mathbb{R}$ be a scalar field, and assume that $f$ is differentiable in $A$. Let $\vec{v} \in \mathbb{R}^n \setminus \{0\}$ be such that $x + t\,\vec{v} \in A$ for sufficiently small $t \in \mathbb{R}$. Then the *directional derivative of $f$ in the direction of $\vec{v}$ at the point* $x$ is defined and denoted by

$$D_{\vec{v}} f(x) = \lim_{t \to 0} \frac{f(x + t\vec{v}) - f(x)}{t}.$$

**Theorem 2.6.2**   Let $A \subseteq \mathbb{R}^n$ be open and let $f : A \to \mathbb{R}$ be a scalar field, and assume that $f$ is differentiable in $A$. Let $\vec{v} \in \mathbb{R}^n \setminus \{\vec{0}\}$ be such that $\vec{x} + t\,\vec{v} \in A$ for sufficiently small $t \in \mathbb{R}$. Then the *directional derivative of $f$ in the direction of $\vec{v}$ at the point* $\vec{x}$ is given by

$$\nabla f(x) \cdot \vec{v}.$$

**EXAMPLE 2.6.10**

Find the directional derivative of $f(x,y,z) = x^3 + y^3 - z^2$ in the direction

of $\begin{bmatrix} 1 \\ 2 \\ 3 \end{bmatrix}$.

*Solution:*

We have

$$\nabla f(x,y,z) = \begin{bmatrix} 3x^2 \\ 3y^2 \\ -2z \end{bmatrix}$$

and so

$$\nabla f(x,y,z) \bullet \vec{v} = 3x^2 + 6y^2 - 6z.$$

**EXAMPLE 2.6.11**

Find the directional derivative of $f(x,y,z) = xy^3 + yz^2$ at the point $\begin{pmatrix} 1 \\ -1 \\ 2 \end{pmatrix}$ in the direction of the vector $\vec{a} = 2\,\vec{i} + \vec{j} + 3\,\vec{k}$.

*Solution:*

$$\nabla f(x,y,z) = \frac{\partial f}{\partial x}\,\vec{i} + \frac{\partial f}{\partial y}\,\vec{j} + \frac{\partial f}{\partial z}\,\vec{k}$$

$$\nabla f(x,y,z) = y^3\,\vec{i} + (3xy^2 + z^2)\,\vec{j} + (2yz)\,\vec{k}$$

$$\nabla f(1,-1,2) = -\vec{i} + 7\,\vec{j} - 4\,\vec{k}$$

$$\frac{\vec{a}}{\sqrt{4+1+9}} = \frac{2\,\vec{i}}{\sqrt{14}} + \frac{\vec{j}}{\sqrt{14}} + \frac{3\,\vec{k}}{\sqrt{14}}$$

$$\nabla f(1,-1,2) \bullet \left( \frac{2\,\vec{i}}{\sqrt{14}} + \frac{\vec{j}}{\sqrt{14}} + \frac{3\,\vec{k}}{\sqrt{14}} \right) = \left( -\vec{i} + 7\,\vec{j} - 4\,\vec{k} \right) \bullet \left( \frac{2\,\vec{i}}{\sqrt{14}} + \frac{\vec{j}}{\sqrt{14}} + \frac{3\,\vec{k}}{\sqrt{14}} \right)$$

$$= -\frac{2}{\sqrt{14}} + \frac{7}{\sqrt{14}} - \frac{12}{\sqrt{14}} = -\frac{7}{\sqrt{14}}.$$

## EXERCISES 2.6

**2.6.1**  Let $g(x,y) = (\ln x)(e^y)$. Find $\nabla g(x,y)$.

**2.6.2**  Let $f(x,y,z) = x^4 - xy + z^2$. Find $\nabla f(x,y,z)$.

**2.6.3**  Let $f(x,y) = x^3 - xy + y^2$. Find $(\nabla f)(1,1)$.

**2.6.4**  Let $f(x,y,z) = xe^{yz}$. Find $(\nabla f)(2,1,1)$.

**2.6.5**  Let $f(x,y,z) = x^2 y \sin(yz)$. Find $\nabla f(x,y,z)$.

**2.6.6**  Let $f(x,y,z) = \begin{bmatrix} xz \\ e^{xy} \\ z \end{bmatrix}$. Find $(\nabla \times f)(2,1,1)$.

**2.6.7**  If $f(x,y,z) = x^2 e^y$ and $k(x,y,z) = y^2 e^{xz}$. Find $\nabla f(x,y,z), \nabla k(x,y,z)$, and $\nabla(fk)$. Verify that $\nabla(fk) = f\nabla k + k\nabla f$.

**2.6.8**  Find the point on the surface $x^2 + y^2 - 5xy + xz - yz = -3$ for which the tangent plane is $x - 7y = -6$.

**2.6.9**  Use a linear approximation of the function $f(x,y) = e^{x\cos 2y}$ at $(0,0)$ to estimate $f(0.1, 0.2)$.

**2.6.10**  Find the directional derivative of $f(x,y) = x^2 - 5xy + 3y^2$ at the point $(2,-1)$ in the direction $\theta = \pi/4$.

**2.6.11**  Prove that the gradient $\nabla f$ points in the direction of most rapid increase for $f$.

**2.6.12**  Find the angles made by the gradient of $f(x,y) = x^{\sqrt{3}} + y$ at the point $(1,1)$ with the coordinate axes.

**2.6.13**  Find the directional derivative of $f(x,y) = \dfrac{(x-y)}{(x+y)}$ at the point $(2,-1)$ in the direction $\vec{v} = 3\vec{i} + 4\vec{j}$.

**2.6.14**  Find the directional derivative of $f(x,y,z) = z^2 \arctan(x+y)$ at the point $(0,0,4)$ in the direction $\vec{v} = \begin{bmatrix} 6 \\ 0 \\ 1 \end{bmatrix}$.

**2.6.15**  Find the directional derivative of $f(x,y,z) = \left(\dfrac{x}{y}\right) - \left(\dfrac{y}{z}\right)$ at point $P_1(0,-1,2)$ in the direction from $P_1(0,-1,2)$ to $P_2(3,1,-4)$. Also determine the direction in which $f(x,y,z)$ increases most rapidly at $P_1(0,-1,2)$ and find the maximum rate of increase.

**2.6.16**  Prove that
$$\nabla \bullet (u \times v) = v \bullet (\nabla \times u) - u \bullet (\nabla \times v).$$

**2.6.17**  Let $\phi : \mathbb{R}^3 \to \mathbb{R}$ be a scalar field, and let $\vec{U}, \vec{V} : \mathbb{R}^3 \to \mathbb{R}^3$ be vector fields. Prove that

1.  $\nabla \bullet \left(\phi \vec{U}\right) = \phi \nabla \bullet \vec{U} + \vec{U} \bullet \nabla \phi$

2.  $\nabla \times \left(\phi \vec{U}\right) = \phi \nabla \times \vec{U} + \nabla \phi \times \vec{U}$

3.  $\nabla \bullet \left(\vec{U} \times \vec{V}\right) = \vec{V} \bullet \nabla \times \vec{U} - \vec{U} \bullet \nabla \times \vec{V}$

**2.6.18**  Find the tangent plane equation and the normal line equation to the graph of the equation $4x^2 - y^2 + 3z^2 = 10$ at the point $(2,-3,1)$.

**2.6.19**  Find the points on the hyperboloid $x^2 - 2y^2 - 4z^2 = 16$ at which the tangent plane is parallel to the plane $4x - 2y + 4z = 5$.

**HINT**  *Parallel planes have proportional gradients.*

**2.6.20**  Find the gradient vector of the function $f(x,y) = \cos \pi x \sin \pi\, y$ $+ \sin 2\pi\, y$ at point $(-1, 1/2)$. Then find the equation of the tangent plane.

**2.6.21**  In what direction $\vec{v}$ does $f(x,y) = 1 - x^2 - y^2$ decrease most rapidly at point $(-1,2)$?

**2.6.22** Find the equation of the tangent plane to the surface $z = xe^{-2y}$ at the point $(1,0,1)$.

**2.6.23** Verify that $\nabla \times \left( \nabla \times \vec{f} \right) = \nabla \left( \nabla \bullet \vec{f} \right) - \nabla^2 \vec{f}$ for the vector field

$$\vec{f} = \begin{bmatrix} 3xz^2 \\ -yz \\ x + 2z \end{bmatrix}.$$

**2.6.24** Find the directional derivative of $f(x,y) = 4x + xy^2 - 5y$ at the point $(2,-1)$ in the direction of a unit vector whose angle $\theta$ with positive $x$-axis is $\pi / 4$.

## 2.7 LEVI-CIVITTA AND EINSTE

In this section, unless otherwise noted, we are dealing in the space $\mathbb{R}^3$ and so, subscripts are not in the set $\{1,2,3\}$.

**Definition 2.7.1 (Einstein's Summation Convention):** In any expression containing subscripted variables appearing twice (and only twice) in any term, the subscripted variables are assumed to be summed over.

**NOTE** *In order to emphasize that we are using Einstein's convention, we will enclose any terms under consideration with* $\prec . \succ$.

### EXAMPLE 2.7.1

Using Einstein's Summation convention, the dot product of two vectors $\vec{x} \in \mathbb{R}^n$ and $\vec{y} \in \mathbb{R}^n$ can be written as

$$\vec{x} \bullet \vec{y} = \sum_{i=1}^{n} x_i y_i = \prec x_t y_t \succ.$$

### EXAMPLE 2.7.2

Given that $a_i, b_j, c_k, d_l$ are the components of vectors in $\mathbb{R}^3$, $\vec{a}, \vec{b}, \vec{c}, \vec{d}$, respectively, what is the meaning of $\prec a_i b_i c_k d_k \succ$?

*Solution:*

We have

$$\prec a_i b_i c_k d_k \succ = \sum_{i=1}^{3} a_i b_i \prec c_k d_k \succ = \vec{a} \cdot \vec{b} \prec c_k d_k \succ = \vec{a} \cdot \vec{b} \sum_{i=1}^{3} c_k d_k = \left(\vec{a} \cdot \vec{b}\right)\left(\vec{c} \cdot \vec{d}\right).$$

## EXAMPLE 2.7.3

Using Einstein's Summation convention, the $ij$–th entry $(AB)_{ij}$ of the product of two matrices $A \in M_{m \times n}(\mathbb{R})$ and $B \in M_{n \times r}(\mathbb{R})$ can be written as

$$(AB)_{ij} = \sum_{k=1}^{n} A_{ik} B_{kj} = \prec A_{it} B_{kj} \succ = \prec A_{it} B_{tj} \succ.$$

## EXAMPLE 2.7.4

Using Einstein's Summation convention, the trace $\text{tr}(A)$ of a square matrix $A \in M_{n \times n}(\mathbb{R})$ is $\text{tr}(A) = \sum_{t=1}^{n} A_{tt} = \prec A_{tt} \succ.$

## EXAMPLE 2.7.5

Demonstrate, via Einstein's Summation convention, that if $A, B$ are two $n \times n$ matrices, then

$$\text{tr}(AB) = \text{tr}(BA).$$

*Solution:*

We have

$$\text{tr}(AB) = \text{tr}\left((AB)_{ij}\right) = \text{tr}\left(\prec A_{ik} B_{kj} \succ\right) = \prec \prec A_{tk} B_{kt} \succ \succ,$$

and

$$\text{tr}(BA) = \text{tr}\left((BA)_{ij}\right) = \text{tr}\left(\prec B_{kj} A_{ik} \succ\right) = \prec \prec B_{tk} A_{kt} \succ \succ,$$

from where the assertion follows, since the indices are dummy variables and can be exchanged.

**Definition 2.7.2 (Kroenecker's Delta):** The symbol $\delta_{i,j}$ is defined as follows:

$$\delta_{ij} = \begin{cases} 0 \text{ if } i \neq j \\ 1 \text{ if } i = j. \end{cases}$$

## EXAMPLE 2.7.6

It is easy to see that $\prec \delta_{ik}\delta_{kj} \succ = \sum_{k=1}^{3} \delta_{ik}\delta_{kj} = \delta_{ij}$.

## EXAMPLE 2.7.7

We have that

$$\prec \delta_{ij}a_ib_j \succ = \sum_{i=1}^{3}\sum_{j=1}^{3} \delta_{ij}a_ib_j = \sum_{k=1}^{3} a_kb_k = \vec{a}\cdot\vec{b}.$$

Recall that a *permutation* of distinct objects is a reordering of them. The $3! = 6$ permutations of the index set $\{1, 2, 3\}$ can be classified into even or odd. We start with the identity permutation 123 and say it is even. Now, for any other permutation, we will say that it is even if it takes an even number of transpositions (switching only two elements in one move) to regain the identity permutation and odd if it takes an odd number of transpositions to regain the identity permutation. Since

$$231 \to 132 \to 123, \quad 312 \to 132 \to 123,$$

the permutations 123 (identity), 231, and 312 are even. Since

$$132 \to 123, \quad 321 \to 123, \quad 213 \to 123,$$

the permutations 132, 321, and 213 are odd.

**Definition 2.7.3 (Levi-Civitta's Alternating Tensor):**   The symbol $\varepsilon_{jkl}$ is defined as follows:

$$\varepsilon_{jkl} = \begin{cases} 0 & \text{if} & \{j,k,l\} \neq \{1,2,3\} \\ -1 & \text{if} & \begin{pmatrix} 1 & 2 & 3 \\ j & k & l \end{pmatrix} \text{ is an odd permutation} \\ +1 & \text{if} & \begin{pmatrix} 1 & 2 & 3 \\ j & k & l \end{pmatrix} \text{ is an even permutation} \end{cases}$$

**NOTE**   *In particular, if one subindex is repeated we have* $\varepsilon_{rrs} = \varepsilon_{rsr} = \varepsilon_{srr}$. *Also,*

$$\varepsilon_{123} = \varepsilon_{231} = \varepsilon_{312} = 1, \quad \varepsilon_{132} = \varepsilon_{321} = \varepsilon_{213} = -1.$$

## EXAMPLE 2.7.8

Using the Levi-Civitta, alternating tensor and Einstein's summation convention, the cross-product can also be expressed, if $\vec{i} = \vec{e_1}$, $\vec{j} = \vec{e_2}$, $\vec{k} = \vec{e_3}$,

Then

$$\vec{x} \times \vec{y} = \prec \varepsilon_{jkl} \left( a_k b_l \right) \vec{e_j} \succ.$$

## EXAMPLE 2.7.9

If $A = \left[ a_{ij} \right]$ is a $3 \times 3$ matrix, then, using the Levi-Civitta alternating tensor,

$$\det A = \prec \varepsilon_{ijk} a_{1i} a_{2j} a_{3k} \succ.$$

## EXAMPLE 2.7.10

Let $\vec{x}, \vec{y}, \vec{z}$ be vectors in $\mathbb{R}^3$. Then

$$\vec{x} \bullet \left( \vec{y} \times \vec{z} \right) = \prec x_i \left( \vec{y} \times \vec{z} \right)_i \succ = \prec x_i \varepsilon_{ikl} \left( y_k z_l \right) \succ.$$

# EXERCISES 2.7

**2.7.1**    Use the Einstein's summation convention and the Levi-Civitta's alternating tensor to show that $\vec{x} \times \left( \vec{y} \times \vec{z} \right) = \left( \vec{x} \bullet \vec{z} \right) \vec{y} - \left( \vec{x} \bullet \vec{y} \right) \vec{z}$.

**2.7.2**    Show that $\nabla \times \left( \nabla f \right) = 0$ using the Einstein's summation convention and the Levi-Civitta's alternating tensor.

**2.7.3**    Show that $\nabla \bullet \left( \nabla \times \vec{u} \right) = 0$ using the Einstein's summation convention and the Levi-Civitta's alternating tensor.

**2.7.4**    Show that $\nabla \times \left( \nabla \times \vec{u} \right) = \nabla \left( \nabla \bullet \vec{u} \right) - \nabla^2 \vec{u}$ using the Einstein's summation convention and the Levi-Civitta's alternating tensor.

**2.7.5**    Show that $\vec{v} \cdot \nabla \vec{v} = \nabla \left( \dfrac{|\vec{v}|^2}{2} \right) + \left( \nabla \times \vec{v} \right) \times \vec{v}$ using the Einstein's summation convention and the Levi-Civitta's alternating tensor.

**2.7.6**    Write True or false for the following statements:

**1.**    $\varepsilon_{ijk} = -\varepsilon_{ikj}$

**2.**    $\varepsilon_{ijk} \varepsilon_{ilm} = \delta_{jl} \delta_{km} - \delta_{jm} \delta_{kl}$

**3.** $\vec{a} \cdot \vec{b} = a_i b_i$

**4.** $\left(\vec{a} \times \vec{b}\right)_i \neq \varepsilon_{ijk} a_j b_k$

## 2.8   EXTREMA

We now turn to the problem of finding maxima and minima for vector functions. As in the one-variable case, the derivative will provide us with information about the extrema, and the "second derivative" will provide us with information about the nature of these extreme points.

To define an analog for the second derivative, let us consider the following. Let $A \subseteq \mathbb{R}^n$ and $f : A \to \mathbb{R}^m$ be differentiable on $A$. We know that for fixed $x_0 \in A, D_{x_0}(f)$, Dx0 (f) is a linear transformation from $\mathbb{R}^n$ to $\mathbb{R}^m$. This means that we have a function

$$T : \begin{array}{c} A \to l\left(\mathbb{R}^n, \mathbb{R}^m\right) \\ x \to D_x\left(f\right) \end{array},$$

Where $l\left(\mathbb{R}^n, \mathbb{R}^m\right)$ denotes the space of linear transformation from $\mathbb{R}^n$ to $\mathbb{R}^m$. Hence, if we differentiate $T$ at $x_0$ again, we obtain a linear transformation $D_{x_0}(T) = D_{x_0}(D_{x_0}(f)) = D_{x_0}^2(f)$ from $\mathbb{R}^n$ to $l\left(\mathbb{R}^n, \mathbb{R}^m\right)$. Hence, given $D_{x_0}^2(f)\left(x_1\right) \in l\left(\mathbb{R}^n, \mathbb{R}^m\right)$. Again, this means that given $x_2 \in \mathbb{R}^n, D_{x_0}^2(f)\left(x_1\right)\left(x_2\right) \in \mathbb{R}^m$, thus the function

$$T : \begin{array}{c} \mathbb{R}^n \times \mathbb{R}^n \to l\left(\mathbb{R}^n, \mathbb{R}^m\right) \\ \left(x_1, x_2\right) \to D_{x_0}^2(f)\left(x_1, x_2\right) \end{array}$$

is well defined, and linear in each variable $x_1$ and $x_2$, that is, it is a *bilinear* function. Just as the Jacobi matrix was a handy tool for finding a matrix representation of $D_x(f)$ in the natural bases, when $f$ maps into $\mathbb{R}$, we have the following analogue representation of the second derivative.

**Theorem 2.8.1**   Let $A \subseteq \mathbb{R}^n$ be an open set, and $f : A \to \mathbb{R}$ be twice differentiable on $A$. Then the matrix of $D_x^2(f) : \mathbb{R}^n \times \mathbb{R}^n \to \mathbb{R}$ with respect to the standard basis is given by the *Hessian matrix*:

$$H_x f = \begin{bmatrix} \dfrac{\partial^2 f}{\partial x_1 \partial x_1}(\mathbf{x}) & \dfrac{\partial^2 f}{\partial x_1 \partial x_2}(\mathbf{x}) & \cdots & \dfrac{\partial^2 f}{\partial x_1 \partial x_n}(\mathbf{x}) \\[2ex] \dfrac{\partial^2 f}{\partial x_2 \partial x_1}(\mathbf{x}) & \dfrac{\partial^2 f}{\partial x_2 \partial x_2}(\mathbf{x}) & \cdots & \dfrac{\partial^2 f}{\partial x_2 \partial x_n}(\mathbf{x}) \\[2ex] \vdots & \vdots & \vdots & \vdots \\[2ex] \dfrac{\partial^2 f}{\partial x_n \partial x_1}(\mathbf{x}) & \dfrac{\partial^2 f}{\partial x_n \partial x_2}(\mathbf{x}) & \cdots & \dfrac{\partial^2 f}{\partial x_n \partial x_n}(\mathbf{x}) \end{bmatrix}.$$

**EXAMPLE 2.8.1**

Let $f : \mathbb{R}^3 \to \mathbb{R}$ be given by

$$f(x,y,z) = xy^2 z^3.$$

Then

$$H_{(x,y,z)} f = \begin{bmatrix} 0 & 2yz^3 & 3y^2 z^2 \\ 2yz^3 & 2xz^3 & 6xyz^2 \\ 3y^2 z^2 & 6xyz^2 & 6xy^2 z \end{bmatrix}.$$

From the preceding example, we notice that the Hessian is symmetric, as the mixed partial derivatives $\dfrac{\partial^2}{\partial x \partial y} f = \dfrac{\partial^2}{\partial y \partial x} f$, etc., are equal. This is no coincidence, as guaranteed by the following theorem.

**Theorem 2.8.2**  Let $A \subseteq \mathbb{R}^n$ be an open set and $f : A \to \mathbb{R}$ be twice differentiable on $A$. If $D_{x_0}^2 (f)$ is continuous, then $D_{x_0}^2 (f)$ is symmetric, that is, $\forall (\mathbf{x}_1, \mathbf{x}_2) \in \mathbb{R}^n \times \mathbb{R}^n$ have

$$D_{x_0}^2 (f)(\mathbf{x}_1, \mathbf{x}_2) = D_{x_0}^2 (f)(\mathbf{x}_2, \mathbf{x}_1).$$

We are now ready to study extrema in several variables. The basic theorems resemble those of one-variable calculus. First, we make some analogous definitions.

**Definition 2.8.1**  Let $A \subseteq \mathbb{R}^n$ be an open set, and $f : A \to \mathbb{R}$. If there is some open ball $B_{x_0}(r)$ on which $\forall \mathbf{x} \in B_{x_0}(r), f(\mathbf{x}_0) \geq f(\mathbf{x})$, we say that $f(\mathbf{x}_0)$

is a *local maximum* of $f$. Similarly, if there is some open ball $B_{x_1}(r)$ on which $\forall x \in B_{x_0}(r'), f(x_1) \le f(x)$ we say that $f(x_1)$ is a *local minimum* of $f$. A point is called an *extreme point* if it is either a local minimum or local maximum. A point t is called a *critical point* if $f$ is differentiable at $t$ and $D_t(f) = 0$. A critical point which is neither a maxima nor a minima is called a *saddle point*.

**Theorem 2.8.3**   Let $A \subseteq \mathbb{R}^n$ be an open set, and $f : A \to \mathbb{R}$ be differentiable on $A$. If $x_0$ is an extreme point, then $D_{x_0}(f) = 0$, that is, $x_0$ is a critical point. Moreover, if $f$ is twice differentiable with continuous second derivative and $x_0$ is a critical point such that $H_{x_0} f$ is negative definite, then $f$ has a local maximum at $x_0$. If $H_{x_0} f$ is positive definite, then $f$ has a local minimum at $x_0$. If $H_{x_0} f$ is indefinite, then $f$ has a saddle point. If $H_{x_0} f$ is semi-definite (positive or negative), the test is inconclusive.

**EXAMPLE 2.8.2**

Find the critical points of

$$f : \begin{array}{ccc} \mathbb{R}^2 & \to & \mathbb{R} \\ (x,y) & \mapsto & x^2 + xy + y^2 + 2x + 3y \end{array}.$$

and investigate their nature.

**Solution:**

We have

$$(\nabla f)(x,y) = \begin{bmatrix} 2x + y + 2 \\ x + 2y + 3 \end{bmatrix},$$

and so to find the critical points, we solve

$$2x + y + 2 = 0,$$
$$x + 2y + 3 = 0,$$

which yields $x = -\dfrac{1}{3}$, $y = -\dfrac{4}{3}$. Now, $H_{(x,y)}f = \begin{bmatrix} 2 & 1 \\ 1 & 2 \end{bmatrix}$, which is positive definite, since $\Delta_1 = 2 > 0$ and $\Delta_2 = \det \begin{bmatrix} 2 & 1 \\ 1 & 2 \end{bmatrix} = 3 > 0$. Thus $x_0 = \left( -\dfrac{1}{3}, -\dfrac{4}{3} \right)$ is a relative minimum and we have $-\dfrac{7}{3} = f\left( -\dfrac{1}{3}, -\dfrac{4}{3} \right) \le f(x,y) = x^2 + xy + y^2 + 2x + 3y$.

## EXAMPLE 2.8.3

Find the extrema of

$$f : \begin{array}{ccc} \mathbb{R}^3 & \to & \mathbb{R} \\ (x,y,z) & \mapsto & x^2 + y^2 + 3z^2 - xy + 2xz + yz \end{array}.$$

**Solution:**

We have

$$(\nabla f)(x,y,z) = \begin{bmatrix} 2x - y + 2z \\ 2y - x + z \\ 6z + 2x + y \end{bmatrix},$$

which vanishes when $x = y = z = 0$. Now,

$$H_r f = \begin{bmatrix} 2 & -1 & 2 \\ -1 & 2 & 1 \\ 2 & 1 & 6 \end{bmatrix},$$

which is positive definite, since $\Delta_1 = 2 > 0$ and $\Delta_2 = \det \begin{bmatrix} 2 & -1 \\ -1 & 2 \end{bmatrix} = 3 > 0$, and

$\Delta_3 = \det \begin{bmatrix} 2 & -1 & 2 \\ -1 & 2 & 1 \\ 2 & 1 & 6 \end{bmatrix} = 4 > 0$. Thus $f$ has a relative minimum at $(0,0,0)$ and

$0 = f(0,0,0) \le f(x,y,z) = x^2 + y^2 + 3z^2 - xy + 2xz + yz$.

**EXAMPLE 2.8.4**

Let $f(x,y) = x^3 - y^3 + axy$, with $a \in \mathbb{R}$ a parameter. Determine the nature of the critical point of $f$.

**Solution:**

We have

$$(\nabla f)(x,y) = \begin{bmatrix} 3x^2 + ay \\ -3y^2 + ax \end{bmatrix} = \begin{bmatrix} 0 \\ 0 \end{bmatrix} \Rightarrow 3x^2 = -ay, \ \ 3y^2 = ax.$$

If $a = 0$, then $x = y = 0$ and so $(0,0)$ is a critical point. If $a \neq 0$, then

$$3\left( 3\frac{y^2}{a} \right)^2 = -ay \Rightarrow 27y^4 = -a^2y$$

$$\Rightarrow y\left( 27y^3 + a^3 \right) = 0$$

$$\Rightarrow y\left( 3y + a \right)\left( 9y^2 - 3ay + a^2 \right) = 0$$

$$\Rightarrow y = 0 \ \text{ or } y = -\frac{a}{3}.$$

If $y = 0$ $x = 0$, so again $(0,0)$ is a critical point. If $y = -\dfrac{a}{3}$ $x = \dfrac{3}{a} \times \left( -\dfrac{a}{3} \right)^2 = \dfrac{a}{3}$ so $\left( \dfrac{a}{3}, -\dfrac{a}{3} \right)$ is a critical point.

Now,

$$H_{f(x,y)} = \begin{bmatrix} 6x & a \\ a & -6y \end{bmatrix} \Rightarrow \Delta_1 = 6x, \ \Delta_2 = -36xy - a^2.$$

At $(0,0), \Delta_1 = 0, \Delta_2 = -a^2$. If $a \neq 0$, then there is a saddle point. At $\left( \dfrac{a}{3}, -\dfrac{a}{3} \right)$, $\Delta_1 = 2a, \Delta_2 = 3a^2$, hence $\left( \dfrac{a}{3}, -\dfrac{a}{3} \right)$ will be a local minimum if $a > 0$ and a local maximum if $a < 0$.

We can use the $Maple^{\text{Tm}}$ commands to find the Hessian of functions. For example, to find the Hessian $H_{(x,y)}f$ for $f(x,y) = \dfrac{5xy}{x^2 + y^2}$, we obtain

```
>  with(linalg) :
>  hessian( (5·x·y)/(x² + y²) , [x, y]);
```

$$\left[\left[-\frac{30\,yx}{\left(x^2+y^2\right)^2}+\frac{40\,x^3 y}{\left(x^2+y^2\right)^3},\; \frac{5}{x^2+y^2}-\frac{10\,x^2}{\left(x^2+y^2\right)^2}\right.\right.$$

$$\left.-\frac{10\,y^2}{\left(x^2+y^2\right)^2}+\frac{40\,x^2 y^2}{\left(x^2+y^2\right)^3}\right],$$

$$\left[\frac{5}{x^2+y^2}-\frac{10\,x^2}{\left(x^2+y^2\right)^2}-\frac{10\,y^2}{\left(x^2+y^2\right)^2}+\frac{40\,x^2 y^2}{\left(x^2+y^2\right)^3},\right.$$

$$\left.\left.-\frac{30\,yx}{\left(x^2+y^2\right)^2}+\frac{40\,xy^3}{\left(x^2+y^2\right)^3}\right]\right]$$

## EXERCISES 2.8

**2.8.1**  Determine the critical points of $f(x,y)=xy-x-y$.

**2.8.2**  Determine the nature of the critical points of
$f(x,y)=x^4+y^4-2(x-y)^2$.

**2.8.3**  Determine the nature of the critical points of
$f(x,y,z)=4x^2 z-2xy-4x^2-z^2+y$.

**2.8.4**  Find the extreme of $f(x,y,z)=x^2+y^2+z^2+xyz$.

**2.8.5**  Find the extreme of $f(x,y,z)=x^2 y+y^2 z+2x-z$.

**2.8.6**  Determine the nature of the critical points of
$f(x,y,z)=4xyz-x^4-y^4-z^2$.

**2.8.7**  Determine the nature of the critical points of
$f(x,y,z)=xyz(4-x-y-z)$.

**2.8.8**  Determine the nature of the critical points of
$g(x,y,z)=xyze^{-x^2-y^2-z^2}$.

**2.8.9**  Let $f(x,y)=\int_{y^2-x}^{x^2+y} g(t)\,dt$, where $g$ is a continuously differentiable
function defined over all real numbers and $g(0)=0$, $g'(0)\neq 0$.
Prove that $(0,0)$ is a saddle point of $f$.

**2.8.10** Find the minimum of $F(x,y) = (x-y)^2 + \left( \dfrac{\sqrt{144 - 16x^2}}{3} - \sqrt{4-y^2} \right)^2$,

for $-3 \leq x \leq 3, -2 \leq y \leq 2$.

**2.8.11** Find the extreme of $f(x,y) = x^2 - 3xy - y^2 + 2y - 6x$.

**2.8.12** Find the extreme of $f(x,y) = 4x^3 - 2x^2y + y^2$.

**2.8.13** Find the extreme of $f(x,y) = 5 + 4x - 2x^2 + 3y - y^2$.

**2.8.14** Find the extreme of $f(x,y) = \dfrac{x}{(x+y)}$.

**2.8.15** Show that the critical points $P_c$ for a function $g(x,y)$ correspond to points $P$ on the graph of $g$ where the normal is vertical.

**2.8.16** Find the critical points of $f(x,y) = x^2 + 2x + xy + 2y + 1$.

**2.8.17** Find the maximum and minimum of $f(x,y,z) = xy + yz$ on the set of points which satisfy $y^2 = 1 - x^2$ and $y = \dfrac{x}{z}$.

**2.8.18** Find the lowest and highest points on the ellipse of intersection of the $x^2 + y^2 = 1$ (cylinder) and the plane $x + y + z = 1$.

**2.8.19** Let a point $P$ within a triangle in which the sum of the squares of the distances to the sides is a minimum. Determine this minimum in terms of the lengths of sides and area.

**2.8.20**

$f_x = 8x^3 - 2x = 0 \Rightarrow x(8x^2 - 2) = 0 \Rightarrow x = 0 \text{ or } x = \pm\dfrac{1}{2}$,

$f_y = 6y = 0 \Rightarrow y = 0$;

$D = \det \begin{bmatrix} f_{xx} & f_{xy} \\ f_{yx} & f_{yy} \end{bmatrix} = \det \begin{bmatrix} 24x^2 - 2 & 0 \\ 0 & 6 \end{bmatrix} \Rightarrow D(0,0) \det \begin{bmatrix} -2 & 0 \\ 0 & 6 \end{bmatrix} = -12 < 0 \Rightarrow$

So $(0,0)$ is saddle point.

$D\left(\pm\dfrac{1}{2}, 0\right) \det \begin{bmatrix} 4 & 0 \\ 0 & 6 \end{bmatrix} = 24 > 0 \Rightarrow \left(\pm\dfrac{1}{2}, 0\right)$ is local minimum point.

## 2.9    LAGRANGE MULTIPLIERS

In some situations, we wish to optimize a function given a set of constraints. For such cases, we have the following.

**Theorem 2.9.1**    Let $A \subseteq \mathbb{R}^n$ and let $f : A \to \mathbb{R}$, $g : A \to \mathbb{R}$ be functions whose respective derivatives are continuous. Let $g(x_0) = c_0$ and let $S = g^{-1}(c_0)$ be the level set for g with value $c_0$, and assume $\nabla g(x_0) \neq 0$. If the restriction of $f$ to $S$ has an extreme point at $x_0$, then $\exists \lambda \in \mathbb{R}$ such that

$$\nabla f(x_0) = \lambda \nabla g(x_0).$$

**NOTE**    *Theorem 2.9.1 only locates extrema, it does not say anything concerning the nature of the critical points found.*

### EXAMPLE 2.9.1

Optimize $f : \mathbb{R}^2 \to \mathbb{R}, f(x,y) = x^2 - y^2$ given that $x^2 + y^2 = 1$.

**Solution:**

Let $g(x,y) = x^2 + y^2 - 1$. We solve

$$\nabla f \begin{bmatrix} x \\ y \end{bmatrix} = \lambda \nabla g \begin{bmatrix} x \\ y \end{bmatrix} \text{ for } x, y, \lambda. \text{ This requires}$$

$$\begin{bmatrix} 2x \\ -2y \end{bmatrix} = \begin{bmatrix} 2x\lambda \\ 2y\lambda \end{bmatrix}.$$

From $2x = 2x\lambda$, we get either $x = 0$ or $\lambda = 1$. If $x = 0$, then $y = \pm 1$ and $\lambda = -1$. If $\lambda = 1$, then $y = 0$, $x = \pm 1$. Thus the potential critical points are $(\pm 1, 0)$ and $(0, \pm 1)$. If $x^2 + y^2 = 1$, then

$$f(x,y) = x^2 - (1 - x^2) = 2x^2 - 1 \geq -1,$$

and

$$f(x,y) = 1 - y^2 - y^2 = 1 - 2y^2 \leq 1.$$

Thus $(\pm 1, 0)$ are maximum points and $(0, \pm 1)$ are minimum points.

## EXAMPLE 2.9.2

Find the maximum and the minimum points of $f(x,y) = 4x + 3y$, subject to the constraint $x^2 + 4y^2 = 4$, using Lagrange multipliers.

**Solution:**

Putting $g(x,y) = x^2 + 4y^2 - 4$, we have

$$\nabla f(x,y) = \lambda \nabla g(x,y) \Rightarrow \begin{bmatrix} 4 \\ 3 \end{bmatrix} = \lambda \begin{bmatrix} 2x \\ 8y \end{bmatrix}.$$

Thus $4 = 2\lambda x$, $3 = 8\lambda y$, clearly then $\lambda \neq 0$. Upon division, we find $\dfrac{x}{y} = \dfrac{16}{3}$. Hence

$$x^2 + 4y^2 = 4 \Rightarrow \frac{256}{9} y^2 + 4y^2 = 4 \Rightarrow y = \pm\frac{3}{\sqrt{73}}, x = \pm\frac{16}{\sqrt{73}}.$$

The maximum is clearly then

$$4\left(\frac{16}{\sqrt{73}}\right) + 3\left(\frac{3}{\sqrt{73}}\right) = \sqrt{73},$$

and the minimum is $-\sqrt{73}$.

## EXAMPLE 2.9.3

Let $a > 0, b > 0, c > 0$. Determine the maximum and minimum values of $f(x,y) = \dfrac{x}{a} + \dfrac{y}{b} + \dfrac{z}{c}$ and the ellipsoid $\dfrac{x^2}{a^2} + \dfrac{y^2}{b^2} + \dfrac{z^2}{c^2} = 1$.

**Solution:**

We use Lagrange multipliers. Put $g(x,y) = \dfrac{x^2}{a^2} + \dfrac{y^2}{b^2} + \dfrac{z^2}{c^2} - 1$. Then

$$\nabla f(x,y,z) = \lambda \nabla g(x,y,z) \Leftrightarrow \begin{bmatrix} 1/a \\ 1/b \\ 1/c \end{bmatrix} = \lambda \begin{bmatrix} 2x/a^2 \\ 2y/b^2 \\ 2z/c^2 \end{bmatrix}.$$

It follows that $\lambda \neq 0$. Hence $x = \dfrac{a}{2\lambda}, y = \dfrac{b}{2\lambda}, z = \dfrac{c}{2\lambda}$. Since $\dfrac{x^2}{a^2} + \dfrac{y^2}{b^2} + \dfrac{z^2}{c^2} = 1$,

we deduce $\dfrac{3}{4\lambda^2} = 1$ $\lambda = \pm\dfrac{\sqrt{3}}{2}$. Since $a, b, c$ are positive, $f$ will have a maximum when all $x, y, z$ are positive and a minimum which all $x, y, z$ are negative. Thus the maximum is when

$$x = \frac{a}{\sqrt{3}}, y = \frac{b}{\sqrt{3}}, z = \frac{c}{\sqrt{3}},$$

and

$$f(x, y, z) \le \frac{3}{\sqrt{3}} = \sqrt{3}$$

and the minimum is when

$$x = -\frac{a}{\sqrt{3}}, y = -\frac{b}{\sqrt{3}}, z = -\frac{c}{\sqrt{3}},$$

and

$$f(x, y, z) \ge -\frac{3}{\sqrt{3}} = -\sqrt{3} \ .$$

## Alternative Method: Using the CBS Inequality

$$\left| \frac{x}{a} \cdot 1 + \frac{y}{b} \cdot 1 + \frac{z}{c} \cdot 1 \right| \le \left( \frac{x^2}{a^2} + \frac{y^2}{b^2} + \frac{z^2}{c^2} \right)^{1/2} \left( 1^2 + 1^2 + 1^2 \right)^{1/2}$$

$$= (1)\sqrt{3} \Rightarrow -\sqrt{3} \le \frac{x}{a} + \frac{y}{b} + \frac{z}{c} \le \sqrt{3} \ .$$

## EXAMPLE 2.9.4

Let $a > 0, b > 0, c > 0$. Determine the maximum volume of the parallelepiped with sides parallel to the axes that can be enclosed inside the ellipsoid $\dfrac{x^2}{a^2} + \dfrac{y^2}{b^2} + \dfrac{z^2}{c^2} = 1$.

*Solution:*

Let $2x, 2y, 2z$, be the dimensions of the box. We must maximize $f(x,y,z) = 8xyz$ subject to the constraint $g(x,y,z) = \dfrac{x^2}{a^2} + \dfrac{y^2}{b^2} + \dfrac{z^2}{c^2} - 1$. Using Lagrange multipliers,

$$\nabla f(x,y,z) = \lambda \nabla g(x,y,z) \Leftrightarrow \begin{bmatrix} 8yz \\ 8xz \\ 8xy \end{bmatrix}$$

$$= \lambda \begin{bmatrix} 2x/a^2 \\ 2y/b^2 \\ 2z/c^2 \end{bmatrix} \Rightarrow 4yz = \lambda \frac{x}{a^2}, 4xz = \lambda \frac{y}{b^2}, 4xy = \lambda \frac{z}{c^2}.$$

Multiplying the first inequality by $x$, the second by $y$, the third by $z$, and adding,

$$4xyz = \lambda \frac{x^2}{a^2}, 4xyz = \lambda \frac{y^2}{b^2}, 4xyz = \lambda \frac{z^2}{c^2} \Rightarrow 12xyz = \lambda \left( \frac{x^2}{a^2} + \frac{y^2}{b^2} + \frac{z^2}{c^2} \right) = \lambda.$$

Hence

$$\frac{\lambda}{3} = \lambda \frac{x^2}{a^2} = \lambda \frac{y^2}{b^2} = \lambda \frac{z^2}{c^2}.$$

If $\lambda = 0$, $8xyz = 0$, which minimizes the volume. If $\lambda \neq 0$, then

$$x = \frac{a}{\sqrt{3}}, \quad y = \frac{b}{\sqrt{3}}, \quad z = \frac{c}{\sqrt{3}},$$

and the maximum value is

$$8xyz \leq 8 \frac{abc}{3\sqrt{3}}.$$

## Alternative Method: Using the AM-GM Inequality

$$\left( x^2 y^2 z^2 \right)^{1/3} = (abc)^{2/3} \left( \frac{x^2}{a^2} \cdot \frac{y^2}{b^2} \cdot \frac{z^2}{c^2} \right)^{1/3} \leq (abc)^{2/3} \cdot \frac{\dfrac{x^2}{a^2} + \dfrac{y^2}{b^2} + \dfrac{z^2}{c^2}}{3} = \frac{1}{3}$$

$$\Rightarrow 8xyz \leq \frac{8}{3\sqrt{3}} (abc).$$

## EXERCISES 2.9

**2.9.1** A closed box (with six outer faces) has fixed surface area of S square units. Find its maximum volume using Lagrange multipliers. That is, subject to the constraint $2ab + 2bc + 2ca = S$, you must maximize $abc$.

**2.9.2** Consider the problem of finding the closest point $P'$ on the plane $\Pi : ax + by + cz = d$, $a$, $b$, $c$ non-zero constants with $a + b + c \neq d$ to the point $P(1, 1, 1)$. In this exercise, you will do this in three essentially different ways.

1. Do this by a geometric argument, arguing the point $P'$ closest to $P$ on $\Pi$ is on the perpendicular passing through $P$ and $P'$.

2. Do this by means of Lagrange multipliers, by minimizing a suitable function $f(x, y, z)$ subject to the constraint $g(x, y, z) = ax + by + cz - d$.

3. Do this considering the unconstrained extrema of a suitable function $h\left(x, y, \dfrac{d - ax - by}{c}\right)$.

**2.9.3** Given that $x$, $y$ are positive real numbers such that $x^4 + 81y^4 = 36$, find the maximum of $x + 3y$.

**2.9.4** If x, y, z are positive real numbers such that $x^2 y^3 z = \dfrac{1}{6^2}$, what is the minimum value of $f(x, y, z) = 2x + 3y + z$?

**2.9.5** Find the maximum and the minimum values of $f(x, y) = x^2 + y^2$ subject to the constraint $5x^2 + 6xy + 5y^2 = 8$.

**2.9.6** Let $a > 0$, $b > 0$, $p > 1$. Maximize $f(x, y) = ax + by$ subject to the constraint $x^p + y^p = 1$.

**2.9.7** Find the extrema of $f(x, y, z) = x^2 + y^2 + z^2$, subject to the constraint $(x - 1)^2 + (y - 2)^2 + (z - 3)^2 = 4$.

**2.9.8** Find the axes of the ellipse $5x^2 + 8xy + 5y^2 = 9$.

**2.9.9**   Optimize $f(x, y, z) = x + y + z$ subject to $x^2 + y^2 = 2$, and $x + z = 1$.

**2.9.10**   Let $x, y$ be strictly positive real numbers with $x + y = 1$. What is the maximum value of $x + \sqrt{xy}$?

**2.9.11**   Let $a, b$ be positive real constants. Maximize $f(x, y) = x^a e^{-x} y^b e^{-y}$ on the triangle in $\mathbb{R}^2$ bounded by the lines $x \geq 0, y \geq 0, x + y \leq 1$.

**2.9.12**   Does there exist a polynomial in two variables with real coefficients $p(x, y)$ such that $p(x, y) > 0$ for all $x$ and $y$, and that for all real numbers $c > 0$ there exists $(x_0, y_0) \in \mathbb{R}^2$ such that $p(x_0, y_0) = c$?

**2.9.13**   Determine the maximum volume of the rectangular solid in the first octant with one vertex at the origin and the opposite vertex lying in the plane $\dfrac{x}{k_1} + \dfrac{y}{k_2} + \dfrac{z}{k_2} = 1$ where $k_1, k_2$, and $k_3$ are positive constants, using Language multipliers.

**HINT**   *First octant* $\Rightarrow (x \geq 0, y \geq 0, z \geq 0)$.

**2.9.14**   Find the local extreme of the function $f(x, y, z) = x + y + z$ subject to constraint $x^2 + y^2 + z^2 = 25$.

# *INTEGRATION*

In this chapter, we focus on differentiation forms, zero-manifolds, one-manifolds, closed and exact forms, two-manifolds, change of variables in double integrals, change to polar coordinates, three-manifolds, change of variables in triple integrals, surface integrals, and Green's, Stokes', and Gauss' Theorems.

## 3.1    DIFFERENTIAL FORMS

We will now consider integration in several variables. In order to begin our discussion, we need to consider the concept of differential forms.

**Definition 3.1.1**    Consider $n$ variables $x_1, x_2, \ldots, x_n$ in $n$-dimensional space (used as the names of the axes) and let

$$\mathbf{a}_j = \begin{bmatrix} a_{1j} \\ a_{2j} \\ \vdots \\ a_{nj} \end{bmatrix} \in \mathbb{R}^n, \ \ 1 \le j \le k,$$

be $k \le n$ vectors in $\mathbb{R}^n$. Moreover, let $\{j_1, j_2, \ldots, j_k\} \subseteq \{1, 2, \ldots, n\}$ be a collection of $k$ sub-indices.

An *elementary k-differential form* $(k > 1)$ acting on the vectors $a_j$, $1 \le j \le k$ is defined and denoted by

$$dx_{j_1} \wedge dx_{j_2} \wedge \ldots \wedge dx_{j_k} (a_1, a_2, \ldots, a_k) = \det \begin{bmatrix} ax_{j_1 1} & ax_{j_1 2} & \ldots & ax_{j_1 k} \\ ax_{j_2 1} & ax_{j_2 2} & \ldots & ax_{j_2 k} \\ \vdots & \vdots & \ldots & \\ ax_{j_k 1} & ax_{j_k 2} & \ldots & ax_{j_k k} \end{bmatrix}.$$

In other words, $dx_{j_1} \wedge dx_{j_2} \wedge \ldots \wedge dx_{j_k} (a_1, a_2, \ldots, a_k)$ is the $x_{j_1}, x_{j_2}, \ldots, x_{jk}$ component of the signed $k$-parallelotope in $\mathbb{R}^n$ spanned by $a_1, a_2, \ldots, a_k$.

**NOTE**   *By virtue of being a determinant, the wedge product $\wedge$ of differential forms has the following properties:*

1. **Anti-commutative:** $da \wedge db = -db \wedge da$.
2. **Linearity:** $d(a + b) = da + db$.
3. **Scalar homogeneity:** if $\lambda \in \mathbb{R}$, $d\lambda a = \lambda da$.
4. **Associative:** $(da \wedge db) \wedge dc = da \wedge (db \wedge dc)$, notice that associative does not hold for the wedge product of vectors.

**NOTE**   *Anti-commutative yields, $da \wedge da = 0$.*

### EXAMPLE 3.1.1
Consider

$$a = \begin{bmatrix} 1 \\ 0 \\ -1 \end{bmatrix} \in \mathbb{R}^3.$$

Then

$$dx(a) = \det(1) = 1,$$

$$dy(a) = \det(0) = 0,$$

$$dz(a) = \det(-1) = -1,$$

Are the (signed) 1-volumes (that is, the length) of the projections of a onto the coordinate axes.

## EXAMPLE 3.1.2

In $\mathbb{R}^3$, we have $dx \wedge dy \wedge dx = 0$, since we have a repeated variable.

## EXAMPLE 3.1.3

In $\mathbb{R}^3$, we have

$$dx \wedge dz + 5dz \wedge dx + 4dx \wedge dy - dy \wedge dx + 12dx \wedge dx = -4dx \wedge dz + 5dx \wedge dy.$$

**NOTE**  *In order to avoid redundancy, we will make the convention that if a sum of two or more terms have the same differential form up to the permutation of the variables, we will simplify the summands and express the other differential forms in terms of the one differential form whose indices appear in increasing order.*

**Definition 3.1.2**  A *0-differential form in* $\mathbb{R}^n$ is simply a differentiable function in $\mathbb{R}^n$.

**Definition 3.1.3**  A *k-differential form field in* $\mathbb{R}^n$ is an expression of the form

$$\omega = \sum_{1 \leq j_1 \leq j_2 \leq \ldots \leq j_k \leq n} a_{j_1 j_2 \cdots j_k} \, dx_{j_1} \wedge dx_{j_2} \wedge \ldots \wedge dx_{j_k},$$

where the $a_{j_1, j_2, \ldots, j_k}$ are differentiable function in $\mathbb{R}^n$.

## EXAMPLE 3.1.4

$$g(x, y, z, w) = x + y^2 + z^3 + w^4$$

is a 0- form in $\mathbb{R}^4$.

## EXAMPLE 3.1.5

An example of a 1-form field in $\mathbb{R}^3$ is

$$\omega = xdx + y^2dy + xyz^3dz.$$

**EXAMPLE 3.1.6**

An example of a 2-form field in $\mathbb{R}^3$ is

$$\omega = x^2 dx \wedge dy + y^2 dy \wedge dz + dz \wedge dx.$$

**EXAMPLE 3.1.7**

An example of a 3-form field in $\mathbb{R}^3$ is

$$\omega = (x+y+z) dx \wedge dy \wedge dz.$$

We show now how to multiply differential forms.

**EXAMPLE 3.1.8**

The product of the 1-form fields in $\mathbb{R}^3$

$$\omega_1 = y dx + x dy,$$

$$\omega_2 = -2x dx + 2y dy,$$

is

$$\omega_1 \wedge \omega_2 = (2x^2 + 2y^2) dx \wedge dy.$$

**Definition 3.1.4**   Let $f(x_1, x_2, \ldots, x_n)$ be a 0-form in $\mathbb{R}^n$. The *exterior derivative* $df$ of $f$

$$df = \sum_{i=1}^{n} \frac{\partial f}{\partial x_i} dx_i.$$

Furthermore, if

$$\omega = f(x_1, x_2, \ldots, x_n) dx_{j_1} \wedge dx_{j_2} \wedge \ldots \wedge dx_{j_k}$$

is a k-form in $\mathbb{R}^n$, the *exterior derivative* $d\omega$ of $\omega$ is the $(k+1)$-form

$$d\omega = df(x_1, x_2, \ldots, x_n) dx_{j_1} \wedge dx_{j_2} \wedge \ldots \wedge dx_{j_k}.$$

### EXAMPLE 3.1.9

If in $\mathbb{R}^2$, $\omega = x^3 y^4$, then

$$d\left(x^3 y^4\right) = 3x^2 y^4 dx + 4x^3 y^3 dy \cdot$$

### EXAMPLE 3.1.10

If in $\mathbb{R}^2$, $\omega = x^2 y dx + x^3 y^4 dy$, then

$$d\omega = d\left(x^2 y dx + x^3 y^4 dy\right)$$

$$= \left(2xy dx + x^2 dy\right) \wedge dx + \left(3x^2 y^4 dx + 4x^3 y^3 dy\right) \wedge dy$$

$$= x^2 dy \wedge dx + 3x^2 y^4 dx \wedge dy$$

$$= \left(3x^2 y^4 - x^2\right) dx \wedge dy.$$

### EXAMPLE 3.1.11

Consider the change of variables $x = u + v, y = uv$. Then

$$dx = du + dv,$$
$$dy = v du + u dv,$$

hence

$$dx \wedge dy = \left(u - v\right) du \wedge dv.$$

### EXAMPLE 3.1.12

Consider the transformation of coordinates $xyz$ into $uvw$ coordinates given by

$$u = x + y + z, \quad v = \frac{z}{y + z}, \quad w = \frac{y + z}{x + y + z} \cdot$$

Then

$$du = dx + dy + dz,$$

$$dv = -\frac{z}{(y+z)^2}dy + \frac{y}{(y+z)^2}dz,$$

$$dw = -\frac{y+z}{(x+y+z)^2}dx + \frac{x}{(x+y+z)^2}dy + \frac{x}{(x+y+z)^2}dz.$$

Multiplication gives

$$du \wedge dv \wedge dw = \left( -\frac{zx}{(y+z)^2(x+y+z)^2} - \frac{y(y+z)}{(y+z)^2(x+y+z)^2} \right.$$

$$\left. +\frac{z(y+z)}{(y+z)^2(x+y+z)^2} - \frac{xy}{(y+z)^2(x+y+z)^2} \right) dx \wedge dy \wedge dz$$

$$= \frac{z^2 - y^2 - zx - xy}{(y+z)^2(x+y+z)^2}dx \wedge dy \wedge dz.$$

## EXERCISES 3.1

**3.1.1**    Match the differential forms types with the elements forms to make the statement true.

    **1.**  0-forms                   A.  Surface elements

    **2.**  1-forms                   B.  Volume forms

    **3.**  2-forms                   C.  Functions forms

    **4.**  3-forms                   D.  Line elements

**3.1.2**    Let $\omega_1 = y dx \wedge dz + dx \wedge dt$ and $\omega_2 = (x+1)dy \wedge dt$. Find $\omega_1 \wedge \omega_2$.

**3.1.3**    Let $\omega = xy dx - xy dy + xy^2 z^3 dz$. Find the exterior derivative $d\omega$.

**3.1.4**   Let $\omega = x^2\left(y + z^2\right)dx \wedge dy + z\left(x^3 + y\right)dy \wedge dz$. Find the exterior derivative $d\omega$.

**3.1.5**   Express the 2-form $dx \wedge dy$ in polar coordinates.

**3.1.6**   Consider $f : \mathbb{R}^2 \to \mathbb{R}^2$ and $(x,y) \mapsto (u,v)$, where $u = x^2 - y^2$ and $v = 2xy$. Find $du \wedge dv$ in terms of $dx \wedge dy$.

**3.1.7**   Let the 1-form, $w = fdx + gdy + hdz$. Find the exterior derivative $dw$.

**3.1.8**   Let the 2-form, $\omega = fdy \wedge dz - gdx \wedge dz + hdx \wedge dy$. Find the exterior derivative $d\omega$.

**3.1.9**   Let $\omega = \left(x + z^2\right)dx \wedge dy$. Find the exterior derivative $d\omega$.

**3.1.10**   Let the 2-form, $\omega = (x + 2z)dx \wedge dy + ydx \wedge dz$ and the vectors

$$\vec{v} = \begin{bmatrix} 1 \\ 5 \\ 5 \end{bmatrix}, \vec{r} = \begin{bmatrix} -1 \\ 0 \\ 3 \end{bmatrix}. \text{ Find } \omega(\vec{v}, \vec{r}).$$

**HINT**   $\omega(\vec{v}, \vec{r}) = \sum_{1 \le i < j \le n} F_{ij}(\vec{a})dx_i \wedge dx_j(\vec{v}, \vec{r}) = \sum_{i, j > i} F_{ij}(\vec{a})\det\begin{bmatrix} dx_i(\vec{v}) & dx_i(\vec{r}) \\ dx_j(\vec{v}) & dx_j(\vec{r}) \end{bmatrix}.$

## 3.2   ZERO-MANIFOLDS

**Definition 3.2.1**   A 0-*dimensional oriented manifold of* $\mathbb{R}^n$ is simply a point $x \in \mathbb{R}^n$, with a choice of the $+$ or $-$ sign. A general oriented 0-manifold is a union of oriented points.

**Definition 3.2.2**   Let $M = +\{b\} \cup -\{a\}$ be an oriented 0-manifold, and let $\omega$ be a 0-form. Then

$$\int_M \omega = \omega(b) - \omega(a).$$

**NOTE**   $-x$ *has opposite orientation to* $+x$ *and*

$$\int_{-x} \omega = -\int_{+x} \omega.$$

### EXAMPLE 3.2.1

Let $M = -\{(1,0,0)\} \cup +\{(1,2,3)\} \cup -\{(0,-2,0)\}$ be an oriented 0-manifold, and let $\omega = x + 2y + z^2$. Then

$$\int_M \omega = -\omega(1,0,0) + \omega(1,2,3) - \omega(0,0,3) = -(1) + (14) - (-4) = 17.$$

**NOTE** *Do not confuse, say,* $-\{(1,0,0)\}$ *with* $-(1,0,0) = (-1,0,0)$. *The first one means that the point* $(1,0,0)$ *is given negative orientation, the second means that* $(-1,0,0)$ *is the additive inverse of* $(1,0,0)$.

## EXERCISES 3.2

**3.2.1** Let $M = -\{(1,1,0)\} \cup +\{(0,2,0)\} \cup -\{(1,-1,2)\}$ be an oriented 0-manifold, and let $\omega = 3x - 2y + z^2$. Find $\int_M \omega$.

**3.2.2** Let $M = +\{(1,2,0)\} \cup +\{(1,0,3)\} \cup -\{(7,-1,0)\}$ be an oriented 0-manifold, and let $\omega = -3x + y^2 + z$. Find $\int_M \omega$.

**3.2.3** Let $M = -\{(1,1,1)\} \cup -\{(0,-2,5)\} \cup -\{(1,7,5)\}$ be an oriented 0-manifold, and let $\omega = 3x - y + 4z$. Find $\int_M \omega$.

**3.2.4** Let $M = +\{(-1,2,1)\} \cup -\{(0,2,1)\} \cup -\{(-3,1,6)\}$ be an oriented 0-manifold, and let $\omega = -2x^2 + y - z$. Find $\int_M \omega$.

**3.2.5** Let $M = +\{(0,1,0)\} \cup +\{(-2,0,3)\} \cup +\{(1,-5,0)\}$ be an oriented 0-manifold, and let $\omega = x + y - z^2$. Find $\int_M \omega$.

**3.2.6** Let $M = -\{(1,0,1)\} \cup +\{(-3,1,3)\} \cup -\{(1,2,0)\}$ be an oriented 0-manifold, and let $\omega = -7x - y^2 + 3z$. Find $\int_M \omega$.

**3.2.7** Let $M = +\{(0,0,1)\} \cup -\{(1,3,2)\} \cup +\{(1,-2,1)\}$ be an oriented 0-manifold, and let $\omega = x - 2y + z^3$. Find $\int_M \omega$.

**3.2.8** Let $M = -\{(1,1,0)\} \cup -\{(1,3,1)\} \cup +\{(6,2,0)\}$ be an oriented 0-manifold, and let $\omega = 3x - y + z$. Find $\int_M \omega$.

## 3.3 ONE MANIFOLD

**Definition 3.3.1** A *one-dimensional oriented manifold of* $\mathbb{R}^n$ is simply an oriented smooth curve $\Gamma \in \mathbb{R}^n$, with a choice of a + orientation if the curve traverses in the direction of increasing $t$, or with a choice of a − sign if the curve traverses in the direction of decreasing $t$. A general oriented 1-manifold is a union of oriented curves.

**NOTE** *The curve* $-\Gamma$ *has opposite orientation to* $\Gamma$ *and*

$$\int_{-\Gamma} \omega = -\int_{\Gamma} \omega.$$

*If* $\vec{f} : \mathbb{R}^2 \to \mathbb{R}^2$ *and if* $d\vec{r} = \begin{bmatrix} dx \\ dy \end{bmatrix}$, *the classical way of writing this is*

$$\int_{\Gamma} \vec{f} \cdot d\vec{r}.$$

We now turn to the problem of integrating 1-forms.

### EXAMPLE 3.3.1

Calculate

$$\int_{\Gamma} xy\, dx + (x+y)\, dy$$

where $\Gamma$ is the parabola $y = x^2, x \in [-1;2]$ oriented in the positive direction.

***Solution:***

We parameterize the curve as $x = t, y = t^2$. Then

$$xy\, dx + (x+y)\, dy = t^3 dt + (t+t^2) dt^2 = (3t^3 + 2t^2) dt,$$

hence

$$\int_{\Gamma} \omega = \int_{-1}^{2} (3t^3 + 2t^2)\, dt$$

$$= \left[ \frac{2}{3} t^3 + \frac{3}{4} t^4 \right]_{-1}^{2}$$

$$= \frac{69}{4}.$$

What would happen if we had given the curve above a different parameterization? First observe that the curve travels from $(-1, 1)$ to $(2, 4)$ on the parabola $y = x^2$. These conditions are met with parameterization. Then

$$xy\,dx + (x+y)\,dy = \left(\sqrt{t}-1\right)^3 d\left(\sqrt{t}-1\right) + \left(\left(\sqrt{t}-1\right)+\left(\sqrt{t}-1\right)^2\right)d\left(\sqrt{t}-1\right)^2$$

$$= \left(3\left(\sqrt{t}-1\right)^3 + 2\left(\sqrt{t}-1\right)^2\right)d\left(\sqrt{t}-1\right)$$

$$= \frac{1}{2\sqrt{t}}\left(3\left(\sqrt{t}-1\right)^3 + 2\left(\sqrt{t}-1\right)^2\right)dt,$$

hence

$$\int_\Gamma \omega = \int_0^9 \frac{1}{2\sqrt{t}}\left(3\left(\sqrt{t}-1\right)^3 + 2\left(\sqrt{t}-1\right)^2\right)dt$$

$$= \left[\frac{3t^2}{4} - \frac{7t^{3/2}}{3} + \frac{5t}{2} - \sqrt{t}\right]_0^9$$

$$= \frac{69}{4},$$

as before.

To solve this problem using *Maple*<sup>Tm</sup> commands, you may use the following code.

```
> with(Student[VectorCalculus]) :
> LineInt( VectorField( ⟨x * y, x + y⟩ ), Path( ⟨t, t^2⟩, t =-1 ..2));
                          69
                          --
                           4
```

**NOTE** *It turns out that if two different parameterizations of the same curve have the same orientation, then their integrals are equal. Hence, we only need to worry about finding a suitable parameterization.*

## EXAMPLE 3.3.2

Calculate the line integral

$$\int_{\Gamma} y \sin x \, dx + x \cos y \, dy,$$

where $\Gamma$ is the line segment from $(0,0)$ to $(1,1)$ in the positive direction.

### Solution:

This line has equation $y = x$, so we choose the parameterization $x = y = t$. The integral is thus

$$\int_{\Gamma} y \sin x \, dx + x \cos y \, dy = \int_{0}^{1} \left( t \sin t + t \cos t \right) dt$$

$$= \left[ t \left( \sin t - \cos t \right) \right]_{0}^{1} - \int_{0}^{1} \left( \sin t - \cos t \right) dt$$

$$= 2 \sin 1 - 1,$$

upon integrating by parts.

To solve this problem using *Maple*$^{Tm}$ you may use the following code.

```
>  with(Student[VectorCalculus]) :
>  LineInt( VectorField( ⟨y * sin(x), x * cos(y)⟩ ), Line( ⟨0, 0⟩, ⟨1, 1⟩ ));
                        -1 + 2 sin(1)
```

## EXAMPLE 3.3.3

Calculate the path integral

$$\int_{\Gamma} \frac{x+y}{x^2+y^2} \, dy + \frac{x-y}{x^2+y^2} \, dx$$

around the closed square $\Gamma = ABCD$ with $A = (1,1), B = (-1,1), C = (-1,-1)$, and $D = (1,-1)$ in the direction $ABCDA$.

*Solution:*

On $AB, y = 1, dy = 0$, on $BC, x = -1, dx = 0$, on $CD, y = -1, dy = 0$, and on $DA, x = 1, dx = 0$. The integral is thus

$$\int_\Gamma \omega = \int_{AB} \omega + \int_{BC} \omega + \int_{CD} \omega + \int_{DA} \omega$$

$$= \int_1^{-1} \frac{x-1}{x^2+1} \, dx + \int_1^{-1} \frac{y-1}{y^2+1} \, dy + \int_{-1}^1 \frac{x+1}{x^2+1} \, dx + \int_{-1}^1 \frac{y+1}{y^2+1} \, dy$$

$$= 4 \int_{-1}^1 \frac{1}{x^2+1} \, dx$$

$$= 4 \arctan x \Big|_{-1}^1$$

$$= 2\pi.$$

**NOTE** *When the integral is along a closed path, like in the preceding example, it is customary to use the symbol $\oint_\Gamma$ rather than $\int_\Gamma$. The positive direction of integration is that sense that when traversing the path, the area enclosed by the curve is to the left of the curve.*

### EXAMPLE 3.3.4

Calculate the path integral

$$\oint_\Gamma x^2 dy + y^2 dx,$$

where $\Gamma$ is the ellipse $9x^2 + 4y^2 = 36$ traversed once in the positive sense.

*Solution:*

Parameterize the ellipse as $x = 2\cos t, y = 3\sin t, t \in [0; 2\pi]$. Observe that when traversing this closed curve, the area of the ellipse is on the left-hand side of the path, so this parameterization traverses the curve in the positive sense. We have

$$\oint_{\Gamma} \omega = \int_0^{2\pi} \left( \left( 4\cos^2 t \right) \left( 3\cos t \right) + \left( 9\sin t \right) \left( -2\sin t \right) \right) dt$$

$$= \int_0^{2\pi} \left( 12\cos^3 t - 18\sin^3 t \right) dt$$

$$= 0.$$

To solve this problem, using *Maple*$^{Tm}$ you may use the following code.

```
> with(Student[VectorCalculus]) :
> LineInt( VectorField( ⟨y^2, x^2⟩ ), Ellipse(9*x^2 + 4*y^2 -36));
                              0
```

**Definition 3.3.2**  Let $\Gamma$ be a smooth curve. The integral

$$\int_{\Gamma} f(\mathbf{x}) \|d\mathbf{x}\|$$

is called the *path integral of f along* $\Gamma$.

**EXAMPLE 3.3.5**

Find $\int_{\Gamma} x \|d\mathbf{x}\|$ where $\Gamma$ is the triangle starting at $A:(-1,-1)$ to $B:(2,-2)$, and ending in $C:(1,2)$, see Figure 3.3.1.

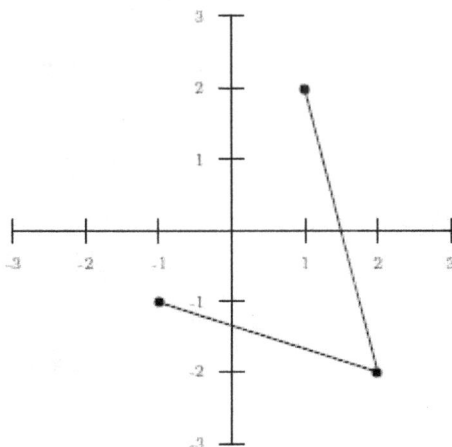

**FIGURE 3.3.1** Example 3.3.5.

*Solution:*

The lines passing through the given points have equations $L_{AB} : y = \dfrac{-x-4}{3}$

and $L_{BC} : y = -4x + 6$. On $L_{AB}$.

$$x\|dx\| = x\sqrt{(dx)^2 + (dy)^2} = x\sqrt{1 + \left(-\frac{1}{3}\right)^2}\, dx = \frac{x\sqrt{10}\ dx}{3},$$

and on $L_{BC}$

$$x\|dx\| = x\sqrt{(dx)^2 + (dy)^2} = x\sqrt{1 + (-4)^2}\, dx = x\sqrt{17}\ dx.$$

Hence

$$\int_{\Gamma} x\|dx\| = \int_{L_{AB}} x\|dx\| + \int_{L_{BC}} x\|dx\|$$

$$= \int_{-1}^{2} \frac{x\sqrt{10}\ dx}{3} + \int_{2}^{1} x\sqrt{17}\ dx$$

$$= \frac{\sqrt{10}}{2} - \frac{3\sqrt{17}}{2}.$$

## EXERCISES 3.3

**3.3.1**   Consider $\int_C x\,dx + y\,dy$. Evaluate $\int_C x\,dx + y\,dy$ where $C$ is the straight line path that starts at $(-1,0)$ goes to $(0,1)$ and ends at $(1,0)$, by parameterizing this path.

**3.3.2**   Consider $\int_C xy\|dx\|$. Calculate $\int_C xy\|dx\|$ where $C$ is the straight line path that starts at $(-1,0)$ goes to $(0,1)$ and ends at $(1,0)$, by parameterizing this path.

**3.3.3**   Evaluate $\int_C x\,dx + y\,dy$ where $C$ is the semicircle that starts at $(-1,0)$ goes to $(0,1)$ and ends at $(1,0)$, by parameterizing this path.

**3.3.4**   Calculate $\int_C xy\|dx\|$ where $C$ is the semicircle that starts at $(-1,0)$ goes to $(0,1)$ and ends at $(1,0)$, by parameterizing this path.

**3.3.5**  Find $\int_{\Gamma} x\,dx + y\,dy$ where $\Gamma$ is the path shown in Figure 3.3.2, starting at $O(0,0)$ going on a straight line to $A\left(4\cos\dfrac{\pi}{6}, 4\sin\dfrac{\pi}{6}\right)$ and continuing on an arc a circle to $B\left(4\cos\dfrac{\pi}{5}, 4\sin\dfrac{\pi}{5}\right)$.

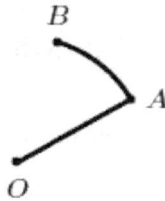

**FIGURE 3.3.2**  Exercises 3.3.5 and 3.3.7.

**3.3.6**  Solve Exercise 3.3.5 using Maple commands.

**3.3.7**  Find $\int_{\Gamma} x\|dx\|$ where $\Gamma$ is the path as shown in Figure 3.3.2.

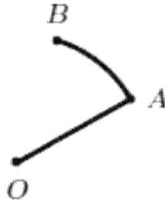

**FIGURE 3.3.2**  Exercises 3.3.5 and 3.3.7.

**3.3.8**  Find $\oint_{\Gamma} z\,dx + x\,dy + y\,dz$ where $\Gamma$ is the intersection of the sphere $x^2 + y^2 + z^2 = 1$ and the plane $x + y = 1$, traversed in the positive direction.

**3.3.9**  Solve Exercise 3.3.7 using Maple commands.

**3.3.10**  Evaluate $\int_{C}\left(x^2 + y^2\right) dx + 2x\,dy$.

    **1.**  Where $C$ consists of line segment from $(1,2)$ to $(1,8)$ and from $(1,8)$ to $(-2,8)$.

    **2.**  Where $C$ is the graph of $y = 2x^2$ form $(1,2)$ to $(-2,8)$.

**3.3.11** Evaluate $\int_C \left( x^2 + y^2 \right) \mathrm{d}x + 2xy \; \mathrm{d}y$ along the curve

$C : x = t, y = 2t^2, 0 \le t \le 1$, where $t = \sin u$ for $0 \le u \le \pi/2$.

**3.3.12** Compute the curve integral $\int_C \left( x^2 - 2xy \right) \mathrm{d}x + \left( y^2 - 2xy \right) \mathrm{d}y$, along the parabola $y = x^2$ from $(-2, 4)$ to $(1, 1)$.

## 3.4 CLOSED AND EXACT FORMS

**Lemma 3.4.1 (Poincare Lemma):** If $\omega$ is a $p$-differential form of continuously differentiable functions in $\mathbb{R}^n$ then

$$\mathrm{d}(\mathrm{d}\omega) = 0.$$

**Proof:**

We will prove this by induction on $p$. For $p = 0$ if

$$\omega = f\left( x_1, x_2, \ldots, x_n \right)$$

then

$$\mathrm{d}\omega = \sum_{k=1}^{n} \frac{\partial f}{\partial x_k} \; \mathrm{d}x_k$$

and

$$\mathrm{d}\omega = \sum_{k=1}^{n} \mathrm{d}\left( \frac{\partial f}{\partial x_k} \right) \wedge \mathrm{d}x_k$$

$$= \sum_{k=1}^{n} \left( \sum_{j=1}^{n} \left( \frac{\partial^2 f}{\partial x_j \partial x_k} \right) \wedge \mathrm{d}x_j \right) \wedge \mathrm{d}x_k$$

$$= \sum_{1 \le j \le k \le n} \left( \frac{\partial^2 f}{\partial x_j \partial x_k} - \frac{\partial^2 f}{\partial x_k \partial x_j} \right) \mathrm{d}x_j \wedge \mathrm{d}x_k$$

$$= 0,$$

since $\omega$ is continuously differentiable and so the mixed partial derivatives are equal. Consider now an arbitrary $p$-form, $p > 0$. Since such a form can be written as

$$\omega = \sum_{1 \le j_1 \le j_2 \le \ldots \le j_p \le n} a_{j_1 j_2 \ldots j_p} \, dx_{j_1} \wedge dx_{j_2} \wedge \ldots dx_{j_p},$$

where $a_{j_1, j_2 \ldots j_p}$ is the continuous differentiable functions in $\mathbb{R}^n$, we have

$$d\omega = \sum_{1 \le j_1 \le j_2 \le \ldots \le j_p \le n} da_{j_1 j_2 \ldots j_p} \, dx_{j_1} \wedge dx_{j_2} \wedge \ldots dx_{j_p}$$

$$= \sum_{1 \le j_1 \le j_2 \le \ldots \le j_p \le n} \left( \sum_{i=1}^{n} \frac{\partial a_{j_1 j_2 \ldots j_p}}{\partial x_i} \, dx_i \right) dx_{j_1} \wedge dx_{j_2} \wedge \ldots dx_{j_p},$$

it is enough to prove that for each summand

$$d\left( da \wedge dx_{j_1} \wedge dx_{j_2} \wedge \ldots dx_{j_p} \right) = 0.$$

But

$$d\left( da \wedge dx_{j_1} \wedge dx_{j_2} \wedge \ldots dx_{j_p} \right) = dda \wedge \left( dx_{j_1} \wedge dx_{j_2} \wedge \ldots dx_{j_p} \right)$$

$$+ da \wedge d\left( dx_{j_1} \wedge dx_{j_2} \wedge \ldots dx_{j_p} \right)$$

$$= da \wedge d\left( dx_{j_1} \wedge dx_{j_2} \wedge \ldots dx_{j_p} \right),$$

Since $dda = 0$ from the case $p = 0$. But an independent induction argument proves that

$$d\left( dx_{j_1} \wedge dx_{j_2} \wedge \ldots dx_{j_p} \right) = 0,$$

completing the proof.

**Definition 3.4.1**  A differential form $\omega$ is said to be *exact* if there is a continuously differentiable function $F$ such that

$$dF = \omega.$$

### EXAMPLE 3.4.1

The differential form

$$x\,dx + y\,dy$$

is exact, since

$$x\,dx + y\,dy = d\left(\frac{1}{2}\left(x^2 + y^2\right)\right).$$

### EXAMPLE 3.4.2

The differential form

$$y\,dx + x\,dy$$

is exact, since

$$y\,dx + x\,dy = d\left(xy.\right)$$

### EXAMPLE 3.4.3

The differential form

$$\frac{x}{x^2 + y^2}\,dx + \frac{y}{x^2 + y^2}\,dy$$

is exact, since

$$\frac{x}{x^2 + y^2}\,dx + \frac{y}{x^2 + y^2}\,dy = d\left(\frac{1}{2}\log_e\left(x^2 + y^2\right)\right).$$

**NOTE**  *Let $\omega = dF$ be an exact form. By the Poincare Lemma 3.4.1, $d\omega = ddF = 0$. A result of Poincare says that for certain domains (called star-shaped domains) the converse is also true, that is, if $d\omega = 0$ on a star-shaped domain then $\omega$ is exact.*

### EXAMPLE 3.4.4

Determine whether the differential form

$$\omega = \frac{2x\left(1 - e^y\right)}{\left(1 + x^2\right)^2}\,dx + \frac{e^y}{1 + x^2}\,dy$$

is exact.

### Solution:

Assume there is a function $F$ such that

$$\mathrm{d}F = \omega.$$

By the Chain Rule

$$\mathrm{d}F = \frac{\partial F}{\partial x}\,\mathrm{d}x + \frac{\partial F}{\partial y}\,\mathrm{d}y\cdot$$

This demands that

$$\frac{\partial F}{\partial x} = \frac{2x\left(1-e^{y}\right)}{\left(1+x^{2}\right)^{2}},$$

$$\frac{\partial F}{\partial y} = \frac{e^{y}}{1+x^{2}}\cdot$$

We have a choice here of integrating either the first, or the second expression. Since integrating the second expression (with respect to $y$) is easier, we find

$$F\left(x,y\right) = \frac{e^{y}}{1+x^{2}} + \phi\left(x\right),$$

where $\phi\left(x\right)$ is a function depending only on $x$. To find it, we differentiate the obtained expression for $F$ with respect to $x$ and find

$$\frac{\partial F}{\partial x} = -\frac{2xe^{y}}{\left(1+x^{2}\right)^{2}} + \phi'\left(x\right).$$

Comparing this with our first expression for $\dfrac{\partial F}{\partial x}$, we find

$$\phi'\left(x\right) = \frac{2x}{\left(1+x^{2}\right)^{2}},$$

that is

$$\phi\left(x\right) = -\frac{1}{1+x^{2}} + c,$$

where $c$ is a constant. We then take

$$F(x,y) = \frac{e^y - 1}{1 + x^2} + c.$$

## EXAMPLE 3.4.5

Is there a continuously differentiable function such that

$$dF = \omega = y^2 z^3 dx + 2xyz^3 dy + 3xy^2 z^2 dz \ ?$$

### *Solution:*

We have

$$d\omega = \left(2yz^3 dy + 3y^2 z^2 dz\right) \wedge dx$$

$$+ \left(2yz^3 dx + 2xz^3 dy + 6xyz^2 dz\right) \wedge dy$$

$$+ \left(3y^2 z^2 dx + 6xyz^2 dy + 6xy^2 z dz\right) \wedge dz$$

$$= 0,$$

so this form is exact in a star-shaped domain. So put

$$dF = \frac{\partial F}{\partial x} dx + \frac{\partial F}{\partial y} dy + \frac{\partial F}{\partial z} dz = y^2 z^3 dx + 2xyz^3 dy + 3xy^2 z^2 dz.$$

Then

$$\frac{\partial F}{\partial x} = y^2 z^3 \Rightarrow F = xy^2 z^3 + a(y,z),$$

$$\frac{\partial F}{\partial y} = 2xyz^3 \Rightarrow F = xy^2 z^3 + b(x,z),$$

$$\frac{\partial F}{\partial z} = 3xy^2 z^2 \Rightarrow F = xy^2 z^3 + c(x,y).$$

Comparing these three expressions for $F$, we obtain $F(x,y,z) = xy^2z^3$.

We have the following equivalent of the Fundamental Theorem of Calculus.

**Theorem 3.4.1** Let $U \subseteq \mathbb{R}^n$ be an open set. Assume $w = dF$ is an exact form, and $\Gamma$ a path in $U$ with starting point $A$ and endpoint $B$. Then

$$\int_\Gamma \omega = \int_A^B dF = F(B) - F(A).$$

In particular, if $\Gamma$ is a simple closed path, then

$$\oint_\Gamma \omega = 0.$$

### EXAMPLE 3.4.6

Evaluate the integral

$$\oint_\Gamma \frac{2x}{x^2 + y^2}\, dx + \frac{2y}{x^2 + y^2}\, dy$$

where $\Gamma$ is the closed polygon with vertices at $A = (0,0), B = (5,0), C = (7,2),$ $D = (3,2), E = (1,1)$, traversed in the order $ABCDEA$.

### *Solution:*

Observe that

$$d\left( \frac{2x}{x^2 + y^2}\, dx + \frac{2y}{x^2 + y^2}\, dy \right) = -\frac{4xy}{\left(x^2 + y^2\right)^2}\, dy \wedge dx - \frac{4xy}{\left(x^2 + y^2\right)^2}\, dx \wedge dy = 0,$$

and so the form is exact in a start-shaped domain. By virtue of Theorem 3.4.1, the integral is 0.

### EXAMPLE 3.4.7

Calculate the path integral

$$\oint_\Gamma \left(x^2 - y\right) dx + \left(y^2 - x\right) dy,$$

where $\Gamma$ is a loop of $x^2 + y^3 - 2xy = 0$ traversed once in the positive sense.

*Solution:*

Since

$$\frac{\partial}{\partial y}\left(x^2 - y\right) = -1 = \frac{\partial}{\partial x}\left(y^2 - x\right),$$

the form is exact, and since this is a closed simple path, the integral is 0.

## EXERCISES 3.4

**3.4.1**   Are the following statements true or false?

1. A form $\alpha$ is closed if $d\alpha = 0$.

2. If $\omega$ is exact and $C$ is closed, then $\int_C \omega = 0$.

**3.4.2**   Are the following statements true or false?

1. Every exact form is closed, since $d(d\omega) = 0$. On the other hand, there are closed but not exact forms.

2. A form $\alpha$ is exact if $\alpha = d\beta$ for some form $\beta$.

**3.4.3**   Show that the differential form $x^2 dx + y dy$ is exact or not exact.

**3.4.4**   Show that the 1-form $\omega = d\theta$ on $\mathbb{R}^2 \setminus \{0\}$, where $\theta$ is the polar angle. In standard Cartesian coordinates:

$$\omega = \frac{x dy - y dx}{x^2 + y^2}.$$ Is the form exact or not exact.

**3.4.5**   Prove that on a rectangular parallelepiped, $\Pi$, all closed forms are exact.

**3.4.6**   Consider the differential 1-form

$$\omega = -\frac{y}{x^2 + y^2} dx + \frac{x}{x^2 + y^2} dy.$$ Defined on $\Omega = \mathbb{R}^2 - (0,0)$, and be closed curve at

$$C: \begin{array}{ccc} [0, 2\pi] & \to & \mathbb{R}^2 \\ \theta & \mapsto & (\cos\theta, \sin\theta). \end{array}$$

Show that $w$ is exact or not exact. Also, show that $w$ is a closed 1-form.

## 3.5   TWO-MANIFOLDS

**Definition 3.5.1**   A *two-dimensional oriented manifold of* $\mathbb{R}^2$ is simply an open set (region) $D \in \mathbb{R}^2$, where the + orientation is counter-clockwise and the − orientation is clockwise. A general oriented 2-manifold is a union of open sets.

**NOTE**   *The region* $-D$ *has opposite orientation to* $D$ *and*

$$\int_{-D} \omega = -\int_D \omega.$$

*We will often write*

$$\int_D f(x,y)\, \mathrm{d}A$$

*where* $\mathrm{d}A$ *denotes the area element.*

**NOTE**   *In this section, unless otherwise noted, we will choose the positive orientation for the regions considered. This corresponds to using the area form* $\mathrm{d}x\mathrm{d}y$.

Let $U \subseteq \mathbb{R}^2$. Given a function $f : D \to \mathbb{R}$, the integral

$$\int_D f\mathrm{d}A$$

Is the sum of all the values of $f$ restricted to $D$. In particular,

$$\int_D \mathrm{d}A$$

is the area of $D$.

In order to evaluate double integrals, we need the following.

**Theorem 3.5.1 (Fubini's Theorem):**   Let $D = [a;b] \times [c;d]$, and let $f : A \to \mathbb{R}$ be continuous. Then

$$\int_D f\mathrm{d}A = \int_a^b \left( \int_c^d f(x,y)\mathrm{d}y \right) \mathrm{d}x = \int_c^d \left( \int_a^b f(x,y)\mathrm{d}x \right) \mathrm{d}y$$

Fubini's Theorem allows us to convert the double integral into iterated (single) integrals.

**EXAMPLE 3.5.1**

$$\int_{[0;1] \times [2;3]} xy\, dA = \int_0^1 \left( \int_2^3 xy\, dy \right)\, dx$$

$$= \int_0^1 \left( \left[ \frac{xy^2}{2} \right]_2^3 \right)\, dx$$

$$= \int_0^1 \left( \frac{9x}{2} - 2x \right)\, dx$$

$$= \left[ \frac{5x^2}{4} \right]_0^1$$

$$= \frac{5}{4}.$$

Notice that if we had integrated first with respect to $x$ we would have obtained the same result:

$$\int_2^3 \left( \int_0^1 xy\, dx \right)\, dy = \int_2^3 \left( \left[ \frac{x^2 y}{2} \right]_0^1 \right)\, dy$$

$$= \int_2^3 \left( \frac{y}{2} \right)\, dx$$

$$= \left[ \frac{y^2}{4} \right]_2^3$$

$$= \frac{5}{4}.$$

Also, this integral is "factorable into $x$ and $y$ pieces" meaning that

$$\int_{[0;1] \times [2;3]} xy\,dA = \left(\int_0^1 x\,dx\right)\left(\int_2^3 y\,dy\right)$$
$$= \left(\frac{1}{2}\right)\left(\frac{5}{2}\right)$$
$$= \frac{5}{4}.$$

To solve this problem using $Maple^{Tm}$, you may use the following code.

```
> with(Student[VectorCalculus]) :
> int(x*y, [x, y] = Region(0..1, 2..3));
5
-
4
```

To solve this problem using *MATLAB*, you may use the following code.

```
>> syms x y
>> firstint = int(x*y,x,0,1)

firstint =

1/2*y

>> answer = int(firstint,y,2,3)

answer =

5/4
```

## EXAMPLE 3.5.2

We have

$$\int_3^4 \int_0^1 (x+2y)(2x+y)\,dx\,dy = \int_3^4 \int_0^1 \left(2x^2 + 5xy + 2y^2\right)\,dx\,dy$$
$$= \int_3^4 \left(\frac{2}{3} + \frac{5}{2}y + 2y^2\right)\,dy$$
$$= \frac{409}{12}.$$

To solve this problem using *Maple*$^{Tm}$, you may use the following code.

```
>  with( Student[ VectorCalculus]) :
>  int( (x + 2*y) * (2*x + y), [x, y] = Region(3 ..4, 0 ..1));
         409
         ───
          12
```

To solve this problem using *MATLAB*, you may use the following code.

```
>> syms x y
>> firstans = int(int((x+2*y)*(2*x+y),x,0,1),y,3,4)

firstans =

409/12
```

In the cases when the domain of integration is not a rectangle, we decompose so that, one variable is kept constant.

### EXAMPLE 3.5.3

Find $\int_D xy \, dxdy$ in the triangle with vertices $A:(-1,-1), B:(2,-2), C:(1,2)$.

**Solution:**

The lines passing through the given points have equations $L_{AB} : y = \dfrac{-x-4}{3}$, $L_{BC} : y = -4x+6, L_{CA} : y = \dfrac{3x+1}{2}$.

Now, we draw the region carefully. If we integrate first with respect to $y$, we must divide the region as in Figure 3.5.1, because there are two upper lines, which the upper value of $y$ might be. The lower point of the dashed line is $(1, -5/3)$.

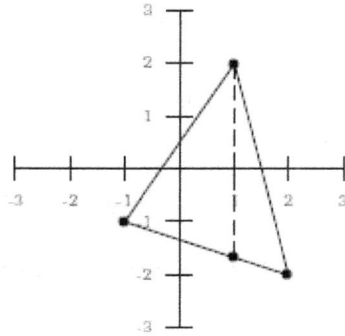

*FIGURE 3.5.1* Example 3.5.3, integration order dydx.

The integral is thus

$$\int_{-1}^{1} x\left(\int_{(-x-4)/3}^{(3x+1)/2} y \ dy\right)dx + \int_{1}^{2} x\left(\int_{(-x-4)/3}^{-4x+6} y \ dy\right)dx = -\frac{11}{8}.$$

If we integrate first with respect to $x$, we must divide the region as in Figure 3.5.2, because there are two left-most lines, which the left value of $x$ might be. The right point of the dashed line is $(7/4, -1)$.

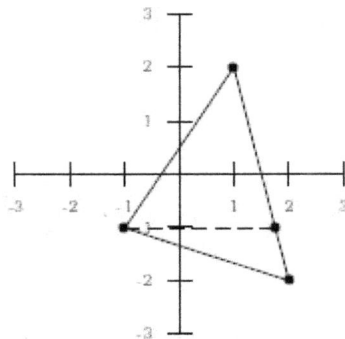

*FIGURE 3.5.2* Example 3.5.3, integration order dxdy.

The integral is thus

$$\int_{-2}^{-1} y\left(\int_{-4-3y}^{(6-y)/4} x \ dx\right)dy + \int_{-1}^{2} y\left(\int_{(2y-1)/3}^{(6-y)/4} x \ dx\right)dy = -\frac{11}{8}.$$

To solve this problem using $Maple^{Tm}$, you may use the following code.

```
>  with(Student[VectorCalculus]) :
>  int(x·y, [x, y] = Triangle(⟨-1,-1⟩, ⟨2,-2⟩, ⟨1, 2⟩));
      11
    - ──
       8
```

To solve this problem using *MATLAB*, you may use the following code.

```
>> syms x y

>> firstans = int(int(x*y,x,-4-3*y,(6-y)/4),y,-2,-1)+int(int(x*y,x,(2*y-1)/3,(6-
y)/4),y,-1,2)

firstans =

-11/8
```

### EXAMPLE 3.5.4

Consider the region inside the parallelogram $P$ with vertices at $A:(6,3)$, $B:(8,4),C:(9,6),D:(7,5)$, as in Figure 3.5.3.

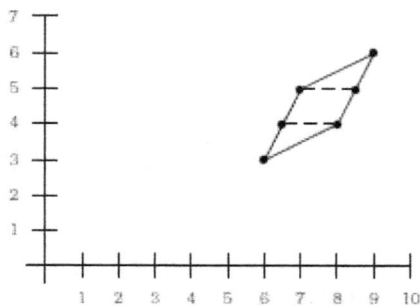

**FIGURE 3.5.3** Example 3.5.4.

Find

$$\int_P xy \; dxdy.$$

### Solution:

The lines joining the points have equations

$$L_{AB} : y = \frac{x}{2},$$

$$L_{BC} : y = 2x - 12,$$

$$L_{CD} : y = \frac{x}{2} + \frac{3}{2},$$

$$L_{DA} : y = 2x - 9.$$

The integral is thus

$$\int_3^4 \int_{(y+9)/2}^{2y} xy \; dxdy + \int_4^5 \int_{(y+9)/2}^{(y+12)/2} xy \; dxdy + \int_5^6 \int_{2y-3}^{(y+12)/2} xy \; dxdy = \frac{409}{4}.$$

To solve this problem using *Maple*$^{Tm}$, you may use the following code.

> *with(Student[VectorCalculus]) :*
> *int(x\*y, [x, y] = Triangle( ⟨6, 3⟩, ⟨8, 4⟩, ⟨7, 5⟩)) + int(x\*y, [x, y]*
>       *= Triangle( ⟨8, 4⟩, ⟨9, 6⟩, ⟨7, 5⟩));*

$$\frac{409}{4}$$

To solve this problem using *MATLAB*, you may use the following code.

>> syms x y

>> firstans = int(int(x*y,x,(y+9)/2,2*y),y,3,4)+int(int(x*y,x,(y+9)/2,(y+12)/2),y,4,5)+ int(int(x*y,x,2*y-3, (y+12)/2),y,5,6)

firstans =

409/4

**EXAMPLE 3.5.5**

Find

$$\int_D \frac{y}{x^2+1} \, dxdy$$

where

$$D = \left\{ (x,y) \in \mathbb{R}^2 \mid x \geq 0, x^2 + y^2 \leq 1 \right\}.$$

*Solution:*

The integral is 0. Observe that if $(x,y) \in D$, then $(x,-y) \in D$. Also, $f(x,-y) = -f(x,y)$.

**EXAMPLE 3.5.6**

Find

$$\int_0^4 \left( \int_{y/2}^{\sqrt{y}} e^{y/x} \, dx \right) dy.$$

See Figure 3.5.5.

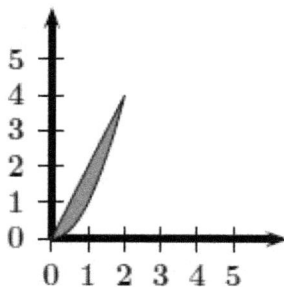

**FIGURE 3.5.4** Example 3.5.6.

*Solution:*

We have

$$0 \leq y \leq 4, \ \frac{y}{2} \leq x \leq \sqrt{y} \Rightarrow 0 \leq x \leq 2, \ x^2 \leq y \leq 2x.$$

We then have

$$\int_0^4 \left( \int_{y/2}^{\sqrt{y}} e^{y/x} dx \right) dy = \int_0^2 \left( \int_{x^2}^{2x} e^{y/x} dy \right) dx$$

$$= \int_0^2 \left( xe^{y/x} \Big|_{x^2}^{2x} \right) dx$$

$$= \int_0^2 \left( xe^2 - xe^x \right) dx$$

$$= 2e^2 - \left( 2e^2 - e^2 + 1 \right)$$

$$= e^2 - 1.$$

## EXAMPLE 3.5.7

Find the area of the region

$$R = \left\{ (x,y) \in \mathbb{R}^2 : \sqrt{x} + \sqrt{y} \geq 1, \sqrt{1-x} + \sqrt{1-y} \geq 1 \right\}.$$

See Figure 3.5.5.

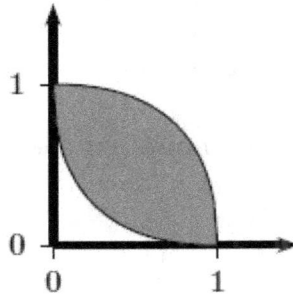

**FIGURE 3.5.5** Example 3.5.7.

***Solution:***

The area is given by

$$\int_D dA = \int_0^1 \left( \int_{(1-\sqrt{x})^2}^{1-(1-\sqrt{1-x})^2} dy \right) dx$$

$$= 2\int_0^1 \left( \sqrt{1-x} + \sqrt{x} - 1 \right) dx$$

$$= \frac{2}{3}.$$

### EXAMPLE 3.5.8

Evaluate $\int_R \llbracket x^2 + y^2 \rrbracket \, dA$, where $R$ is the rectangle $\left[ 0; \sqrt{2} \right] \times \left[ 0; \sqrt{2} \right]$.
See Figure 3.5.6.

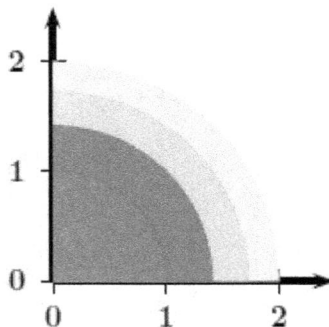

**FIGURE 3.5.6** Example 3.5.8.

***Solution:***

The function $(x,y) \mapsto \llbracket x^2 + y^2 \rrbracket$ jumps every time $x^2 + y^2$ is an integer. For $(x,y) \in R$, we have $0 \le x^2 + y^2 \le \left( \sqrt{2} \right)^2 + \left( \sqrt{2} \right)^2 = 4$. Thus we decompose $R$ as the union of the

$$R_k = \left\{ (x,y) \in \mathbb{R}^2 : x \geq 0, y \geq 0, k \leq x^2 + y^2 \leq k+1 \right\}, \quad k \in \{1,2,3\}.$$

$$\int_R \left[\!\left[ x^2 + y^2 \right]\!\right] dA = \sum_{1 \leq k \leq 3} \int_{R_k} \left[\!\left[ x^2 + y^2 \right]\!\right] dA$$

$$= \iint_{1 \leq x^2+y^2 < 2, x \geq 0, y \geq 0} 1 dA + \iint_{2 \leq x^2+y^2 < 3, x \geq 0, y \geq 0} 2 dA + \iint_{3 \leq x^2+y^2 < 4, x \geq 0, y \geq 0} 3 dA.$$

Now the integrals can be computed by realizing that they are areas of quarter annuli, and so,

$$\iint_{k \leq x^2+y^2 < k+1, x \geq 0, y \geq 0} k dA = k \cdot \frac{1}{4} \cdot \pi \left(k+1-k\right) = \frac{\pi k}{4}.$$

Hence

$$\int_R \left[\!\left[ x^2 + y^2 \right]\!\right] dA = \frac{\pi}{4}(1+2+3) = \frac{3\pi}{2}.$$

## EXERCISES 3.5

**3.5.1**   Evaluate the iterated integral

$$\int_1^3 \int_0^x \frac{1}{x} \, dy \, dx.$$

**3.5.2**   Let $S$ be the interior and boundary of the triangle with vertices $(0,0)$, $(2,1)$, and $(2,0)$. Find $\int_S y \, dA$.

**3.5.3**   Let $S = \left\{ (x,y) \in \mathbb{R}^2 : x \geq 0, y \geq 0, 1 \leq x^2 + y^2 \leq 4 \right\}$. Find $\int_S x^2 \, dA$.

**3.5.4**   Find $\int_D xy \, dx \, dy$

where

$$D = \left\{ (x,y) \in \mathbb{R}^2 \mid y \geq x^2, x \geq y^2 \right\}.$$

**3.5.5**    Find $\int_D (x+y)(\sin x)(\sin y)\,dA$

where

$$D = [0;\pi]^2.$$

**3.5.6**    Find $\int_0^1 \int_0^1 \min(x^2 + y^2)\,dxdy$.

**3.5.7**    Find $\int_D xy\,dxdy$

where

$$D = \{(x,y) \in \mathbb{R}^2 : x > 0, y > 0, 9 < x^2 + y^2 < 16, 1 < x^2 - y^2 < 16\}.$$

**3.5.8**    Evaluate $\int_R x\,dA$ where $R$ is the (un-oriented) circular segment in Figure 3.5.7, which is created by the intersection of regions

$$\{(x,y) \in \mathbb{R}^2 : x^2 + y^2 \le 16\}$$

and

$$\left\{(x,y) \in \mathbb{R}^2 : y \ge -\frac{\sqrt{3}}{3}x + 4\right\}.$$

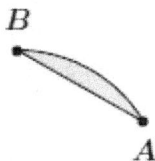

**FIGURE 3.5.7** Exercise 3.5.8.

**3.5.9**    Find $\int_0^1 \int_y^1 2e^{x^2}\,dxdy$.

**3.5.10**    Evaluate $\int_{[0;1]^2} \min(x, y^2)\,dA$.

**3.5.11**    Find $\int_R xy\,dA$, where $R$ is the (un-oriented) $\triangle OAB$ in Figure 3.5.8 with $O(0,0), A(3,1)$, and $B(4,4)$.

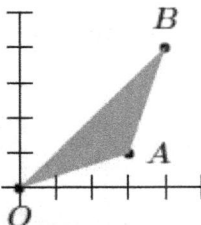

**FIGURE 3.5.8** Exercise 3.5.11.

**3.5.12** Solve Exercise 3.5.11 using Maple commands.

**3.5.13** Find $\int_D \log_e (1 + x + y) \, dA$

where

$$D = \left\{ (x, y) \in \mathbb{R}^2 \mid x \geq 0, y \geq 0, x + y \leq 1 \right\}.$$

**3.5.14** Evaluate $\int_{[0;2]^2} \llbracket x + y^2 \rrbracket \, dA$.

**3.5.15** Evaluate $\int_R \llbracket x + y \rrbracket \, dA$,

where $R$ is the rectangle $[0;1] \times [0;2]$.

**3.5.16** Evaluate $\int_R x \, dA$ where $R$ is the quarter annulus in Figure 3.5.9, which is formed by the area between the circles $x^2 + y^2 = 1$ and $x^2 + y^2 = 4$ in the first quadrant.

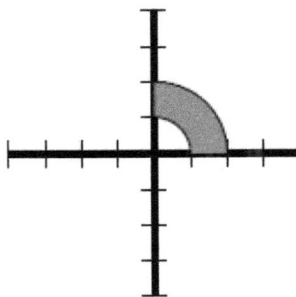

**FIGURE 3.5.9** Exercise 3.5.16.

**3.5.17**   Evaluate $\int_R x\,dA$, where $R$ is the E-shape figure in Figure 3.5.10.

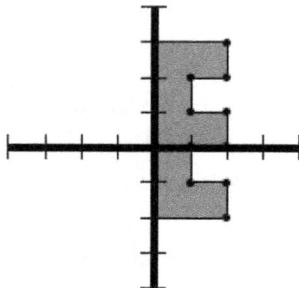

**FIGURE 3.5.10** Exercise 3.5.17.

**3.5.18**   Evaluate $\int_0^{\pi/2} \int_x^{\pi/2} \dfrac{\cos y}{y}\, dy\,dx$.

**3.5.19**   Find $\int_1^2 \left( \int_{\sqrt{x}}^x \sin\dfrac{\pi x}{2y}\, dy \right) dx + \int_2^4 \left( \int_{\sqrt{x}}^2 \sin\dfrac{\pi x}{2y}\, dy \right) dx$.

**3.5.20**   Find $\int_D 2x\left(x^2 + y^2\right) dA$

where

$$D\left\{(x,y) \in \mathbb{R}^2 : x^4 + y^4 + x^2 - y^2 \le 1\right\}.$$

**3.5.21**   Find the area bounded by the ellipses $x^2 + \dfrac{y^2}{4} = 1$ and $\dfrac{x^2}{4} + y^2 = 1$. as in Figure 3.5.11.

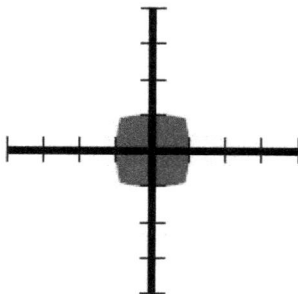

**FIGURE 3.5.11** Exercise 3.5.21.

**3.5.22** Find $\int\limits_{D} xy dA$

where

$$D\left\{(x,y) \in \mathbb{R}^2 : x \geq 0, y \geq 0, xy + y + x \leq 1\right\}.$$

**3.5.23** Find $\int\limits_{D} \log_e\left(1 + x^2 + y\right) dA$

where

$$D\left\{(x,y) \in \mathbb{R}^2 : x \geq 0, y \geq 0, x^2 + y \leq 1\right\}.$$

**3.5.24** Evaluate $\int_R x dA$, where $R$ is the region between the circles $x^2 + y^2 = 4$ and $x^2 + y^2 = 2y$, as shown in Figure 3.5.12.

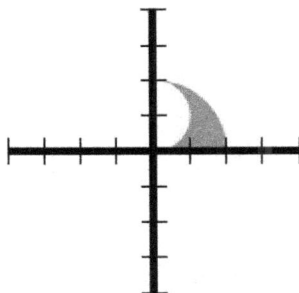

**FIGURE 3.5.12** Exercise 3.5.24.

**3.5.25** Find $\int\limits_{D} |x - y| \, dA$

where

$$D\left\{(x,y) \in \mathbb{R}^2 : |x| \leq 1, |y| \leq 1\right\}.$$

**3.5.26** Find $\int\limits_{D} (2x + 3y + 1) \, dA.$

Where $D$ is the triangle with vertices at $A(-1,-1), B(2,-4)$, and $C(1,3).$

**3.5.27** Let $f:[0;1] \to ]0;+\infty]$ be a decreasing function. Prove that

$$\frac{\int_0^1 x f^2(x)\,dx}{\int_0^1 x f(x)\,dx} \leq \frac{\int_0^1 f^2(x)\,dx}{\int_0^1 f(x)\,dx}.$$

**3.5.28** Find $\int_D \big(xy(x+y)\big)\,dA$. Where

$$D\big\{(x,y) \in \mathbb{R}^2 : x \geq 0, y \geq 0, x+y \leq 1\big\}.$$

**3.5.29** Let $f,g:[0;1] \to [0;1]$ be continuous, with $f$ increasing. Prove that

$$\int_0^1 (f \circ g)(x)\,dx \leq \int_0^1 f(x)\,dx + \int_0^1 g(x)\,dx.$$

**3.5.30** Compute $\int_S (xy + y^2)\,dA$, where $S = \big\{(x,y) \in \mathbb{R}^2 : |x|^{1/2} + |y|^{1/2} \leq 1\big\}$.

**3.5.31** Evaluate $\int_0^a \int_0^b e^{\max(b^2 x^2, a^2 y^2)}\,dy\,dx$, where $a$ and $b$ are positive.

**3.5.32** Find $\int_D \sqrt{xy}\,dA$.

Where

$$D\big\{(x,y) \in \mathbb{R}^2 : y \geq 0, (x+y)^2 \leq 2x\big\}.$$

**3.5.33** A rectangle $R$ on the plane is the disjoint union $R = \bigcup_{k=1}^N R_k$ of rectangles $R_k$. It is known that at least one side of each of the rectangles $R_k$ is an integer. Show that at least one side of $R$ is an integer.

**3.5.34** Evaluate $\int_0^1 \int_0^1 \cdots \int_0^1 (x_1 x_2 \cdots x_n)\,dx_1\,dx_2 \cdots dx_n$.

**3.5.35** Evaluate $\int_0^1 \int_0^1 \cdots \int_0^1 (x_1 + x_2 + \cdots + x_n)\,dx_1\,dx_2 \cdots dx_n$.

**3.5.36** Find $\int_D \frac{1}{(x+y)^4}\,dA$, where

$$D\big\{(x,y) \in \mathbb{R}^2 \mid x \geq 1, y \geq 1, x+y \leq 4\big\}.$$

**3.5.37** Find $\int\limits_D x\,dA$. Where

$$D\left\{(x,y)\in\mathbb{R}^2\mid y\geq 0, x-y+1\geq 0, x+2y-4\leq 0\right\}.$$

**3.5.38** Evaluate $\lim\limits_{n\to+\infty}\int_0^1\int_0^1\cdots\int_0^1\cos^2\left(\frac{\pi}{2n}(x_1+x_2+\cdots+x_n)\right)dx_1dx_2\cdots dx_n$.

## 3.6   CHANGE OF VARIABLES IN DOUBLE INTEGRALS

We now perform a multidimensional analogue of the change of variables theorem in double integrals.

**Theorem 3.6.1**   Let $(D,\Delta)\in(\mathbb{R}^n)^2$ open, bounded sets in $\mathbb{R}^n$ with volume and let $g:\Delta\to D$ be a continuously differentiable bijective mapping such that $\det g'(u)\neq 0$, both $\left|\det g'(u)\right|, \dfrac{1}{\left|\det g'(u)\right|}$ bounded on $\Delta$. For $f:D\to\mathbb{R}$ bounded and integrable. $f\circ g\left|\det g'(u)\right|$ is integrable on $\Delta$ and

$$\int\cdots\int_D f = \int\cdots\int_\Delta (f\circ g)\left|\det g'(u)\right|,$$

that is

$$\int\cdots\int_D f(x_1,x_2,\ldots,x_n)\,dx_1\wedge dx_2\wedge\ldots\wedge dx_n$$

$$=\int\cdots\int_\Delta f(g(u_1,u_2,\ldots,u_n))\left|\det g'(u)\right|\,du_1\wedge du_2\wedge\ldots\wedge du_n.$$

One normally chooses changes of variables that map into rectangular regions, or that simplify the integrand. Let us start with a rather trivial example.

**EXAMPLE 3.6.1**

Evaluate the integral

$$\int_3^4\int_0^1(x+2y)(2x+y)\,dx\,dy.$$

### Solution:

Observe that we have already computed this integral in Example 3.5.2. Put

$$u = x + 2y \Rightarrow du = dx + 2dy,$$
$$v = 2x + y \Rightarrow dv = 2dx + dy,$$

giving

$$du \wedge dv = -3dx \wedge dy.$$

Now,

$$(u, v) = \begin{bmatrix} 1 & 2 \\ 2 & 1 \end{bmatrix} \begin{bmatrix} x \\ y \end{bmatrix}$$

is a linear transformation, and hence it maps quadrilaterals into quadrilaterals. The corners of the rectangle in the area of integration in the $xy$-plane are $(0, 3)$, $(1, 3)$, $(1, 4)$, and $(0, 4)$ (traversed counter-clockwise; see Figure 3.6.1). They map into $(6, 3)$, $(7, 5)$, $(9, 6)$, and $(8, 4)$, respectively, in the $uv$-plane (see Figure 3.6.2).

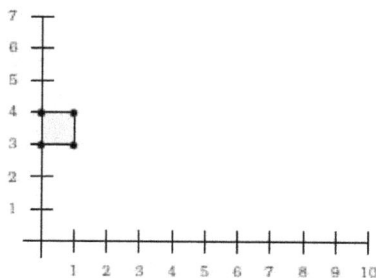

**FIGURE 3.6.1** Example 3.6.1, $xy$-plane.

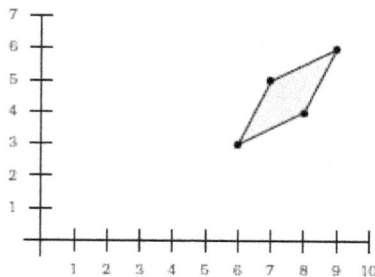

**FIGURE 3.6.2** Example 3.6.1, $uv$-plane.

The form $dx \wedge dy$ has opposite orientation to $du \wedge dv$ so we use

$$dv \wedge du = 3dx \wedge dy$$

instead. The integral sought is

$$\frac{1}{3}\int_P uv \ dv du = \frac{409}{12},$$

from Example 3.5.4.

### EXAMPLE 3.6.2

The integral

$$\int_{[0;1]^2} \left(x^4 - y^4\right) dA = \int_0^1 \left(\frac{1}{5} - y^4\right) dy = 0.$$

Evaluate it using the change of variables $u = x^2 - y^2, v = 2xy$.

***Solution:***

First we find

$$du = 2xdx - 2ydy,$$

$$dv = 2ydx + 2xdy,$$

and so

$$du \wedge dv = \left(4x^2 + 4y^2\right)dx \wedge dy.$$

We now determine the region $\Delta$ into which the square $D = [0;1]^2$ is mapped. We use the fact that the boundaries will be mapped into boundaries. Put

$$AB = \{(x,0) : 0 \leq x \leq 1\},$$

$$BC = \{(1,y) : 0 \leq y \leq 1\},$$

$$CD = \{(1-x,y) : 0 \leq x \leq 1\},$$

$$DA = \{(0,1-y) : 0 \leq y \leq 1\}.$$

On $AB$, we have $u = x, v = 0$. Since $0 \le x \le 1$, $AB$ is thus mapped into the line segment $0 \le u \le 1$, $v = 0$.

On $BC$, we have $u = 1 - y^2, v = 2y$. Thus $u = 1 - \dfrac{v^2}{4}$. Hence, $BC$ is mapped to the portion of the parabola $u = 1 - \dfrac{v^2}{4}$, $0 \le v \le 2$

On $CD$ we have $u = (1-x)^2 - 1, v = 2(1-x)$. This means that $u = \dfrac{v^2}{4} - 1, 0 \le v \le 2$. Finally, on $DA$, we have $u = -(1-y)^2, v = 0$. Since $0 \le y \le 1$, $DA$ is mapped into the line segment $-1 \le u \le 0, v = 0$. The region $\Delta$ is thus the area in the $uv$ plane enclosed by the parabolas $u \le \dfrac{v^2}{4} - 1, u \le 1 - \dfrac{v^2}{4}$ with $-1 \le u \le 1, 0 \le v \le 2$.

We deduce that

$$\int_{[0;1]^2} \left(x^4 - y^4\right) \mathrm{d}A = \int_\Delta \left(x^4 - y^4\right) \frac{1}{4\left(x^4 + y^4\right)} \mathrm{d}u\mathrm{d}v$$

$$= \frac{1}{4} \int_\Delta \left(x^2 - y^2\right) \mathrm{d}u\mathrm{d}v$$

$$= \frac{1}{4} \int_\Delta u \, \mathrm{d}u\mathrm{d}v$$

$$= \frac{1}{4} \int_0^2 \left( \int_{v^2/4-1}^{1-v^2/4} u \, \mathrm{d}u \right) \mathrm{d}v$$

$$= 0,$$

as before.

## EXAMPLE 3.6.3

Find $\int\limits_D e^{(x^3+y^3)/xy}\,dA$,

where

$$D = \left\{(x,y) \in \mathbb{R}^2 \mid y^2 - 2px \le 0, x^2 - 2py \le 0, p \in \,]0;+\infty[\ \text{fixed}\right\},$$

Using the change of variables $x = u^2v, y = uv^2$.

### *Solution:*

We have

$$dx = 2uv\,du + u^2\,dv,$$

$$dy = v^2\,du + 2uv\,dv,$$

$$dx \wedge dy = 3u^2v^2\,du \wedge dv.$$

The region transforms into

$$\Delta = \left\{(u,v) \in \mathbb{R}^2 \mid 0 \le u \le (2p)^{1/3}, 0 \le v \le (2p)^{1/3}\right\}.$$

The integral becomes

$$\int\limits_D f(x,y)\,dxdy = \int\limits_\Delta \exp\left(\frac{u^6v^3 + u^3v^6}{u^3v^3}\right)\left(3u^2v^2\right)dudv$$

$$= 3\int\limits_\Delta e^{u^3}e^{v^3}u^2v^2\,dudv$$

$$= \frac{1}{3}\left(\int_0^{(2p)^{1/3}} 3u^2 e^{u^3}\,du\right)^2$$

$$= \frac{1}{3}\left(e^{2p} - 1\right)^2.$$

As an exercise, you may try the (more natural) substitution $x^3 = u^2v, y^3 = v^2u$ and verify that the same result is obtained.

## EXAMPLE 3.6.4

In this problem, we will follow an argument of Calabi, Beukers, and Knock to prove that

$$\sum_{n=1}^{+\infty} \frac{1}{n^2} = \frac{\pi^2}{6}.$$

1. Prove that if $S = \sum_{n=1}^{+\infty} \frac{1}{n^2}$, then $\frac{3}{4}S = \sum_{n=1}^{+\infty} \frac{1}{(2n-1)^2}$.

2. Prove that $\sum_{n=1}^{+\infty} \frac{1}{(2n-1)^2} = \int_0^1 \int_0^1 \frac{dxdy}{1-x^2y^2}$.

3. Use the change of variables $x = \dfrac{\sin u}{\cos v}$, $y = \dfrac{\sin v}{\cos u}$ in order to evaluate $\int_0^1 \int_0^1 \frac{dxdy}{1-x^2y^2}$.

### Solution:

1. Observe that the sum of even terms is

$$\sum_{n=1}^{+\infty} \frac{1}{(2n)^2} = \frac{1}{4}\sum_{n=1}^{+\infty} \frac{1}{n^2} = \frac{1}{4}S,$$

a quarter of the sum, hence the sum of the odd terms must be three-quarters of the sum, $\dfrac{3}{4}S$.

2. Observe that

$$\frac{1}{2n-1} = \int_0^1 x^{2n-2}\, dx \Rightarrow \left(\frac{1}{2n-1}\right)^2 = \left(\int_0^1 x^{2n-2}\, dx\right)\left(\int_0^1 y^{2n-2}\, dy\right) = \int_0^1 \int_0^1 (xy)^{2n-2}\, dxdy.$$

Thus

$$\sum_{n=1}^{+\infty} \frac{1}{(2n-1)^2} = \sum_{n=1}^{+\infty} \int_0^1 \int_0^1 (xy)^{2n-2}\, dxdy = \int_0^1 \int_0^1 \sum_{n=1}^{+\infty} (xy)^{2n-2}\, dxdy = \int_0^1 \int_0^1 \frac{dxdy}{1-x^2y^2},$$

as claimed.

3. If $x = \dfrac{\sin u}{\cos v}$, $y = \dfrac{\sin v}{\cos u}$ then

$$dx = (\cos u)(\sec v)\, du + (\sin u)(\sec v)(\tan v)\, dv,$$
$$dy = (\sec u)(\tan v)(\sin v)\, du + (\sec u)(\cos v)\, dv,$$

from where

$$dx \wedge dy = du \wedge dv - \left(\tan^2 u\right)\left(\tan^2 v\right) du \wedge dv = \left(1 - \left(\tan^2 u\right)\left(\tan^2 v\right)\right) du \wedge dv.$$

Also,

$$1 - x^2 y^2 = 1 - \frac{\sin^2 v}{\cos^2 v} \cdot \frac{\sin^2 v}{\cos^2 u} = 1 - \left(\tan^2 u\right)\left(\tan^2 v\right).$$

This gives

$$\frac{dx dy}{1 - x^2 y^2} = du dv.$$

We now have to determine the region that the transformation, $x = \dfrac{\sin u}{\cos v}, y = \dfrac{\sin v}{\cos u}$ forms in the $uv$-plane. Observe that

$$u = \arctan x \sqrt{\frac{1 - y^2}{1 - x^2}}, \quad v = \arctan y \sqrt{\frac{1 - x^2}{1 - y^2}} \quad .$$

This means that the square in the $xy$-plane in Figure 3.6.3 is transformed into the triangle in the $uv$-plane in Figure 3.6.4.

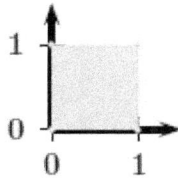

**FIGURE 3.6.3** Example 3.6.4, *xy*-plane.

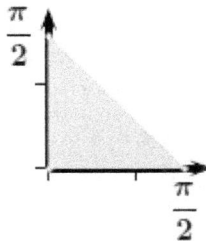

**FIGURE 3.6.4** Example 3.6.4, *uv*-plane.

We deduce,

$$\int_0^1 \int_0^1 \frac{dxdy}{1-x^2y^2} = \int_0^{\pi/2} \int_0^{\pi/2-v} dudv = \int_0^{\pi/2} (\pi/2 - v)dv = \left(\frac{\pi}{2}v - \frac{v^2}{2}\right)\Big|_0^{\pi/2} = \frac{\pi^2}{4} - \frac{\pi^2}{8} = \frac{\pi^2}{8}.$$

Finally,

$$\frac{3}{4}S = \frac{\pi^2}{8} \Rightarrow S = \frac{\pi^2}{6}.$$

## EXAMPLE 3.6.5

Evaluate $\int_0^{1/5} \int_0^{1-3y} e^{\left(\frac{x}{x+3y}\right)} dxdy$ using Maple.

To solve this problem using $Maple^{Tm}$, you may use the following code.

```
> evalf(int(int(exp(x/(x+3*y)),x=0..1-3*y),y=0..1/5));
  0.2552649404
```

Maple automatically performs the change of variables.

## EXERCISES 3.6

**3.6.1**    Let $D' = \{(u,v) \in \mathbb{R}^2 : u \le 1, -u \le v \le u\}$. Consider,

$$\Phi: \begin{array}{ccc} \mathbb{R}^2 & \to & \mathbb{R}^2 \\ (u,v) & \mapsto & \left(\dfrac{u+v}{2}, \dfrac{u-v}{2}\right). \end{array}$$

Find the image of $\Phi$ on $D'$, that is, find $D = \Phi(D')$.

**3.6.2**    Let $D' = \{(u,v) \in \mathbb{R}^2 : u \le 1, -u \le v \le u\}$. Consider,

$$\Phi: \begin{array}{ccc} \mathbb{R}^2 & \to & \mathbb{R}^2 \\ (u,v) & \mapsto & \left(\dfrac{u+v}{2}, \dfrac{u-v}{2}\right). \end{array}$$

Find $\int_D (x+y)^2 e^{x^2-y^2} dA$.

**3.6.3**   Find $\int\int_D f(x,y) \, dA$ where

$$D = \left\{ (x,y) \in \mathbb{R}^2 \mid a \le xy \le b, y \ge x \ge 0, y^2 - x^2 \le 1, (a,b) \in \mathbb{R}^2, 0 < a < b \right\}$$

and $f(x,y) = y^4 - x^4$ by using the change of variables
$u = xy, \quad v = y^2 - x^2$.

**3.6.4**   Use the following steps (due to Tom Apostol), in order to prove that

$$\sum_{n=1}^{\infty} \frac{1}{n^2} = \frac{\pi^2}{6}.$$

1.  Use the series expansion

$$\frac{1}{1-t} = 1 + t + t^2 + t^3 + \cdots, \qquad |t| < 1,$$

in order to prove (formally) that $\int_0^1 \int_0^1 \frac{dx \, dy}{1 - xy} = \sum_{n=1}^{\infty} \frac{1}{n^2}$.

2.  Use the change of variables $u = x + y, \ v = x - y$ to show that

$$\int_0^1 \int_0^1 \frac{dx \, dy}{1 - xy} = 2 \int_0^1 \left( \int_{-u}^u \frac{dv}{4 - u^2 + v^2} \right) du + 2 \int_1^2 \left( \int_{u-2}^{2-u} \frac{dv}{4 - u^2 + v^2} \right) du.$$

3.  Show that the preceding integral reduces to

$$2 \int_0^1 \frac{2}{\sqrt{4 - u^2}} \arctan \frac{u}{\sqrt{4 - u^2}} \, du + 2 \int_1^2 \frac{2}{\sqrt{4 - u^2}} \arctan \frac{2 - u}{\sqrt{4 - u^2}} \, du.$$

4.  Finally, prove that the preceding integral is $\dfrac{\pi^2}{6}$ by using the substitution $\theta = \arcsin \dfrac{u}{2}$.

**3.6.5**   Evaluate the integral $\iint_D \sin^2 (x - y)(x + y)^2 \, dA$ using change of variables, where $D$ is the region bounded by square with vertices $(0,1),(1,2),(2,1)$, and $(1,0)$.

**3.6.6** Evaluate the integral $\iint\limits_D \dfrac{\cos\big((x-y)/2\big)}{3x+y}\, dA$ using change of vari-

ables, where $D$ is the region bounded by the graphs
$y=-3x+3, y=-3x+6, y=x, y=x-\pi; u=x-y$, and $v=3x+y$.

**3.6.7** Evaluate the integral $\iint\limits_D \big(x^2+y^2\big)\sin xy \, dA$ using change of vari-

ables, where $D$ is the region bounded by the graphs
$y=\dfrac{2}{x}, y=-\dfrac{2}{x}, y^2=x^2-1, y^2=x^2-9; u=x^2-y^2$, and $v=xy$.

**3.6.8** Evaluate the integral $\iint\limits_D y \, dA$ using change of variables, where $D$ is

the triangle region with vertices $(0,0),(2,3)$, and $(-4,1)$;
$x=2u-4v$ and $y=3u+v$.

**3.6.9** Evaluate the integral $\iint\limits_D (x+y)^3 \, dA$ using change of variables,

where $D$ is the parallelogram region with sides $x+y=k_1$ and
$x-2y=k_2$ for appropriate choices of $k_1$ and $k_2$; $x=u-y$ and
$x=v+2y$.

**3.6.10** Evaluate the integral $\iint\limits_D \sin\left(\dfrac{-x+y}{x+y}\right) dA$ using change of variables,

where $D$ is the trapezoid region with vertices $(1,1),(2,2),(4,0)$, and
$(2,0)$; $x=y-u$ and $y=v-x$.

**3.6.11** Evaluate the integral $\iint\limits_D (x-y)^2 \cos^2(x+y) \, dA$ using change of

variables, where $D$ is the region by the square with vertices
$(0,1),(1,2),(2,1)$, and $(1,0)$; $x=u+y$ and $y=v-x$.

**3.6.12** Evaluate the integral $\iint\limits_D e^{\left(\frac{x-y}{x+y}\right)} \, dA$ using change of variables, where

$D$ is the region defined by $D=\big\{(x,y): x\geq 0, y\geq 0, x+y\leq 1\big\}$.

## 3.7 CHANGE TO POLAR COORDINATES

One of the most common changes of variable is the passage to polar coordinates where

$$x = \rho \cos\theta \Rightarrow dx = \cos\theta \, d\rho - \rho \sin\theta \, d\theta,$$

$$y = \rho \sin\theta \Rightarrow dy = \sin\theta \, d\rho + \rho \cos\theta \, d\theta,$$

hence

$$dx \wedge dy = \left( \rho \cos^2\theta + \rho \sin^2\theta \right) d\rho \wedge d\theta = \rho \, d\rho \wedge d\theta.$$

### EXAMPLE 3.7.1

Find $\int\limits_D xy \sqrt{x^2 + y^2} \, dA$, where

$$D = \left\{ (x,y) \in \mathbb{R}^2 \mid x \geq 0, y \geq 0, y \leq x, x^2 + y^2 \leq 1 \right\}.$$

See Figure 3.7.1.

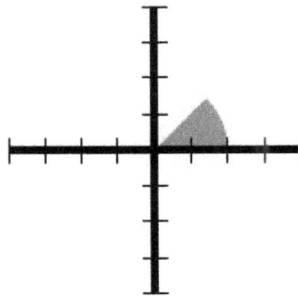

**FIGURE 3.7.1** Example 3.7.1.

### Solution:

We use polar coordinates. The region $D$ transforms into the region

$$\Delta = [0;1] \times \left[ 0; \frac{\pi}{4} \right].$$

Therefore the integral becomes

$$\int_\Delta \rho^4 \cos\theta \sin\theta \, d\rho \, d\theta = \left( \int_0^{\pi/4} \cos\theta \sin\theta \, d\theta \right) \left( \int_0^1 \rho^4 d\rho \right) = \frac{1}{20}.$$

## EXAMPLE 3.7.2

Evaluate $\int_R x \, dA$, where $R$ is the region bounded by the circles $x^2 + y^2 = 4$ and $x^2 + y^2 = 2y$. See Figure 3.7.2.

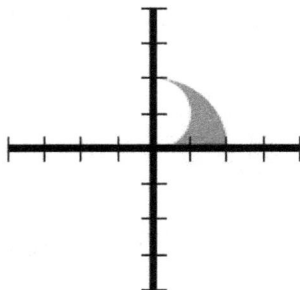

**FIGURE 3.7.2** Example 3.7.2.

## Solution:

Since $x^2 + y^2 = r^2$, the radius sweeps from $r^2 = 2r\sin\theta$ to $r^2 = 4$, that is, from $2\sin\theta$ to 2. The angle clearly sweeps from 0 to $\dfrac{\pi}{2}$. Thus the integral becomes

$$\int_R x \, dA = \int_0^{\pi/2} \int_2^{2\sin\theta} r^2 \cos\theta \, dr \, d\theta = \frac{1}{3} \int_0^{\pi/2} \left( 8\cos\theta - 8\cos\theta \sin^3\theta \right) d\theta = 2.$$

## EXAMPLE 3.7.3

Find $\int_D e^{-x^2 - xy - y^2} \, dA$, where

$$D = \left\{ (x,y) \in \mathbb{R}^2 : x^2 + xy + y^2 \le 1 \right\}.$$

See Figure 3.7.3.

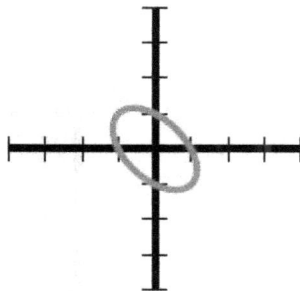

**FIGURE 3.7.3** Example 3.7.3.

### *Solution:*

Completing squares

$$x^2 + xy + y^2 = \left(x + \frac{y}{2}\right)^2 + \left(\frac{\sqrt{3}\,y}{2}\right)^2.$$

Put $U = x + \dfrac{y}{2}, V = \dfrac{\sqrt{3}\,y}{2}$. The integral becomes

$$\int_{\{x^2 + xy + y^2 \le 1\}} e^{-x^2 - xy - y^2}\, \mathrm{d}x\mathrm{d}y = \frac{2}{\sqrt{3}} \int_{\{U^2 + V^2 \le 1\}} e^{-\left(U^2 + V^2\right)}\mathrm{d}U\mathrm{d}V.$$

Passing to polar coordinates, the previous equals

$$\frac{2}{\sqrt{3}} \int_0^{2\pi} \int_0^1 \rho e^{-\rho^2}\, \mathrm{d}\rho\, \mathrm{d}\theta = \frac{2\pi}{\sqrt{3}}\left(1 - e^{-1}\right).$$

### EXAMPLE 3.7.4

Evaluate $\displaystyle\int_R \frac{1}{\left(x^2 + y^2\right)^{3/2}}\, \mathrm{d}A$, over the region

$$\left\{(x,y) \in \mathbb{R}^2 : x^2 + y^2 \le 4, y \ge 1\right\}$$

See Figure 3.7.4.

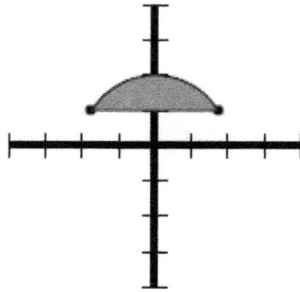

**FIGURE 3.7.4** Example 3.7.4.

### Solution:

The radius sweeps from $r = \dfrac{1}{\sin\theta}$ to $r = 2$ The desired integral is

$$\int_R \frac{1}{\left(x^2 + y^2\right)^{3/2}} \, dA = \int_{\pi/6}^{5\pi/6} \int_{\csc\theta}^{2} \frac{1}{r^2} \, dr d\theta$$

$$= \int_{\pi/6}^{5\pi/6} \left( \sin\theta - \frac{1}{2} \right) d\theta$$

$$= \sqrt{3} - \frac{\pi}{3} \cdot$$

### EXAMPLE 3.7.5

Evaluate $\int_R \left(x^3 + y^3\right) dA$, where $R$ is the region bounded by ellipse $\dfrac{x^2}{a^2} + \dfrac{y^2}{b^2} = 1$, and the first quadrant, $a > 0$ and $b > 0$.

### Solution:

Put $x = ar\cos\theta, y = br\sin\theta$. Then

$$x = ar\cos\theta \Rightarrow dx = a\cos\theta \, dr - ar\sin\theta \, d\theta,$$

$$y = br\sin\theta \Rightarrow dy = b\sin\theta \, dr + br\cos\theta \, d\theta,$$

whence

$$dx \wedge dy = \left( abr \cos^2 \theta + abr \sin^2 \theta \right) dr \wedge d\theta = abrdr \wedge d\theta.$$

Observe that on the ellipse

$$\frac{x^2}{a^2} + \frac{y^2}{b^2} = 1 \Rightarrow \frac{a^2 r^2 \cos^2 \theta}{a^2} + \frac{b^2 r^2 \sin^2 \theta}{b^2} = 1 \Rightarrow r = 1.$$

Thus the required integral is

$$\int_R \left( x^3 + y^3 \right) dA = \int_0^{\pi/2} \int_0^1 abr^4 \left( \cos^3 \theta + \sin^3 \theta \right) drd\theta$$

$$= ab \left( \int_0^1 r^4 dr \right) \left( \int_0^{\pi/2} \left( a^3 \cos^3 \theta + b^3 \sin^3 \theta \right) d\theta \right)$$

$$= ab \left( \frac{1}{5} \right) \left( \frac{2a^3 + 2b^3}{3} \right)$$

$$= \frac{2ab \left( a^3 + b^3 \right)}{15}.$$

## EXERCISES 3.7

**3.7.1**  Evaluate $\int_R xydA$ where $R$ the region is

$$R = \left\{ (x,y) \in \mathbb{R}^2 : x^2 + y^2 \leq 16, x \geq 1, y \geq 1 \right\},$$

as in the Figure 3.7.1. Set up the integral in the Cartesian and polar coordinates.

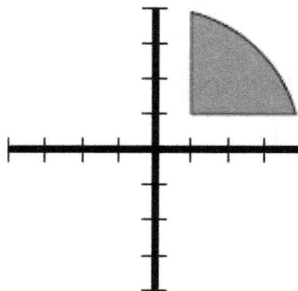

**FIGURE 3.7.5** Exercise 3.7.1.

**3.7.2** Find $\int_D \left( x^2 - y^2 \right) dA$, where

$$D = \left\{ (x,y) \in \mathbb{R}^2 \mid (x-1)^2 + y^2 \le 1 \right\}$$

**3.7.3** Find $\int_D \sqrt{xy} \, dA$, where

$$D = \left\{ (x,y) \in \mathbb{R}^2 \mid \left( x^2 + y^2 \right)^2 \le 2xy \right\}$$

**3.7.4** Find $\int_D f \, dA$, where

$D = \left\{ (x,y) \in \mathbb{R}^2 : b^2 x^2 + a^2 y^2 = a^2 b^2, (a,b) \in \, ]0; +\infty[ \text{ fixed} \right\}$ and $f(x,y) = x^3 + y^3$.

**3.7.5** Find $\int_D \sqrt{x^2 + y^2} \, dA$, where

$$D = \left\{ (x,y) \in \mathbb{R}^2 \mid x \ge 0, y \ge 0, x^2 + y^2 \le 1, x^2 + y^2 - 2y \ge 0 \right\}.$$

**3.7.6** Find $\int_D f \, dA$, where

$D = \left\{ (x,y) \in \mathbb{R}^2 \mid y \ge 0, x^2 + y^2 - 2x \le 0 \right\}$ and $f(x,y) = x^2 y$.

**3.7.7** Find $\int_D f \, dA$, where

$D = \left\{ (x,y) \in \mathbb{R}^2 \mid x \ge 1, x^2 + y^2 - 2x \le 0 \right\}$ and $f(x,y) = \dfrac{1}{\left( x^2 + y^2 \right)^2}$.

**3.7.8** Let $D = \left\{ (x,y) \in \mathbb{R}^2 : x^2 + y^2 - y \le 0, x^2 + y^2 - x \le 0 \right\}$. Find the integral $\int_D (x+y)^2 \, dA$.

**3.7.9**  Let $D = \{(x,y) \in \mathbb{R}^2 \mid y \leq x^2 + y^2 \leq 1\}$. Compute $\int_D \dfrac{dA}{\left(1 + x^2 + y^2\right)^2}$.

**3.7.10**  Evaluate $\displaystyle\int_{\{(x,y)\in\mathbb{R}^2, x\geq 0, y\geq 0, x^4+y^4\leq 1\}} x^3 y^3 \sqrt{1 - x^4 - y^4}\ dA$, using

$x^2 = \rho\cos\theta, y^2 = \rho\sin\theta$.

**3.7.11**  William Thompson (Lord Kelvin) is credited to have said: "A mathematician is someone to whom

$$\int_0^{+\infty} e^{-x^2}\ dx = \frac{\sqrt{\pi}}{2}$$

is as obvious as twice two is four to you. Liouville was a mathematician." Prove that

$$\int_0^{+\infty} e^{-x^2}\ dx = \frac{\sqrt{\pi}}{2}$$

by following these steps.

1.  Let $a > 0$ be a real number and put $D_a = \{(x,y) \in \mathbb{R}^2 \mid x^2 + y^2 \leq a^2\}$.
    Find

    $$I_a = \int_{D_a} e^{-(x^2+y^2)}\ dxdy.$$

2.  Let $a > 0$ be a real number and put $\Delta_a = \{(x,y) \in \mathbb{R}^2 \mid |x| \leq a, |y| \leq a\}$.

    Let $J_a = \int_{\Delta_a} e^{-(x^2+y^2)}\ dxdy$. Prove that

    $$I_a \leq J_a \leq I_{a\sqrt{2}}.$$

3.  Deduce that $\displaystyle\int_0^{+\infty} e^{-x^2}\ dx = \frac{\sqrt{\pi}}{2}$.

**3.7.12**  Let $D = \{(x,y) \in \mathbb{R}^2 : 4 \leq x^2 + y^2 \leq 16\}$ and $f(x,y) = \dfrac{1}{x^2 + xy + y^2}$.
Find $\displaystyle\int_D f(x,y)\ dA$.

**3.7.13**  Prove that every closed convex region in the plane of area $\geq \pi$ has two points which are two units apart.

**3.7.14**   In the $xy$-plane, if $R$ is the set of points inside and on a convex polygon, let $D(x, y)$ be the distance from $(x, y)$ to the nearest point $R$. Show that

$$\int_{-\infty}^{+\infty} \int_{-\infty}^{+\infty} e^{-D(x,y)} \, dxdy = 2\pi + L + A,$$

where $L$ is the perimeter of $R$ and $A$ is the area of $R$.

## 3.8   THREE-MANIFOLDS

**Definition 3.8.1**   A *three-dimensional oriented manifold of* $\mathbb{R}^3$ is simply an open set (body) $V \in \mathbb{R}^3$, where the + orientation is in the direction of the outward pointing normal to the body, and the − orientation is n the direction of the inward-pointing normal to the body. A general oriented three-manifold is a union of pen sets.

**NOTE**   *The region* $-M$ *has opposite orientation to* $M$ *and*

$$\int_{-M} \omega = -\int_{M} \omega.$$

*We will often write*

$$\int_{M} f dV$$

Where $dV$ denotes the volume element.

**NOTE**   *In this section, unless otherwise noticed, we will choose the positive orientation for the regions considered. This corresponds to using the volume form* $dx \wedge dy \wedge dz$.

Let $V \subseteq \mathbb{R}^3$. Given a function $f : V \to \mathbb{R}$, integral

$$\int_{V} f dV$$

is the sum of all the values of $f$ restricted to $V$. In particular,

$$\int_{V} dV$$

is the oriented volume of $V$.

**EXAMPLE 3.8.1**

Find

$$\int_{[0;1]^3} x^2 y e^{xyz}\ dV.$$

*Solution:*

The integral is

$$\int_0^1 \left( \int_0^1 \left( \int_0^1 x^2 y e^{xyz}\ dz \right) dy \right) dx = \int_0^1 \left( \int_0^1 x\left(e^{xy}-1\right) dy \right) dx$$
$$= \int_0^1 \left( e^x - x - 1 \right) dx$$
$$= e - \frac{5}{2}.$$

**EXAMPLE 3.8.2**

Find $\int_R z\ dV$ if

$$R = \left\{ (x,y) \in \mathbb{R}^3 \mid x \geq 0, y \geq 0, z \geq 0, \sqrt{x} + \sqrt{y} + \sqrt{z} \leq 1 \right\}.$$

*Solution:*

The integral is

$$\int_R z\,dx\,dy\,dz = \int_0^1 z \left( \int_0^{(1-\sqrt{z})^2} \left( \int_0^{(1-\sqrt{z}-\sqrt{x})^2} dy \right) dx \right) dz$$
$$= \int_0^1 z \left( \int_0^{(1-\sqrt{z})^2} \left( 1 - \sqrt{z} - \sqrt{x} \right)^2 dx \right) dz$$
$$= \frac{1}{6} \int_0^1 z \left( 1 - \sqrt{z} \right)^4 dz$$
$$= \frac{1}{840}.$$

## EXAMPLE 3.8.3

Prove that $\int_V x\,dV = \dfrac{a^2 bc}{24}$, where $V$ is the tetrahedron

$$V = \left\{ (x,y) \in \mathbb{R}^3 : x \geq 0, y \geq 0, z \geq 0, \frac{x}{a} + \frac{y}{b} + \frac{z}{c} \leq 1 \right\}.$$

**Solution:**

We have

$$\int_V x\,dx\,dy\,dz = \int_0^c \int_0^{b-bz/c} \int_0^{a-ay/b-az/c} x\,dx\,dy\,dz$$

$$= \frac{1}{2} \int_0^c \int_0^{b-bz/c} \left( a - \frac{ay}{b} - \frac{az}{c} \right)^2 dy\,dz$$

$$= \frac{1}{6} \int_0^c \frac{a^2 (-z+c)^3 b}{c^3}\,dx$$

$$= \frac{a^2 bc}{24}.$$

## EXAMPLE 3.8.4

Evaluate the integral, $\int_S x\,dV$ where $S$ is the is the (unoriented) tetrahedron with vertices $(0,0,0), (3,2,0), (0,3,0)$, and $(0,0,2)$. See Figure 3.8.1.

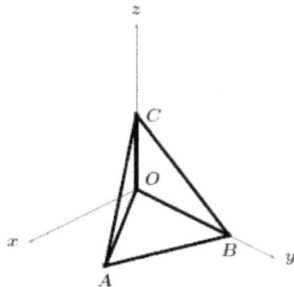

**FIGURE 3.8.1** Example 3.8.4.

### Solution:

A short computation shows that the plane passing through $(3,2,0),(0,3,0)$, and $(0,0,2)$ has equation $2x+6y+9z=18$. Hence, $0 \leq z \leq \dfrac{18-2x-6y}{9}$. We must now figure out the $xy$ limits of integration. In Figure 3.8.2, we draw the projection of the tetrahedron on the $xy$-plane. The line passing through $AB$ has equation $y = -\dfrac{x}{3}+3$. The line passing through $AC$ has equation $y = \dfrac{2}{3}x$.

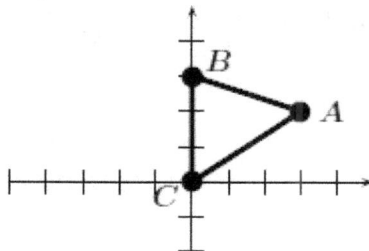

**FIGURE 3.8.2** Example 3.8.4, $xy$-projection.

We find, finally,

$$\int_S x dV = \int_0^3 \int_{2x/3}^{3-x/3} \int_0^{(18-2x-6y)/9} x\, dz\, dy\, dx$$

$$= \int_0^3 \int_{2x/3}^{3-x/3} \frac{18x - 2x^2 - 6yx}{9}\, dy\, dx$$

$$= \int_0^3 \left. \frac{18xy - 2x^2y - 3y^2x}{9}\right|_{2x/3}^{3-x/3}\, dx$$

$$= \int_0^3 \left( \frac{x^3}{3} - 2x^2 + 3x \right) dx$$

$$= \frac{9}{4}.$$

To solve this problem using $Maple^{Tm}$, you may use the following code.

```
> with(Student[VectorCalculus]) :
> int(x, [x, y, z] = Tetrahedron( ⟨0, 0, 0⟩, ⟨3, 2, 0⟩, ⟨0, 3, 0⟩, ⟨0, 0, 2⟩ ));
```

$$\frac{9}{4}$$

To solve this problem using *MATLAB*, you may use the following code.

```
>> syms x y z
>> firstans = int(int(int(x,z,0,(18-2*x-6*y)/9),y,2*x/3,3-x/3),x,0,3)
```

firstans =

9/4

## EXAMPLE 3.8.5

Evaluate $\int_R xyz \, dV$, where $R$ is the solid formed by the intersection of the parabolic cylinder $z = 4 - x^2$, the planes $z = 0, y = x, y = 0$. Use the following orders of integration:

1. $dz\,dx\,dy$
2. $dx\,dy\,dz$

## *Solution:*

We must find the projections of the solid on the coordinate planes.

1. With the order $dz\,dx\,dy$, the limits of integration of $z$ can only depend, if at all, on $x$ and $y$. Given an arbitrary point in the solid, its lowest $z$ coordinate is 0 and its highest one is on the cylinder, so the limits for z are from $z = 0$ to $z = 4 - x^2$. The projection of the solid on the $xy$-plane is the area bounded by the lines $y = x, x = 2$, and the $x$ and $y$ axes.

$$\int_0^2 \int_0^y \int_0^{4-x^2} xyz\,dz\,dx\,dy = \frac{1}{2}\int_0^2 \int_0^y xy\left(4-x^2\right)^2 dx\,dy$$

$$= \frac{1}{2}\int_0^2 \int_0^y y\left(16x - 8x^3 + x^5\right) dx\,dy$$

$$= \int_0^2 \left(4y^3 - y^5 + \frac{y^7}{12}\right) dy$$

$$= 8.$$

2. With the order $dx\,dy\,dz$, the limits of integration of $x$ can only depend, if at all, on $y$ and $z$. Given an arbitrary point in the solid, $x$ sweeps from the plane to $x = 2$, so the limits for $x$ are from $x = y$ to $x = \sqrt{4-z}$. The projection of the solid on the $yz$-plane is the area bounded by $z = 4 - y^2$, and the $z$ and $y$ axes.

$$\int_0^4 \int_0^{\sqrt{4-z}} \int_y^2 xyz\,dx\,dy\,dz = \frac{1}{2}\int_0^4 \int_0^{\sqrt{4-z}} \left(4y - y^3\right) z\,dy\,dz$$

$$= \int_0^4 \left(2z - \frac{z^3}{8}\right) dz$$

$$= 8.$$

## EXERCISES 3.8

**3.8.1**   Compute $\int_E z\,dV$, where $E$ is the region in the first octant bounded by the planes $x + z = 1$ and $y + z = 1$.

**3.8.2**   Evaluate the integrals $\int_R 1\,dV$ and $\int_R x\,dV$, where $R$ is the tetrahedron with vertices at $(0, 0, 0)$, $(1, 1, 1)$, $(1, 0, 0)$, and $(0, 0, 1)$.

**3.8.3**   Compute $\int_E x\,dV$, where $E$ is the region in the first octant bounded by the plane $y = 3z$ and the cylinder $x^2 + y^2 = 9$.

**3.8.4**   Find $\displaystyle\int_D \frac{dV}{\left(1+x^2z^2\right)\left(1+y^2z^2\right)}$ where

$$D=\left\{(x,y,z)\in\mathbb{R}^3:0\le x\le 1,0\le y\le 1,z\ge 0\right\}.$$

**3.8.5**   Write an iterated integral for $\displaystyle\iiint_S f(x,y,z)\,dV$ for the solid region

$$S=\left\{(x,y,z):0\le x\le 1,0\le y\le 3,0\le z\le\left(\frac{12-3x-2y}{6}\right)\right\}.$$

**3.8.6**   Evaluate $\displaystyle\iiint_S\left(3xy^3z^2\right)dV$ for the solid region

$$S=\left\{(x,y,z):-1\le x\le 3,1\le y\le 4,0\le z\le 2\right\}.$$

**3.8.7**   Find the volume of the solid bounded by the cylinders $y=x^2$ and $y=z^2$, and the plane $y=1$.

**3.8.8**   Find the volume of the solid bounded by the graphs $z=2,z=4y^2,x=2$, and $x=0$.

**3.8.9**   Evaluate $\displaystyle\iiint_S\left(e^{x+y+z}\right)dV$ for the solid region $S$ in $\mathbb{R}^3$ bounded by the planes $z=0,z=-x,x=0,y=1$, and $y=-x$.

**3.8.10**   Find the volume of the ellipsoidal solid
$$S=\left\{(x,y,z):4x^2+4y^2+z^2-16=0\right\}.$$

**3.8.11**   Find the volume of the centroid of the tetrahedral defined by
$$S=\left\{(x,y,z):x+y+z\le 1,y\ge 0,z\ge 0\right\}.$$

**3.8.12**   Find the volume of the region $S$ bounded by the parabolic cylinder $z=4-x^2$ and the planes $x=0,y-0,y=6$, and $z=0$.

**3.8.13**   Find the volume $V$ for the solid region $S$ in $\mathbb{R}^3$ bounded by the graphs $z=x^2+y^2,z=0,y=x,y=0$, and $x=2$.

**3.8.14**   Find the volume $V$ for the solid region $S$ in $\mathbb{R}^3$ bounded by the graphs $z=x+y,y^2=-x^2+4$, the coordinate plane, and first octant.

## 3.9   CHANGE OF VARIABLES IN TRIPLE INTEGRALS

We demonstrate in this section change of variables in three integrals through examples.

### EXAMPLE 3.9.1

Find $\int_R (x+y+z)(x+y-z)(x-y-z)dV$, where $R$ is the tetrahedron bounded by the planes $x+y+z=0, x+y-z=0, x-y-z=0$ and $2x-z=1$.

### *Solution:*

We make the charge of variables

$$u = x+y+z \Rightarrow du = dx + dy + dz,$$

$$v = x+y-z \Rightarrow dv = dx + dy - dz,$$

$$w = x-y-z \Rightarrow dw = dx - dy - dz.$$

This gives

$$du \wedge dv \wedge dw = -4dx \wedge dy \wedge dz.$$

These forms have opposite orientations, so we choose, say,

$$du \wedge dw \wedge dv = 4dx \wedge dy \wedge dz$$

which have the same orientation. Also,

$$2x - z = 1 \Rightarrow u + v + 2w = 2.$$

The tetrahedron in the $xyz$-coordinate frame is mapped into a tetrahedron bounded by $u=0, v=0, u+v+2w=1$ in the $uvw$-coordinate frame. The integral becomes

$$\frac{1}{4}\int_0^2 \int_0^{1-v/2} \int_0^{2-v-2w} uvw\, du\, dw\, dv = \frac{1}{180}.$$

Consider a transformation to cylindrical coordinates

$$(x,y,z) = (\rho \cos\theta, \rho \sin\theta, z).$$

From what we know about polar coordinates

$$dx \wedge dy = \rho \, d\rho \wedge d\theta.$$

Since the wedge product of forms is associative,

$$dx \wedge dy \wedge dz = \rho \, d\rho \wedge d\theta \wedge dz.$$

## EXAMPLE 3.9.2

Find $\int_R z^2 \, dx dy dz$, if

$$R = \left\{ (x,y,z) \in \mathbb{R}^3 \mid x^2 + y^2 \leq 1, 0 \leq z \leq 1 \right\}.$$

*Solution:*

The region of integration is mapped into $\Delta = [0; 2\pi] \times [0; 1] \times [0; 1]$ through a cylindrical coordinate change. The integral is therefore

$$\int_R f(x,y,z) \, dx dy dz = \left( \int_0^{2\pi} d\theta \right) \left( \int_0^1 \rho \, d\rho \right) \left( \int_0^1 z^2 dz \right) = \frac{\pi}{3}.$$

## EXAMPLE 3.9.3

Evaluate $\int_D (x^2 + y^2) \, dx dy dz$ over the first octant region bounded by the cylinders $x^2 + y^2 = 1$ and $x^2 + y^2 = 4$ and the planes $z = 0, z = 1, x = 0, x = y$.

*Solution:*

The integral is

$$\int_0^1 \int_{\pi/4}^{\pi/2} \int_1^2 \rho^3 d\rho \, d\theta \, dz = \frac{15\pi}{16}.$$

## EXAMPLE 3.9.4

Three long cylinders of radius $R$ intersect at right angles. Find the volume of their intersection.

### Solution:

Let $V$ be the desired volume. By symmetry, $V = 2^4 V'$, where

$$V' = \int_{D'} dx\,dy\,dz,$$

$$D' = \left\{ (x,y,z) \in \mathbb{R}^3 : 0 \le y \le x, 0 \le z, x^2 + y^2 \le R^2, y^2 + z^2 \le R^2, z^2 + x^2 \le R^2 \right\}.$$

In this case, it is easier to integrate with respect to $z$ first. Using cylindrical coordinates

$$\Delta' = \left\{ (\theta, \rho, z) \in \left[ 0; \frac{\pi}{4} \right] \times [0; R] \times [0; +\infty], 0 \le z \le \sqrt{R^2 - \rho^2 co^2\theta} \right\}.$$

Now,

$$V' = \int_0^{\pi/4} \left( \int_0^R \left( \int_0^{\sqrt{R^2 - \rho^2 \cos^2\theta}} dz \right) \rho\,d\rho \right) d\theta$$

$$= \int_0^{\pi/4} \left( \int_0^R \rho \sqrt{R^2 - \rho^2 \cos^2\theta} \; d\rho \right) d\theta$$

$$= \int_0^{\pi/4} -\frac{1}{3\cos^2\theta} \left[ \left( R^2 - \rho^2 \cos^2\theta \right)^{3/2} \right]_0^R d\theta$$

$$= \frac{R^3}{3} \int_0^{\pi/4} \frac{1 - \sin^3\theta}{\cos^2\theta} \; d\theta, \text{ now let } u = \cos\theta$$

$$= \frac{R^3}{3} \left( \left[ \tan\theta \right]_0^{\pi/4} + \int_1^{\frac{\sqrt{2}}{2}} \frac{1 - u^2}{u^2} \; du \right)$$

$$= \frac{R^3}{3} \left( 1 - \left[ u^{-1} + u \right]_1^{\frac{\sqrt{2}}{2}} \right)$$

$$= \frac{\sqrt{2} - 1}{\sqrt{2}} R^3.$$

Finally,

$$V = 16V' = 8\left(2 - \sqrt{2}\right)R^3.$$

Consider now a change to spherical coordinates

$$x = \rho \cos\theta \sin\phi, \quad y = \rho \sin\theta \sin\phi, \quad z = \rho \cos\phi.$$

We have

$$\mathrm{d}x = \cos\theta \sin\phi \mathrm{d}\rho - \rho \sin\theta \sin\phi \mathrm{d}\theta + \rho \cos\theta \cos\phi \mathrm{d}\phi,$$

$$\mathrm{d}y = \sin\theta \sin\phi \mathrm{d}\rho + \rho \cos\theta \sin\phi \mathrm{d}\theta + \rho \sin\theta \cos\phi \mathrm{d}\phi,$$

$$\mathrm{d}z = \cos\phi \mathrm{d}\rho - \rho \sin\phi \mathrm{d}\phi.$$

This gives

$$\mathrm{d}x \wedge \mathrm{d}y \wedge \mathrm{d}z = -\rho^2 \sin\phi \mathrm{d}\rho \wedge \mathrm{d}\theta \wedge \mathrm{d}\phi.$$

From this derivation, the form $\mathrm{d}\rho \wedge \mathrm{d}\theta \wedge \mathrm{d}\phi$ is negatively oriented, and so we choose

$$\mathrm{d}x \wedge \mathrm{d}y \wedge \mathrm{d}z = \rho^2 \sin\phi \mathrm{d}\rho \wedge \mathrm{d}\phi \wedge \mathrm{d}\theta$$

instead.

## EXAMPLE 3.9.5

Let $(a,b,c) \in \left]0; +\infty\right[^3$ be fixed. Find $\int_R xyz\mathrm{d}V$ if

$$R = \left\{ (x,y,z) \in \mathbb{R}^3 : \frac{x^2}{a^2} + \frac{y^2}{b^2} + \frac{z^2}{c^2} \le 1, x \ge 0, y \ge 0, z \ge 0 \right\}.$$

***Solution:***

We use spherical coordinates, where

$$(x,y,z) = (a\rho \cos\theta \sin\phi, b\rho \sin\theta \sin\phi, c\rho \cos\phi).$$

We have

$$\mathrm{d}x \wedge \mathrm{d}y \wedge \mathrm{d}z = abc\rho^2 \sin\phi \mathrm{d}\rho \wedge \mathrm{d}\phi \wedge \mathrm{d}\rho.$$

The integration region is mapped into

$$\Delta = [0;1] \times \left[0; \frac{\pi}{2}\right] \times \left[0; \frac{\pi}{2}\right].$$

The integral becomes

$$(abc)^2 \left(\int_0^{\pi/2} \cos\theta \, \sin\theta \, d\theta\right) \left(\int_0^1 \rho^5 d\rho\right) \left(\int_0^{\pi/2} \cos^3\phi \sin\phi \, d\phi\right) = \frac{(abc)^2}{48}.$$

**EXAMPLE 3.9.6**

Let $V = \left\{(x,y,z) \in \mathbb{R}^3 : x^2 + y^2 + z^2 \leq 9, 1 \leq z \leq 2\right\}$. Then

$$\int_V dx \, dy \, dz = \int_0^{2\pi} \int_{\pi/2-\arcsin 2/3}^{\pi/2-\arcsin 1/3} \int_{1/\cos\phi}^{2/\cos\phi} \rho^2 \sin\phi \, d\rho \, d\phi \, d\theta = \frac{63\pi}{4}.$$

# EXERCISES 3.9

**3.9.1**   Consider the region $R$ below the cone $z = \sqrt{x^2 + y^2}$ and above the paraboloid $z = x^2 + y^2$ for $0 \leq z \leq 1$. Set up integrals for the volume of this region in Cartesian, cylindrical, and spherical coordinates. Also, find this volume.

**3.9.2**   Consider the integral $\int_R x \, dV$, where $R$ is the region above the paraboloid $z = x^2 + y^2$ and under the sphere $x^2 + y^2 + z^2 = 4$. Set up integrals for the volume of this region in Cartesian, cylindrical, and spherical coordinates. Also, find this volume.

**3.9.3**   Consider the region R bounded by the sphere $x^2 + y^2 + z^2 = 4$ and the plane $z = 1$. Set up integrals for the volume of this region in Cartesian, cylindrical, and spherical coordinates. Also, find this volume.

**3.9.4**   Compute $\int_E y \, dV$ where $E$ is the region between the cylinders $x^2 + y^2 = 1$ and $x^2 + y^2 = 4$, below the plane $x - z = -2$ and above the $xy$-plane.

**3.9.5**    Compute $\int_E y^2 z^2 dV$ where $E$ is bounded by the paraboloid $x = 1 - y^2 - z^2$ and the plane $x = 0$.

**3.9.6**    Compute $\int_E z\sqrt{x^2 + y^2 + z^2}\, dV$ where $E$ is the upper solid hemisphere bounded by the $xy$-plane and the sphere of radius 1 about the origin.

**3.9.7**    Compute the four-dimentional integral
$$\iiint\limits_{x^2+y^2+u^2+v^2 \le 1} e^{x^2+y^2+u^2+v^2}\, dx\, dy\, du\, dv.$$

**3.9.8**    Find $\int\limits_R x^1 y^9 z^8 \left(1 - x - y - z\right)^4 dx\, dy\, dz$, where
$$R = \left\{ (x,y,z) \in \mathbb{R}^3 : x \ge 0, y \ge 0, z \ge 0, x + y + z \le 1 \right\}.$$

**3.9.9**    Find the volume for the unit ball region $S$ defined by
$$S = \left\{ (x,y,z) : x^2 + y^2 + z^2 \le 1 \right\}.$$

**3.9.10**    Evaluate $\iiint\limits_S \left(x^2 + z^2\right)^{\frac{1}{2}} dV$ for the solid region $S$, which is bounded by the paraboloid $y = z^2 + x^2$ and the plane $y = 4$.

**3.9.11**    Find the volume $V$ for the solid region $S$ in $\mathbb{R}^3$ bounded by the graphs $z = -x^2 - y^2 + 10$, and $z = 1$.

**3.9.12**    Find the volume $V$ for the solid region $S$ in $\mathbb{R}^3$ bounded by the graphs $z^2 = -x^2 + y$ and $z^2 = -x^2 + 2y - 4$.

**3.9.13**    Find the volume $V$ for the solid region $S$ in $\mathbb{R}^3$ bounded by the graphs $z = 0, z^2 = -x^2 - y^2 + 4, y = x, y = \sqrt{3}x$, and first octant.

**3.9.14**    Find the volume $V$ for the solid region $S$ in $\mathbb{R}^3$ bounded by the graphs inside $z^2 = -x^2 - y^2 + 1$ and outside $z^2 - x^2 - y^2 = 0$.

## 3.10 SURFACE INTEGRALS

**Definition 3.10.1** A *two-dimensional oriented manifold of* $\mathbb{R}^3$ is simply a smooth surface $D \in \mathbb{R}^3$, where the + orientation is in the direction of the outward normal pointing away from the origin and the − orientation is in the direction of the inward normal pointing toward the origin. A general oriented 2-manifold in $\mathbb{R}^3$ is a union of surfaces.

**NOTE** *The surface* $-\Sigma$ *has opposite orientation on* $\Sigma$ *and*

$$\int_{-\Sigma} \omega = -\int_{\Sigma} \omega.$$

**NOTE** *In this section, unless otherwise noticed, we will choose the positive orientation for the regions considered. This corresponds to using the ordered basis*

$$\{dy \wedge dz, dz \wedge dx, dx \wedge dy\}.$$

**Definition 3.10.2** Let $f : \mathbb{R}^3 \to \mathbb{R}$. The integral of $f$ over the smooth surface $\Sigma$ (oriented in the positive sense) is given by expression

$$\int_{\Sigma} f \|d^2\mathbf{x}\|.$$

Here,

$$\|d^2\mathbf{x}\| = \sqrt{(dx \wedge dy)^2 + (dz \wedge dx)^2 + (dy \wedge dz)^2}$$

is the surface area element.

### EXAMPLE 3.10.1

Evaluate $\int_{\Sigma} z \|d^2\mathbf{x}\|$ where $\Sigma$ is the outer surface of the section of the paraboloid, $z = x^2 + y^2, 0 \le z \le 1$.

*Solution:*

We parameterize the paraboloid as follows. Let $x = u, y = v, z = u^2 + v^2$. Observe that the domain $D$ of $\Sigma$ is the unit disk $u^2 + v^2 \le 1$. We see that

$$dx \wedge dy = du \wedge dv,$$

$$dy \wedge dz = -2u du \wedge dv,$$

$$dz \wedge dx = -2v du \wedge dv,$$

and so

$$\left\| d^2 x \right\| = \sqrt{\left( dx \wedge dy \right)^2 + \left( dz \wedge dx \right)^2 + \left( dy \wedge dz \right)^2}$$

$$= \sqrt{1 + 4u^2 + 4v^2}\, du \wedge dv.$$

Now,

$$\int_\Sigma z \left\| d^2 x \right\| = \int_D \left( u^2 + v^2 \right) \sqrt{1 + 4u^2 + 4v^2}\, du dv.$$

To evaluate this last integral, we use polar coordinates and so

$$\int_D \left( u^2 + v^2 \right) \sqrt{1 + 4u^2 + 4v^2}\, du dv = \int_0^{2\pi} \int_0^1 \rho^3 \sqrt{1 + 4\rho^2}\, d\rho\, d\theta$$

$$= \frac{\pi}{12} \left( 5\sqrt{5} + \frac{1}{5} \right).$$

### EXAMPLE 3.10.2

Find the area of that part of the cylinder $x^2 + y^2 = 2y$ lying inside the sphere $x^2 + y^2 + z^2 = 4$.

### *Solution:*

We have $x^2 + y^2 = 2y \Leftrightarrow x^2 + (y-1)^2 = 1$. We parameterize the cylinder by putting $x = \cos u, y - 1 = \sin u$, and $z = v$. Hence

$$dx = -\sin u du,\ dy = \cos u du,\ dz = dv,$$

whence

$$dx \wedge dy = 0,\ dy \wedge dz = \cos u du \wedge dv, dz \wedge dx = \sin u du \wedge dv,$$

and so

$$\left\| d^2 \mathbf{x} \right\| = \sqrt{\left( dx \wedge dy \right)^2 + \left( dz \wedge dx \right)^2 + \left( dy \wedge dz \right)^2}$$
$$= \sqrt{\cos^2 u + \sin^2 u} \ du \wedge dv$$
$$= du \wedge dv.$$

The cylinder and the sphere intersect when $x^2 + y^2 = 2y$ and $x^2 + y^2 + z^2 = 4$, that is, when $z^2 = 4 - 2y$, i.e. $v^2 = 4 - 2(1 + \sin u) = 2 - 2\sin u$. Also, $0 \le u \le \pi$. The integral is thus

$$\int_{\Sigma} \left\| d^2 \mathbf{x} \right\| = \int_0^\pi \int_{-\sqrt{2-2\sin u}}^{\sqrt{2-2\sin u}} dv du = \int_0^\pi 2\sqrt{2 - 2\sin u} \ du$$

$$= 2\sqrt{2} \int_0^\pi \sqrt{1 - \sin u} \ du$$

$$= 2\sqrt{2} \left( 4\sqrt{2} - 4 \right).$$

## EXAMPLE 3.10.3

Evaluate $\int_{\Sigma} x dy dz + \left( z^2 - zx \right) dz dx - xy dx dy$, where $\Sigma$ is the top side of the triangle with vertices at $(2,0,0), (0,2,0), (0,0,4)$.

## *Solution:*

Observe that the plane passing through the three given points has equation $2x + 2y + z = 4$. We project this plane onto the coordinate axes obtaining

$$\int_{\Sigma} x dy dz = \int_0^4 \int_0^{2-z/2} \left( 2 - y - z/2 \right) dy dz = \frac{8}{3},$$

$$\int_{\Sigma} \left( z^2 - zx \right) dz dx = \int_0^2 \int_0^{4-2x} \left( z^2 - zx \right) dz dx = 8,$$

$$-\int_{\Sigma} xy dx dy = -\int_0^2 \int_0^{2-y} xy dx dy = -\frac{2}{3},$$

and hence,

$$\int_{\Sigma} x\,dy\,dz + \left(z^2 - zx\right)dz\,dx - xy\,dx\,dy = 10.$$

## EXERCISES 3.10

**3.10.1**    Evaluate $\int_{\Sigma} y\left\|d^2\mathbf{x}\right\|$ where $\Sigma$ is the surface

$z = x + y^2, 0 \le x \le 1, 0 \le y \le 2.$

**3.10.2**    Consider the cone $z = \sqrt{x^2 + y^2}$. Find the surface area of the part of the cone which lies between the planes $z = 1$ and $z = 2$.

**3.10.3**    Evaluate $\int_{\Sigma} x^2 \left\|d^2\mathbf{x}\right\|$ where $\Sigma$ is the surface of the unit sphere

$x^2 + y^2 + z^2 = 1.$

**3.10.4**    Evaluate $\int_S z\left\|d^2\mathbf{x}\right\|$ over the conical surface $z = \sqrt{x^2 + y^2}$ between

$z = 0$ and $z = 1$.

**3.10.5**    You put a perfectly spherical egg through an egg slicer, resulting in $n$ slices of identical height, but you forgot to peel it first! Show that the amount of egg shell in any of the slices is the same. Your argument must use surface integrals.

**3.10.6**    Evaluate $\int_{\Sigma} xy\,dy\,dz - x^2\,dz\,dx + (x+z)\,dx\,dy$, where $\Sigma$ is the top of the triangular region of the plane $2x + 2y + z = 6$, bounded by the first octant.

**3.10.7**    Find the surface area of the part of conical $z^2 = x^2 + y^2$ that is directly over the triangle in the $xy$-plane with vertices $(0,0), (4,0)$, and $(0,4)$.

**3.10.8**    Find the surface area of the part of paraboloid $z = x^2 + y^2$ that lies directly under the plane $z = 9$.

**3.10.9** Find the surface integral $\iint\limits_{S}(x+y+z)\,dS$, where $S$ is the portion of the sphere $x^2+y^2+z^2=1$, that lies in the first octant using spherical polar coordinates.

**3.10.10** Find the surface area of the part of the plane $2x+3y+4z-12=0$ that lies directly above the region in the first octant bounded by the graph $\sin 2\theta = r$.

**3.10.11** Find the surface area of the part of the graph of $z-x^2+y^2=0$ that lies directly in the first octant within the cylinder $x^2+y^2-4=0$.

**3.10.12** Find the surface area of the part of the graph of $4z^2-x^2-y^2=0$ that lies directly within the cylinder $(x-1)^2+y^2-1=0$.

## 3.11 GREEN'S, STOKES', AND GAUSS' THEOREMS

We are now in a position to state the general Stokes' Theorem in this section.

**Theorem 3.11.1 (General Stokes' Theorem):** Let $M$ be a smooth oriented manifold, having boundary $\partial M$. If $\omega$ is a differential form, then

$$\int_{\partial M}\omega = \int_{M}d\omega.$$

In $\mathbb{R}^2$, if $\omega$ is a 1-form, this takes the name of *Green's Theorem*.

**EXAMPLE 3.11.1**

Evaluate $\oint_{C}(x-y^3)\,dx+x^3dy$ where $C$ is the circle $x^2+y^2=1$.

**Solution:**

We will first use *Green's Theorem* and then evaluate the integral directly. We have

$$\begin{aligned}
d\omega &= d\left(x-y^3\right)\wedge dx + d\left(x^3\right)\wedge dy\\
&= \left(dx-3y^2dy\right)\wedge dx + \left(3x^2dx\right)\wedge dy\\
&= \left(3y^2+3x^2\right)dx\wedge dy.
\end{aligned}$$

The region $M$ is the area enclosed by the circle $x^2 + y^2 = 1$. Thus by *Green's Theorem*, and using polar coordinates,

$$\oint_C (x - y^3) \, dx + x^3 dy = \int_M (3y^2 + 3x^2) \, dxdy$$

$$= \int_0^{2\pi} \int_0^1 3\rho^2 d\rho \, d\theta$$

$$= \frac{3\pi}{2}.$$

Alternative Method:

We can evaluate this integral directly, again resorting to polar coordinates.

$$\oint_C (x - y^3) \, dx + x^3 dy = \int_0^{2\pi} (\cos\theta - \sin^3\theta)(-\sin\theta) \, d\theta + (\cos^3\theta)(\cos\theta) \, d\theta$$

$$= \int_0^{2\pi} (\sin^4\theta + \cos^4\theta - \sin\theta \cos\theta) \, d\theta.$$

To evaluate the last integral, observe that $1 = (\sin^2\theta + \cos^2\theta)^2 = \sin^4\theta + 2\sin^2\theta \cos^2\theta + \cos^4\theta$, hence the integral equals

$$\int_0^{2\pi} (\sin^4\theta + \cos^4\theta - \sin\theta \cos\theta) \, d\theta = \int_0^{2\pi} (1 - 2\sin^2\theta \cos^2\theta - \sin\theta \cos\theta) \, d\theta$$

$$= \frac{3\pi}{2}.$$

In general, let

$$\omega = f(x,y) \, dx + g(x,y) \, dy$$

be a 1-form in $\mathbb{R}^2$. Then

$$d\omega = df(x,y) \wedge dx + dg(x,y) \wedge dy$$

$$= \left( \frac{\partial}{\partial x} f(x,y) dx + \frac{\partial}{\partial y} f(x,y) dy \right) \wedge dx + \left( \frac{\partial}{\partial x} g(x,y) dx + \frac{\partial}{\partial y} g(x,y) dy \right) \wedge dy$$

$$= \left( \frac{\partial}{\partial x} g(x,y) - \frac{\partial}{\partial y} f(x,y) \right) dx \wedge dy$$

which gives the classical Green's Theorem

$$\int_{\partial M} f(x,y)\, dx + g(x,y) dy = \int_M \left( \frac{\partial}{\partial x} g(x,y) - \frac{\partial}{\partial y} f(x,y) \right) dx dy.$$

In $\mathbb{R}^3$, if $\omega$ is a 2-form, the above theorem takes the name of *Gauss's Theorem* or the *divergence Theorem*.

### EXAMPLE 3.11.2

Evaluate $\int_S (x-y)\, dy dz + z dz dx - y dx dy$ where $S$ is the surface of the sphere $x^2 + y^2 + z^2 = 9$ and the positive direction is the outward normal.

### *Solution:*

The region $M$ is the interior of sphere $x^2 + y^2 + z^2 = 9$. Now,

$$d\omega = (dx - dy) \wedge dy \wedge dz + dz \wedge dz \wedge dx - dy \wedge dx \wedge dy$$

$$= dx \wedge dy \wedge dz.$$

The integral becomes

$$\int_M dx dy dz = \frac{4\pi}{3}(27) = 36\pi.$$

Alternative Method:

We could evaluate this integral directly, we have

$$\int_\Sigma (x-y)\,dy\,dz = \int_\Sigma x\,dy\,dz,$$

since $(x,y,z) \mapsto -y$ is an odd function of $y$ and the domain of integration is symmetric with respect to $y$. Now

$$\int_\Sigma x\,dy\,dz = \int_{-3}^{3}\int_0^{2\pi} |\rho|\sqrt{9-\rho^2}\,d\rho\,d\theta = 36\pi.$$

Also,

$$\int_\Sigma z\,dz\,dx = 0,$$

since $(x,y,z) \mapsto z$ is an odd function of $z$ and the domain of integration is symmetric with respect to $z$. Similarly,

$$\int_\Sigma -y\,dx\,dy = 0,$$

Since $(x,y,z) \mapsto -y$ is an odd function of $y$ and the domain of integration is symmetric with respect to $y$. Now,

In general, let

$$\omega = f(x,y,z)\,dy \wedge dz + g(x,y,z)\,dz \wedge dx + h(x,y,z)\,dx \wedge dy$$

be a 2-form in $\mathbb{R}^3$. Then

$$\begin{aligned}
d\omega &= df(x,y,z)\,dy \wedge dz + dg(x,y,z)\,dz \wedge dx + dh(x,y,z)\,dx \wedge dy \\
&= \left(\frac{\partial}{\partial x}f(x,y,z)\,dx + \frac{\partial}{\partial y}f(x,y,z)\,dy + \frac{\partial}{\partial z}f(x,y,z)\,dz\right) \wedge dy \wedge dz \\
&\quad + \left(\frac{\partial}{\partial x}g(x,y,z)\,dx + \frac{\partial}{\partial y}g(x,y,z)\,dy + \frac{\partial}{\partial z}g(x,y,z)\,dz\right) \wedge dz \wedge dx \\
&\quad + \left(\frac{\partial}{\partial x}h(x,y,z)\,dx + \frac{\partial}{\partial y}h(x,y,z)\,dy + \frac{\partial}{\partial z}h(x,y,z)\,dz\right) \wedge dx \wedge dy \\
&= \left(\frac{\partial}{\partial x}f(x,y,z) + \frac{\partial}{\partial y}g(x,y,z) + \frac{\partial}{\partial z}h(x,y,z)\right) dx \wedge dy \wedge dz,
\end{aligned}$$

which gives the classical Gauss's Theorem

$$\int_{\partial M} f(x,y,z)\,\mathrm{d}y\mathrm{d}z + g(x,y,z)\mathrm{d}z\mathrm{d}x + h(x,y,z)\mathrm{d}x\mathrm{d}y$$

$$= \int_M \left( \frac{\partial}{\partial x} f(x,y,z) + \frac{\partial}{\partial y} g(x,y,z) + \frac{\partial}{\partial z} h(x,y,z) \right) \mathrm{d}x\mathrm{d}y\mathrm{d}z.$$

Using classical notation, if

$$\vec{a} = \begin{bmatrix} f(x,y,z) \\ g(x,y,z) \\ h(x,y,z) \end{bmatrix}, \quad \mathrm{d}\vec{S} = \begin{bmatrix} \mathrm{d}y\mathrm{d}z \\ \mathrm{d}z\mathrm{d}x \\ \mathrm{d}x\mathrm{d}y \end{bmatrix},$$

then

$$\int_M (\nabla \cdot \vec{a})\,\mathrm{d}V = \int_{\partial M} \vec{a} \cdot \mathrm{d}\vec{S}.$$

The classical Strokes' Theorem occurs when $\omega$ is a 1-form in $\mathbb{R}^3$.

## EXAMPLE 3.11.3

Evaluate $\oint_C y\mathrm{d}x + (2x-z)\,\mathrm{d}y + (z-x)\mathrm{d}z$ where $C$ is the intersection of the sphere $x^2 + y^2 + z^2 = 4$ and the plane $z = 1$.

### Solution:

We have

$$\begin{aligned} \mathrm{d}\omega &= (\mathrm{d}y) \wedge \mathrm{d}x + (2\mathrm{d}x - \mathrm{d}z) \wedge \mathrm{d}y + (\mathrm{d}z - \mathrm{d}x) \wedge \mathrm{d}z \\ &= -\mathrm{d}x \wedge \mathrm{d}y + 2\mathrm{d}x \wedge \mathrm{d}y + \mathrm{d}y \wedge \mathrm{d}z + \mathrm{d}z \wedge \mathrm{d}x \\ &= \mathrm{d}x \wedge \mathrm{d}y + \mathrm{d}y \wedge \mathrm{d}z + \mathrm{d}z \wedge \mathrm{d}x. \end{aligned}$$

Since on $C$, $z = 1$, the surface $\Sigma$ on which we are integrating is the inside of the circle $x^2 + y^2 + 1 = 4$, i.e., $x^2 + y^2 = 3$. Also, $z = 1$ implies $\mathrm{d}z = 0$ and so

$$\int_\Sigma \mathrm{d}\omega = \int_\Sigma \mathrm{d}x\mathrm{d}y.$$

Since this is just the area of the circular region $x^2 + y^2 \leq 3$, the integral evaluates to

$$\int_{\Sigma} dx\, dy = 3\pi.$$

In general, let

$$\omega = f(x,y,z)\, dx + g(x,y,z)\, dy + h(x,y,z)\, dz$$

Be a 1-form in $\mathbb{R}^3$. Then

$$d\omega = df(x,y,z) \wedge dx + dg(x,y,z) \wedge dy + dh(x,y,z) \wedge dz$$

$$= \left( \frac{\partial}{\partial x} f(x,y,z)\, dx + \frac{\partial}{\partial y} f(x,y,z)\, dy + \frac{\partial}{\partial z} f(x,y,z)\, dz \right) \wedge dx$$

$$+ \left( \frac{\partial}{\partial x} g(x,y,z)\, dx + \frac{\partial}{\partial y} g(x,y,z)\, dy + \frac{\partial}{\partial z} g(x,y,z)\, dz \right) \wedge dy$$

$$+ \left( \frac{\partial}{\partial x} h(x,y,z)\, dx + \frac{\partial}{\partial y} h(x,y,z)\, dy + \frac{\partial}{\partial z} h(x,y,z)\, dz \right) \wedge dz$$

$$= \left( \frac{\partial}{\partial y} h(x,y,z) - \frac{\partial}{\partial z} g(x,y,z) \right) dy \wedge dz$$

$$+ \left( \frac{\partial}{\partial z} f(x,y,z) - \frac{\partial}{\partial x} h(x,y,z) \right) dz \wedge dx$$

$$+ \left( \frac{\partial}{\partial x} g(x,y,z) - \frac{\partial}{\partial y} f(x,y,z) \right) dx \wedge dy$$

which gives the classical Strokes' Theorem.

$$\int_{\partial M} f(x,y,z)\,dx + g(x,y,z)dy + h(x,y,z)dz$$

$$= \int_M \left( \frac{\partial}{\partial y} h(x,y,z) - \frac{\partial}{\partial z} g(x,y,z) \right) dydz$$

$$+ \left( \frac{\partial}{\partial z} g(x,y,z) - \frac{\partial}{\partial x} f(x,y,z) \right) dxdy$$

$$+ \left( \frac{\partial}{\partial x} h(x,y,z) - \frac{\partial}{\partial y} f(x,y,z) \right) dxdy.$$

Using classical notation, if

$$\vec{a} = \begin{bmatrix} f(x,y,z) \\ g(x,y,z) \\ h(x,y,z) \end{bmatrix}, \ d\vec{r} = \begin{bmatrix} dx \\ dy \\ dz \end{bmatrix}, \ d\vec{S} = \begin{bmatrix} dydz \\ dzdx \\ dxdy \end{bmatrix},$$

then

$$\int_M (\nabla \times \vec{a}) \cdot d\vec{S} = \int_{\partial M} \vec{a} \cdot d\vec{r}.$$

## EXERCISES 3.11

**3.11.1** Evaluate $\oint_C x^3 y\,dx + xy\,dy$ where $C$ is the square with vertices at $(0,0),(2,0),(2,2)$, and $(0,2)$.

**3.11.2** Consider the triangle $\Delta$ with vertices $A:(0,0), B:(1,1), C:(-2,2)$.

1. If $L_{PQ}$ denotes the equation of the line joining $P$ and $Q$, find $L_{AB}, L_{AC}$, and $L_{BC}$.

2. Evaluate $\oint_\Delta y^2 dx + x\,dy$.

3. Find $\int_D (1-2y)\,dx \wedge dy$ where $D$ is the interior of $\Delta$.

**3.11.3**   Find $\oint_\Gamma f(\vec{a})\cdot d\vec{a}$, where $\Gamma$ is the boundary of the rectangle
$D = \{(x,y): x \in [0,2], y \in [0,1]\}$ when $f(\vec{a}) = 5x^3\vec{i} - 6xy\vec{j}$.

**3.11.4**   Prove that $\int_C \vec{f}\cdot d\vec{r} = 3\pi$, where $\vec{f}(x,y,z) = z\vec{i} + x\vec{j} + y\vec{k}$ and $C$ is the
curve $\vec{r} = \cos t\vec{i} + \sin t\vec{j} + t\vec{k}$ when $t \in [0,2\pi]$.

**3.11.5**   Use Green's Theorem to prove that $\int_\Gamma (x^2 + 2y^3)\,dy = 16\pi$, where $\Gamma$
is the circle $(x-2)^2 + y^2 = 4$. Also, prove this directly by using a
path integral.

**3.11.6**   Let $\Gamma$ denote the curve of intersection of the plane $x+y = 2$ and the
sphere $x^2 - 2x + y^2 - 2y + z^2 = 0$, oriented clockwise when viewed
from the origin. Use Stokes' Theorem to prove that
$\int_\Gamma y\,dx + z\,dy + x\,dz = -2\pi\sqrt{2}$. Prove this directly by parametrizing the
boundary of the surface and evaluating the path integral.

**3.11.7**   Let $M \subset \mathbb{R}^2$ be the upper semi-disk of radius $R$. Find
$\int_{\partial M} (x^2 dx + 2xy\,dy)$ using Green's Theorem.

**3.11.8**   Evaluate $\oint \vec{f}\cdot d\vec{r}$ using Green's Theorem, where
$\vec{f}(x,y,z) = (2xy - x^2)\vec{i} + (x + y^2)\vec{j}$ and $C$ is the boundary of the
region $\Gamma$ defined by the curves $y = x^2$ and $y = \sqrt{x}$, $0 \le x \le 1$.

**3.11.9**   Evaluate $\oint_C 4x^2y\,dx + 2y\,dy$ using Green's Theorem, where $C$ is
the boundary of the triangle with vertices in Figure 3.11.1.

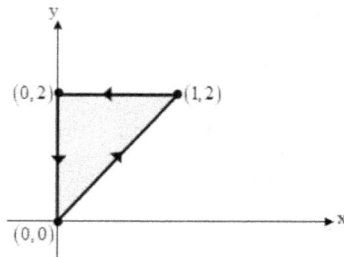

**Figure 3.11.1** Exercise 3.11.7.

**3.11.10** Evaluate $\oint_C \left(x+y^2\right) dx + \left(1+x^2\right) dy$ using Green's Theorem, where $C$ is the closed curve determined by $y = x^3$ and $y = x^2$ form $(0,0)$ to $(1,1)$.

**3.11.11** Evaluate $\oint_C e^x \sin y\ dx + e^x \cos y\ dy$ using Green's Theorem, where $C$ is the ellipse $3x^2 + 8y^2 = 24$.

**3.11.12** Evaluate $\oint_C \left(y - \sin x\right) dx + \cos x\ dy$ using Green's Theorem in the plane, where $C$ is the triangle of the Figure 3.11.2.

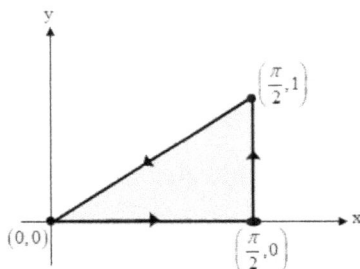

**Figure 3.11.2** Exercise 3.11.10.

**3.11.13** Evaluate $\oint_C \left(2xy - x^2\right) dx + \left(x+y^2\right) dy$ using Green's Theorem in the plane, where $C$ is the closed curve of the region bounded by $y = x^2$ and $y^2 = x$ intersect form $(0,0)$ to $(1,1)$.

**3.11.14** Evaluate $\oint_C \left(\left(2x^2 + 2xy\right) dx + \left(x^2 + xy + y^2\right) dy\right)$ using Green's Theorem in the plane, where $C$ is the square with vertices $(0,0),(1,0),(1,1),$ and $(0,1)$, as in Figure 3.11.3.

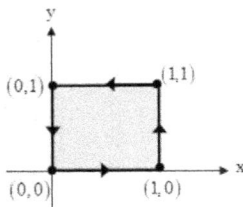

**Figure 3.11.3** Exercise 3.11.12.

**3.11.15** Evaluate $\oint_C (x+y^2)dx+(2x^2-y)dy$ using Green's Theorem, where $C$ is the Boundary of the region determined by the graphs of $y=x^2$ and $y=4$.

**3.11.16** Evaluate $\oint_C e^{x^2}dx+2\arctan x \, dy$ using Green's Theorem, where $C$ is the Triangle with vertices $(0,0),(0,1)$, and $(-1,1)$.

**3.11.17** Evaluate $\oint_C xy^2dx+3\cos y \, dy$ using Green's Theorem, where $C$ is the boundary of the region in the first quadrant determined by the graphs of $y=x^2$ and $y=x^3$.

**3.11.18** Evaluate $\oint_C y^3dx+(x^3+3xy^2) \, dy$ using Green's Theorem, where $C$ is the path from $(0,0)$ to $(1,1)$ along the graph $y=x^3$ and from $(1,1)$ to $(0,0)$ along the graph of $y=x$.

**3.11.19** Use Gauss's Theorem to find the outward flux of the vector field $\vec{f}=\begin{bmatrix} x^2 \\ 2yz \\ 4z^3 \end{bmatrix}$ across the region $M$ bounded by the parallelepiped

$$0\le x\le 1, \quad 0\le y\le 2, \quad 0\le z\le 3.$$

**3.11.20** Use Gauss's Theorem to find the flux of the vector field $\vec{f}=\begin{bmatrix} x^2y \\ 2xz \\ yz^3 \end{bmatrix}$ across the surface of the rectangle solid $M$ as in Figure 3.11.4, where

$$0\le x\le 1, \quad 0\le y\le 2, \quad 0\le z\le 3$$

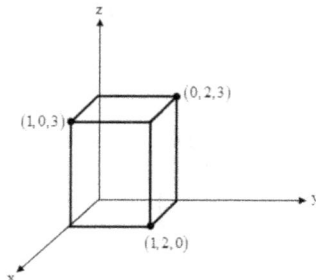

**Figure 3.11.4** Exercise 3.11.18.

**3.11.21** Use Gauss's Theorem to find the outward flux of the vector field

$$\vec{f} = \begin{bmatrix} 4x \\ y \\ 4z \end{bmatrix} \text{ across the region } M \text{ bounded by the sphere}$$

$$x^2 + y^2 + z^2 = 4.$$

**3.11.22** Use Gauss's Theorem to find the outward flux of the vector field

$$\vec{f} = \begin{bmatrix} xy^2 \\ x^2 y \\ 6\sin x \end{bmatrix} \text{ across the region } M \text{ bounded by the cone } z = \sqrt{x^2 + y^2}$$

and the planes $z = 2$ and $z = 4$.

**3.11.23** Use Gauss's Theorem to find the outward flux of the vector field

$$\vec{f} = \begin{bmatrix} 4xz \\ -y^2 \\ yz \end{bmatrix} \text{ across the region } M \text{ of the cube bounded by}$$

$x = 0, x = 1, y = 0, y = 1, z = 0,$ and $z = 1.$

**3.11.24** Use Gauss's Theorem to find the outward flux of the vector field

$$\vec{f} = \begin{bmatrix} y^3 e^z \\ -xy \\ x\tan^{-1} y \end{bmatrix} \text{ across the region } M \text{ bounded by the coordinate}$$

planes $x = 0, y = 0, z = 0$ and the plane $x + y + z = 1$.

**3.11.25** Use Stokes's Theorem to find $\iint\limits_{S} \left( \nabla \times \vec{f} \right) \cdot \vec{n} \, dS$ for the vector field

$$\vec{f} = \begin{bmatrix} y \\ -x \\ yz \end{bmatrix} \text{ across the region } S, \text{ which is paraboloid } z = x^2 + y^2 \text{ with}$$

the circle $x^2 + y^2 = 1$ and $z = 1$ as its bounded.

**3.11.26** Use Stokes's Theorem to find $\oint_{C} \vec{f} \cdot d\vec{r}$ for the vector field

$$\vec{f} = \begin{bmatrix} x^2 \\ y^2 \\ z^2 \end{bmatrix} \text{ across the region } S, \text{ which is part of the cone}$$

$z = \sqrt{x^2 + y^2}$ cutoff by the plane $z = 1$.

# MAPLE

Maple is interactive mathematical and analytical software designed to perform a wide variety of mathematical calculations as well as operations on symbolic, numeric entities, and modeling. In this appendix, we give a general overview of Maple. For more information on Maple, visit the maple website: www.maplesoft.com.

## A.1   GETTING STARTED AND WINDOWS OF MAPLE

When you double-click on the Maple icon, it opens as shown in Figure A.1. This figure shows Maple in the document mode. The worksheet mode is shown in Figure A.2, where the special [> prompt appears. This is the main area in which the user interacts with Maple. For general help, click on **Help** then **Maple Help** in menu bar as shown in Figure A.3. Also, Maple uses the question mark (**?**), followed by the command or topic name, to get help. For example, to get help on solve, you type **?solve**. To terminate the Maple session, from the **File** menu, select **Exit**.

Document mode

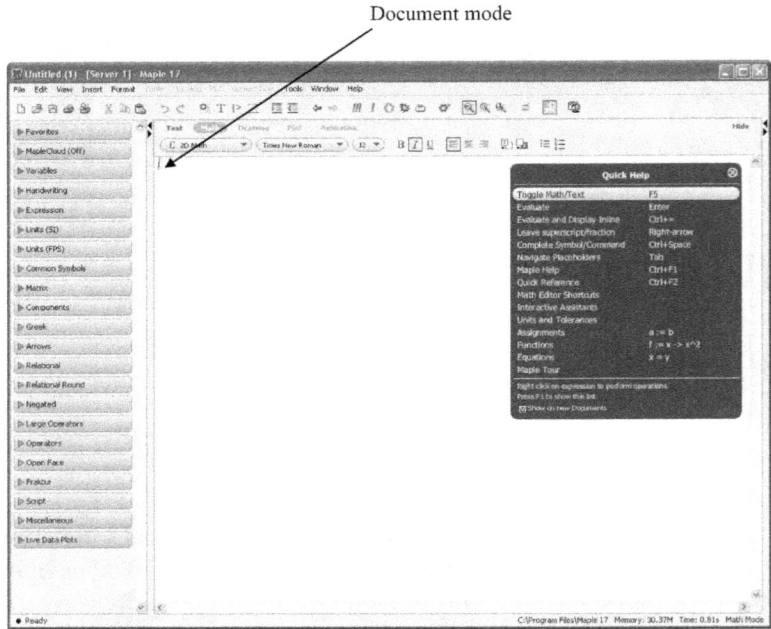

**FIGURE A.1** Default environment (document mode).

By clicking on this icon,
we get the worksheet mode

Worksheet mode

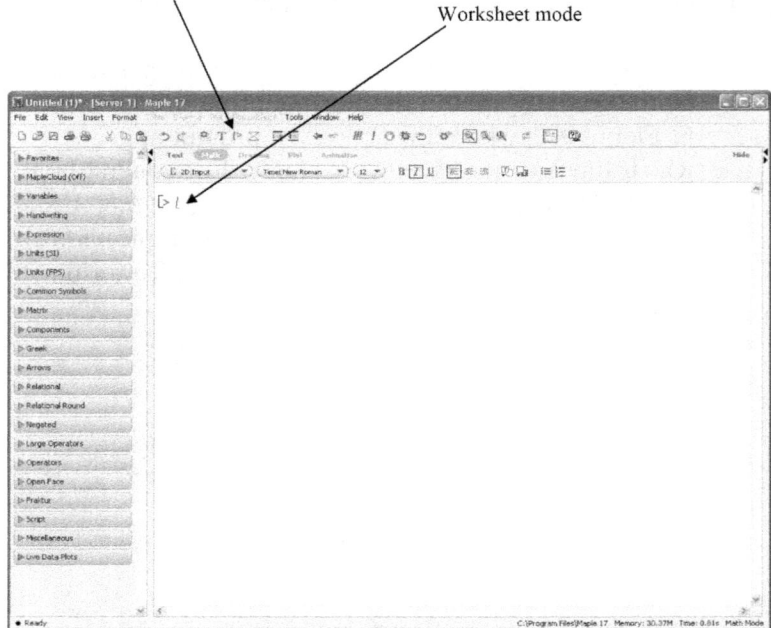

**FIGURE A.2** Worksheet mode.

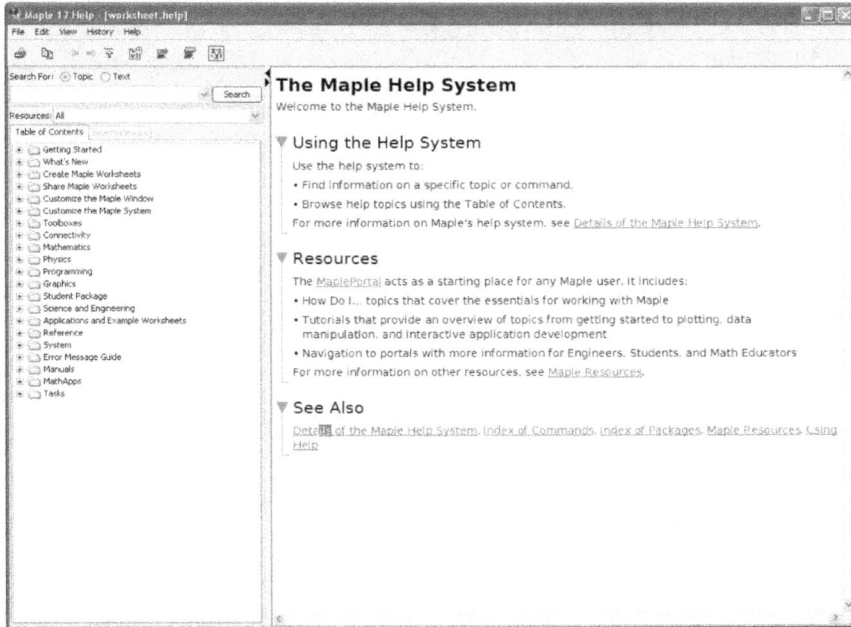

*FIGURE A.3*  Maple help system.

## A.2  ARITHMETIC

Maple can do arithmetic operations like a calculator. Table A.1 provides Maple's common arithmetic operations. To evaluate an arithmetic expression, type the expression and then press the **Enter** key.

**TABLE A.1**  Maple common arithmetic operations.

| Operation | Descriptions |
|---|---|
| + | addition |
| - | subtraction |
| * | multiplication |
| / | division |
| ^ | exponentiation |
| ! | factorial |
| abs (n) | Absolute value of n |
| sqrt (n) | Square root of n |

Maple uses, **pi**, command to present $\pi$ and uses, **exp (1)**, command to present $e$.

### EXAMPLE A.2.1

Calculate $2^5 - \dfrac{(8+6)}{4} + 3 \times 9$.

*Solution:*

> $2^5 - \dfrac{(8+6)}{4} + 3 \cdot 9$

$$\frac{111}{2}$$

### EXAMPLE A.2.2

Simple numerical calculation $5 + |-3|$.

*Solution:*

> $5 + abs(-3)$

$$8$$

## A.3  SYMBOLIC COMPUTATION

Maple can do a variety of symbolic calculations. For example,

> $(x-y)^2 \cdot (x-y)^3$

$$(x-y)^5$$

Maple also makes simplifications to the expression when you use the command **simplify**. For example,

> $simplify\left(15 \cdot \left(\sin(\theta)^2 + \cos(\theta)^2\right)\right)$

$$15$$

The **expand** and **factor** commands are used to expand and factor the expression respectively. For example,

> $expand(\cos(\gamma - \beta))$

$$\cos(\gamma)\cos(\beta) + \sin(\gamma)\sin(\beta)$$

> $factor(2 \cdot x^4 + x^2 + x \cdot y)$

$$x\left(2x^3 + x + y\right)$$

## A.4  ASSIGNMENTS

To assign values to a variable, Maple uses colon equals (:=). For example,

> $x := 3$

$$x := 3$$

> $x^4 + 2 \cdot x \cdot y + z$

$$6y + z + 81$$

To clear the value of the variable $x$, type

> $x := 'x'$

$$x := x$$

## A.5  WORKING WITH OUTPUT

One percent sign (%) refers to the output of the previous command. Two percent signs (%%) refer to the second-to-last output and three percent signs (%%%) to the third-to-last output. Maple remembers the output of the last three statements you entered. For example,

> $6 + 3$

$$9$$

> $\% \cdot 3$

$$27$$

> $\%\% + 3$

$$12$$

> $\%\%\% - 1$

$$8$$

## A.6  SOLVING EQUATIONS

Maple uses **solve** command to solve equations. For example,

> $solve(x^2 + 5 \cdot x - 3 = 0)$

$$-\frac{5}{2} + \frac{1}{2}\sqrt{37}, \ -\frac{5}{2} - \frac{1}{2}\sqrt{37}$$

We can solve equations with more than one variable for a specific variable. For example,

> $solve(x + \cos(y) = 7, y)$

$$\pi - \arccos(x - 7)$$

> $solve(\{y = 2 \cdot x - 1, y = x + 2\}, \{x, y\})$

$$\{x = 3, y = 5\}$$

## A.7 PLOTS WITH MAPLE

Maple uses the basic plotting command, **plot**, to plot functions, expressions, list of points, and parametric functions. For example, to plot the graph of $y = 3x^2 - x + 1$ on the interval -1 to 1, type

> $plot(3 \cdot x^3 - 2 \cdot x, x = -1 ..1, y = -1 ..1)$

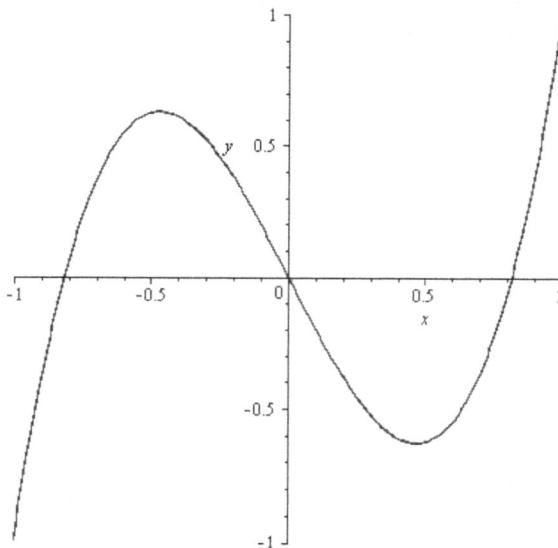

Also, we can plot several functions or expressions on same graph. For example,

> $plot(\{2 \cdot x^3 - 5, \exp(x^2), \cos(x), x + 5\}, x = -4 ..4, y = -3 ..6)$

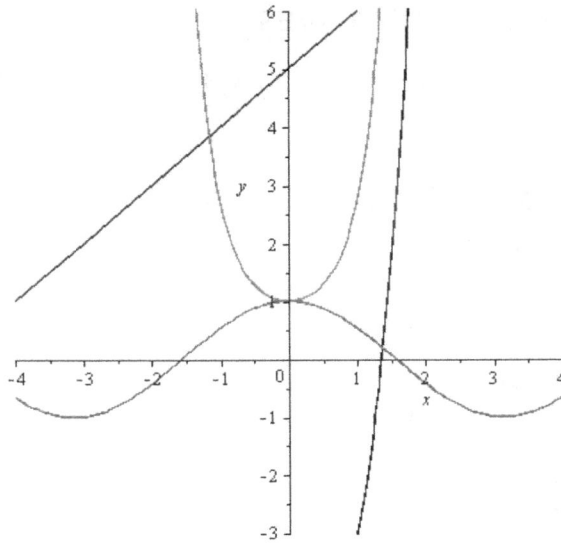

Maple allows you to annotate a plot by adding text and drawings by clicking on the plot. The **Plot** options tool bar will show up. Then click on **Drawing** button and the drawing tool bar will show up. For example,

$plot(\sin(x) + 0.3 \cdot x, x = -5 .. 5);$

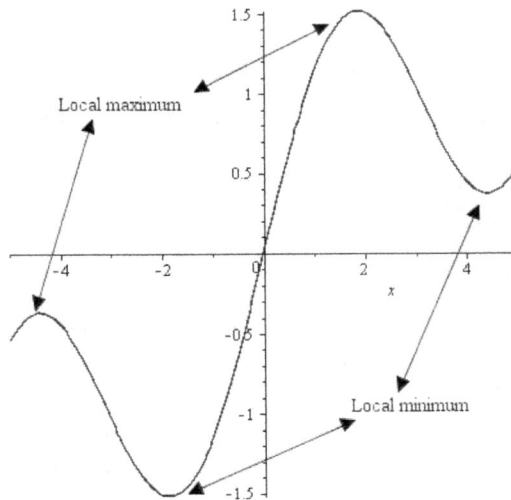

## A.8  LIMITS AND DERIVATIVES

Maple can evaluate limits of $\lim_{x \to a} f(x)$ by using **limit (f(x), x =a);** command. For example,

> $limit\left( \dfrac{(\exp(x) + 2 - x)}{x^3}, x = 0 \right)$

$$undefined$$

Maple uses **diff** command to compute derivatives. For example,

> $diff(2 \cdot x^5, x)$

$$10\, x^4$$

## A.9  INTEGRATION

Maple uses **int** command to compute integrals. For example,

> $int(9 \cdot x^2 + x, x)$

$$3 x^3 + \frac{1}{2} x^2$$

## A.10 MATRIX

Maple uses **Matrix** command to make a matrix. For example,

> $A := Matrix([[0, 1], [2, 5]])$ :
> $B := Matrix([[4, 6], [-3, 8]])$ :
> $A + B$

$$\begin{bmatrix} 4 & 7 \\ -1 & 13 \end{bmatrix}$$

> $C = Matrix([[2, -3], [6, 7]])$

$$C = \begin{bmatrix} 2 & -3 \\ 6 & 7 \end{bmatrix}$$

# B

# *MATLAB*

MATLAB has become a useful and dominant tool for technical professionals around the world. MATLAB is an abbreviation for Matrix Laboratory. It is a numerical computation and simulation tool that uses matrices and vectors. Also, MATLAB enables the user to solve wide analytical problems.

A copy of MATLAB software can be obtained from:

The Mathworks, Inc.

3 Apple Hill Drive
Natick, MA 01760-2098
Phone: 508-647-7000
Website: *http://www.mathworks.com*

This brief introduction of MATLAB (R2010b) is presented here to give a general idea about the software. MATLAB computational applications to science and engineering systems used to solve practical problems.

## B.1 GETTING STARTED AND WINDOWS OF MATLAB

When you double-click on the MATLAB icon, it opens as shown in Figure B.1. The command window, where the special >> prompt appears, is the main area in which the user interacts with MATLAB. To make the Command Window active, you need to click anywhere inside its border. To quit MATLAB, you can select **EXIT MATLAB** from the **File** menu, or by entering *quit* or *exit* at the Command Window prompt. Do not click on the X (close box) in the top right corner of the MATLAB window, because it may cause problems with the operating software. Figure B.1 contains four default windows, which are Command Window, Workplace Window, Command History Window, and Current Folder Window. Table B.1 shows a list of the various windows and their purpose of MATLAB.

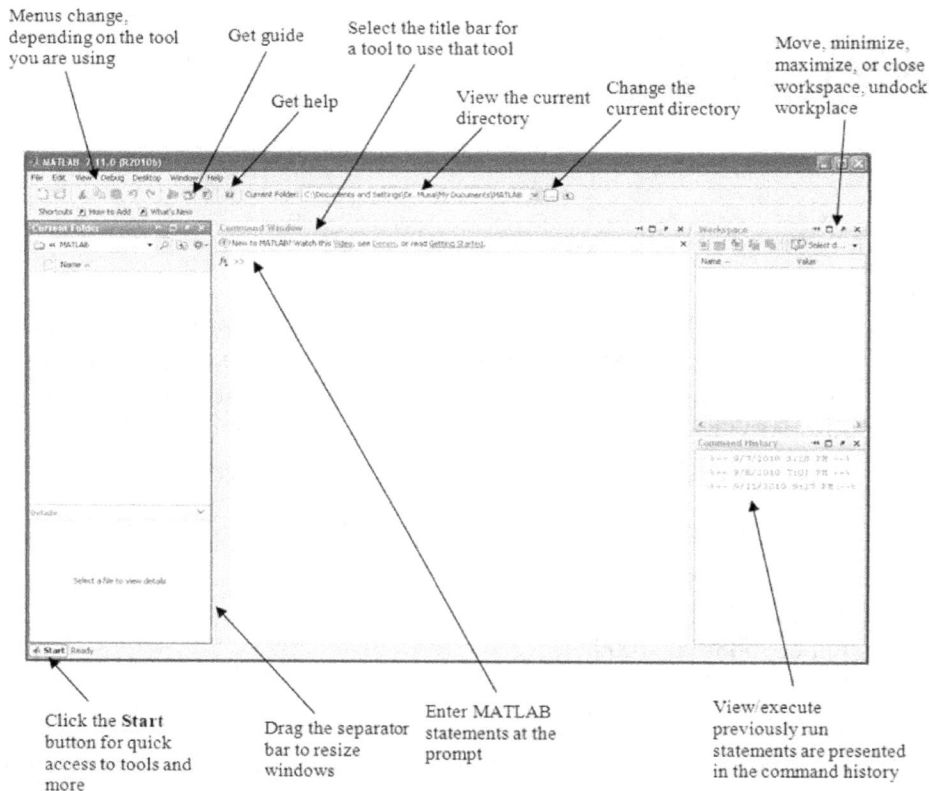

Menus change, depending on the tool you are using

Get guide

Get help

Select the title bar for a tool to use that tool

View the current directory

Change the current directory

Move, minimize, maximize, or close workspace, undock workplace

Click the **Start** button for quick access to tools and more

Drag the separator bar to resize windows

Enter MATLAB statements at the prompt

View/execute previously run statements are presented in the command history

**Figure B.1** MATALB default environment.

**TABLE B.1** MATLAB windows

| Window | Description |
|---|---|
| Command Window | Main window, enter variables, runs programs |
| Workplace Window | Gives information about the variable used |
| Command History Window | Records command entered in the Command Window |

**TABLE B.1** MATLAB windows (Continued)

| Window | Description |
|---|---|
| Current Folder Window | Shows the files in the current directory with details |
| Editor Window | Makes and debugs script and function files |
| Help Window | Gives help information |
| Figure Window | Contains output from the graphic commands |
| Launch Pad window | Provides access to tools, demos, and documentation |

### B.1.1 Using MATLAB in Calculations

Table B.2 shows the MATLAB common arithmetic operators. The order of operations as first, parentheses ( ), the innermost are executed first for nested parentheses; second, exponentiation ^; third, multiplication * and division / (they are equal precedence); fourth, addition + and subtraction -.

**TABLE B.2** MATLAB common arithmetic operators

| Operator symbols | Descriptions |
|---|---|
| + | Addition |
| - | Subtraction |
| * | Multiplication |
| / | Right division (means $\frac{a}{b}$) |
| \ | Left division (means $\frac{b}{a}$) |
| ^ | Exponentiation (raising to a power) |
| ' | Converting to complex conjugate transpose |
| ( ) | Specify evaluation order |

For example,

>> a = 11; b = -3; c = 5;

>> x = 9*a + c^2 - 2

x =

122

>> y = sqrt(x)/6

y =

1.8409

Table B.3 provides a common sample of MATLAB functions. You can obtain more by typing *help* in the Command Window (>> help).

**TABLE B.3** Typical elementary math functions

| Function | Description |
|---|---|
| abs (x) | Absolute value or complex magnitude of x |
| acos (x), acosh (x) | Inverse cosine and inverse hyperbolic cosine of x (in radians) |
| angle (x) | Phase angle (in radians) of a complex number x |
| asin (x), asinh (x) | Inverse sine and inverse hyperbolic sine of x (in radians) |
| atan (x), atanh (x) | Inverse tangent and inverse hyperbolic tangent of x (in radians) |
| conj (x) | Complex conjugate of x (in radians) |
| cos (x), cosh (x) | Cosine and inverse hyperbolic cosine of x (in radians) |
| cot (x), coth (x) | Inverse cotangent and inverse hyperbolic cotangent of x (in radians) |
| exp (x) | Exponential of x |
| Fix | Round toward zero |
| imag (x) | Imaginary part of a complex number x |
| log (x) | Natural logarithm of x |
| log2 (x) | Natural logarithm of x to base 2 |
| log10 (x) | Common logarithms (base 10) of x |
| real (x) | Real part of a complex number of x |
| sin (x), sinh (x) | Sine and inverse hyperbolic sine of x (in radians) |
| sqrt (x) | Square root of x |
| tan (x), tanh (x) | Tangent and inverse hyperbolic tangent of x (in radians) |

For example,

>> 9+3^(log2(4.25))

ans =

18.9077

\>> y=5*cos(pi/4)

y =

   3.5355

\>> z = exp(y+6)

z =

   1.3843e+004

In addition to operating on mathematical functions, MATLAB allows us to work easily with vectors and matrices. A vector (or one-dimensional array) is a special matrix (or two-dimensional array) with one row or one column. Arithmetic operations can apply to matrices and Table B.4 shows extra common operations that can be implemented to matrices.

**TABLE B.4** Matrix operations

| Operations | Descriptions |
| --- | --- |
| $A'$ | Transpose of matrix A |
| det (A) | Determinant of matrix A |
| inv (A) | Inverse of matrix A |
| eig (A) | Eigenvalues of matrix A |
| diag (A) | Diagonal elements of matrix A |

A vector can be created by typing the elements inside brackets [ ] from a known list of numbers.

For example,

\>> A = [1 2 3 6 5 22]

A =

   1   2   3   6   5   22

\>> B = [1 2 6; 8 9 11; 10 14 16]

B =

```
 1   2   6
 8   9   11
10   14  16
```

Also, a vector can be created with constant spacing by using the command *variable-name* = *[a: n: b]*, where a is the first term of the vector; n is spacing; b is the last term.

For example,

\>> x = [1:0.6:5]

x =

```
1.0000   1.6000   2.2000   2.8000   3.4000   4.0000   4.6000
```

Also, a vector can be created with constant spacing by using the command *variable-name* = *linspace (a, b, m)*, where a is the first element of the vector; b is the last element; m is the number of elements.

For example,

\>> x=linspace(0,4*pi,6)

x =

```
0   2.5133   5.0265   7.5398   10.0531   12.5664
```

Examples using Table B.4:

\>> B = [1 2 4; 7 8 9; 3 5 10]

B =

```
1   2   4
7   8   9
3   5   10
```

```
>> D=B^2

D =

    27    38    62
    90   123   190
    68    96   157

>> C= A'

C =

    1
    2
    3
    6
    5
   22

>> E =[4 8;11 25];
>> inv(E)

ans =

    2.0833   -0.6667
   -0.9167    0.3333

>> det(E)

ans =

   12
```

Special constants can be used in MATLAB. Table B.5 provides special constants used in MATLAB.

**TABLE B.5** MATLAB named constants

| Name | Content |
|---|---|
| Pi | $\pi = 3.14159...$ |
| i or j | Imaginary unit, $\sqrt{-1}$ |
| Eps | Floating-point relative precision, $2^{-52}$ |
| Realmin | Smallest floating-point number, $2^{-1022}$ |
| Realmax | Largest floating-point number, $(2\text{-}eps).2^{1023}$ |
| Bimax | Largest positive integer, $2^{53} - 1$ |
| Inf or Inf | Infinity |
| nan or NaN | Not a number |
| Rand | Random element |
| Eye | Identity matrix |
| Ones | An array of 1's |
| Zeros | An array of 0's |

For example,

\>> eye(3)

ans =

$$\begin{array}{ccc} 1 & 0 & 0 \\ 0 & 1 & 0 \\ 0 & 0 & 1 \end{array}$$

\>> ones(3)

ans =

$$\begin{array}{ccc} 1 & 1 & 1 \\ 1 & 1 & 1 \\ 1 & 1 & 1 \end{array}$$

\>> 1/0

ans =

Inf

\>> 0/0

ans =

NaN

Arithmetic operations on arrays are done element by element. Table B.5 provides MATLAB common Arithmetic operations on arrays.

**TABLE B.6** MATLAB common arithmetic operations on arrays

| Operator symbols on arrays | Descriptions |
|---|---|
| + | Addition same as matrices |
| - | Subtraction same as matrices |
| .* | Element-by-element multiplication |
| ./ | Element-by-element right division |
| .\ | Element-by-element left division |
| .^ | Element-by-element power |
| .' | Unconjugated array transpose |

For example,

\>> A=[-1 2 0; 3 5 7; 8 9 9]

A =

  -1   2   0

   3   5   7

   8   9   9

\>> A.*A

ans =

1   4   0

9   25   49

64   81   81

>> C=[0 1;2 4; 7 8];

>> D=[8 11;10 16;14 18];

>> C./D

ans =

0   0.0909

0.2000   0.2500

0.5000   0.4444

>> C.^2

ans =

0   1

4   16

49   64

## B.2   PLOTTING

MATLAB has nice capability to plot in two-dimensional and three-dimensional plots.

### B.2.1   Two-dimensional Plotting

First, we start with two-dimensional plots. The *plot* command is used to create two-dimensional plots. The simplest form of the command is *plot (x,y)*. The arguments x and y are each a vector (one-dimensional array). The vectors x and y must have the same number of elements. When the *plot* command is executed a figure will be created in the Figure Window. The *plot (x, y, 'line specifiers')* command has additional optional arguments that can be used to detail the color and style of the lines. Tables B.7 through B.9 show various types of lines, points, and color types used in MATLAB.

**TABLE B.7** MATLAB various line styles

| Line types | MATLAB symbol |
|---|---|
| Solid (default) | - |
| Dashed | -- |
| Dotted | : |
| Dash-dot | -. |

**TABLE B.8** MATLAB various point styles

| Point type | MATLAB symbol |
|---|---|
| Asterisk | * |
| Plus sign | + |
| x-mark | x |
| Circle | o |
| Point | . |
| Square | s |

**TABLE B.9** MATLAB various line color types

| Color | MATLAB symbol |
|---|---|
| Black | K |
| Blue | B |
| Green | G |
| Red | R |
| Yellow | Y |
| Magenta | M |
| Cyan | C |
| White | W |

For example,

>> x=0:pi/50:2*pi;%x is a vector, 0 <= x <= 2*pi, increments of pi/50

>> y=3*sin(2*pi*x);% y is a vector

>> plot(x,y,'--b')%creates the 2D plot with blue and dashed line

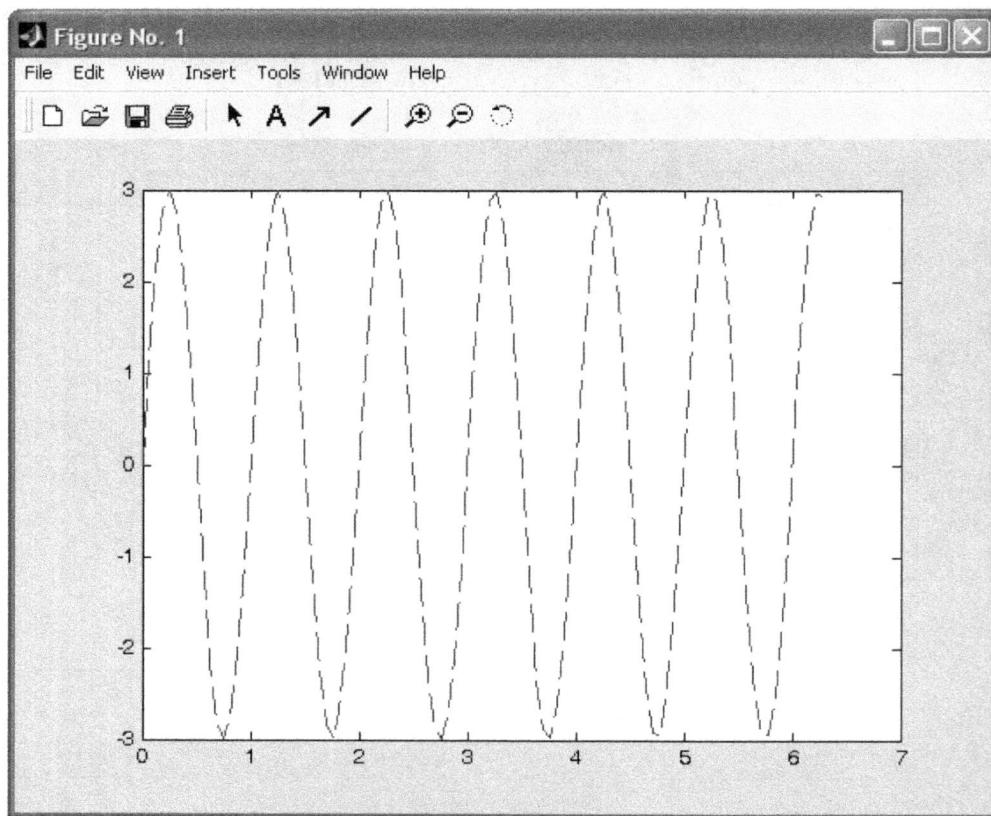

The command *fplot('function', limits, line specifiers)* is used to plot a function with the form y = f (x), where the function can be typed as a string inside the command. The limits are a vector with two elements that specify the domain x [xmin,xmax], or is a vector with four elements that specifies the domain of x and the limits of the y-axis [xmin, xmax, ymin, ymax]. The line specifiers are used the same as in the plot command.

For example,

>> fplot('x^3+2*sin(3*x)-2',[-5,5],'xr')

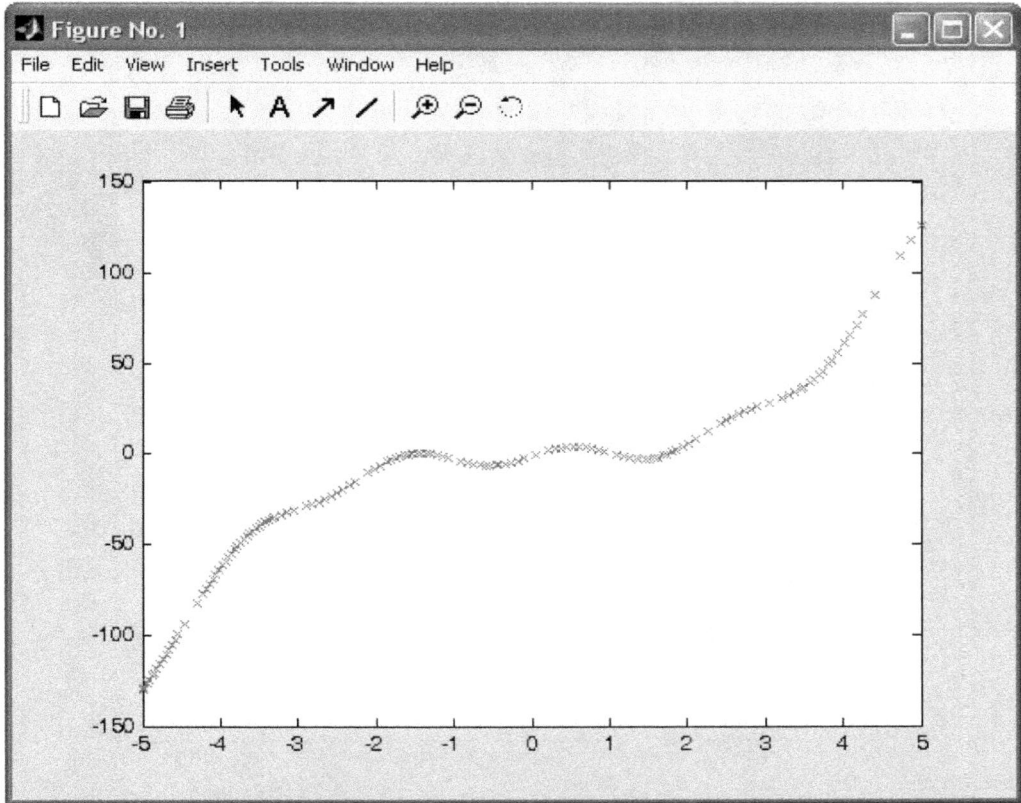

Also, we can create a plot for a function y = f(x) using the command *plot* by creating a vector of values of x for the domain that the function will be plotted, then creating y with corresponding values of f(x).

For example,

```
>> x=[0:0.1:1];
>> y=cos(2*pi*x);
>> plot(x,y,'ro:')
```

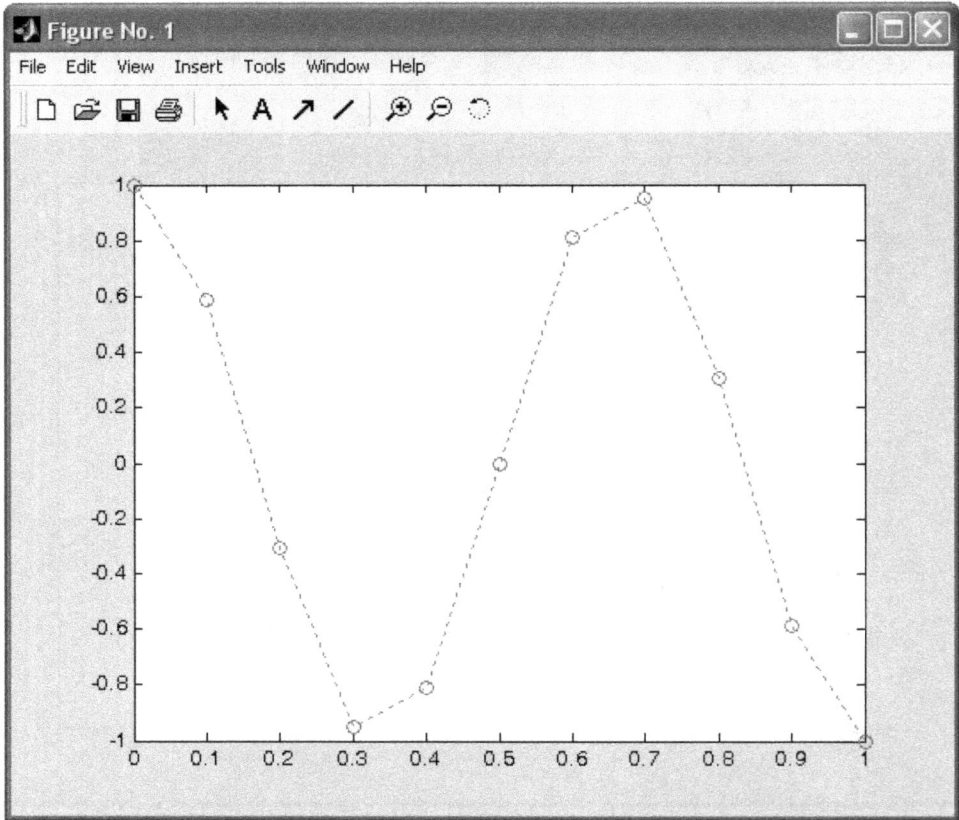

In MATLAB several graphs can be plotted at the same plot in two ways: first, using *plot* command by typing pairs of vectors inside the *Plot* command such as *Plot(x,y, z, t, u, h)*, which will create three graphs: y vs. x, t vs. z, and h vs. u, all in the same plot.

For example, the command to plot the function $y = 2x^3 - 15x + 5$, its first derivative $y' = 6x^2 - 15$, and its second derivative $y'' = 12x$, for domain $-3 \le x \le 6$, all in the same plot, is as follows:

>> x=[-3:0.01:6];% vector x with the domain of the function

>> y=2*x.^3-15*x+5;% vector y with the function value at each x

>> yd=6*x.^2-15;% vector yd with the value of the first derivative

>> ydd=12*x; %vector ydd with the value of the second t derivative

>> plot(x,y,'-r',x,yd,':b',x,ydd,'--k')

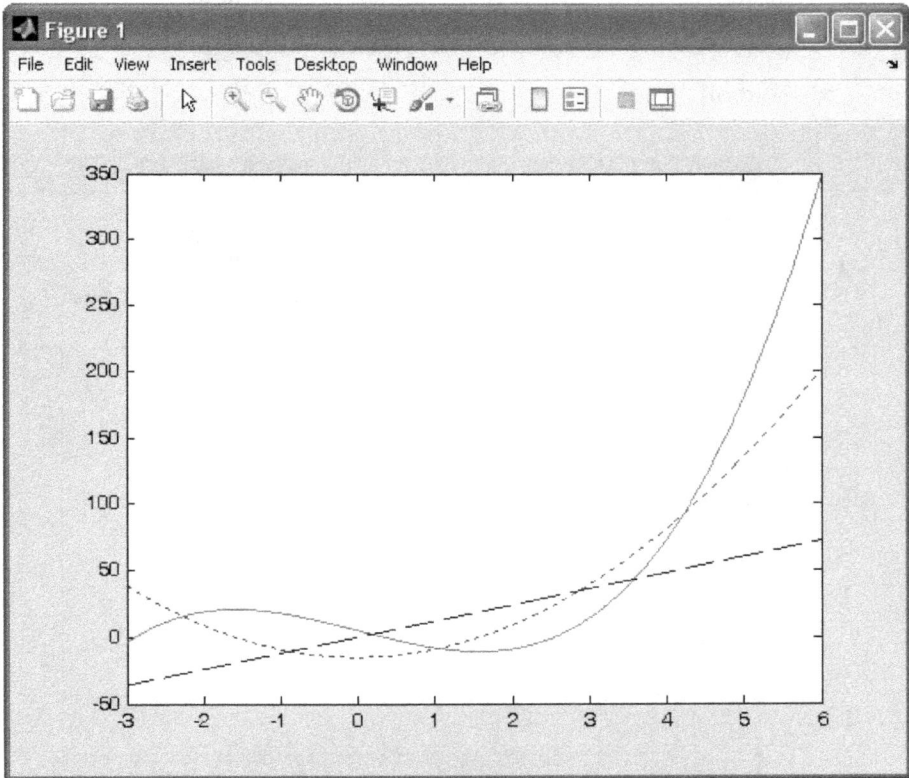

Second, using *hold on, hold off* commands. The *hold on* command will hold the first plotted graph and add to it extra figures for each time the *plot* command is typed. The *hold off* command stops the process of *hold on* command.

For example, if we use the previous example, we get the same result using the following commands:

>> x=[-3:0.01:6];

>> y=2*x.^3-15*x+5;

>> yd=6*x.^2-15;

>> ydd=12*x;

>> plot(x,y,'-r')

>> hold on % the first graph is created

>> plot(x,yd,':b') % second graph is added to the figure

>> plot(x,ydd,'--k') % third graph is added to the figure

>> hold off

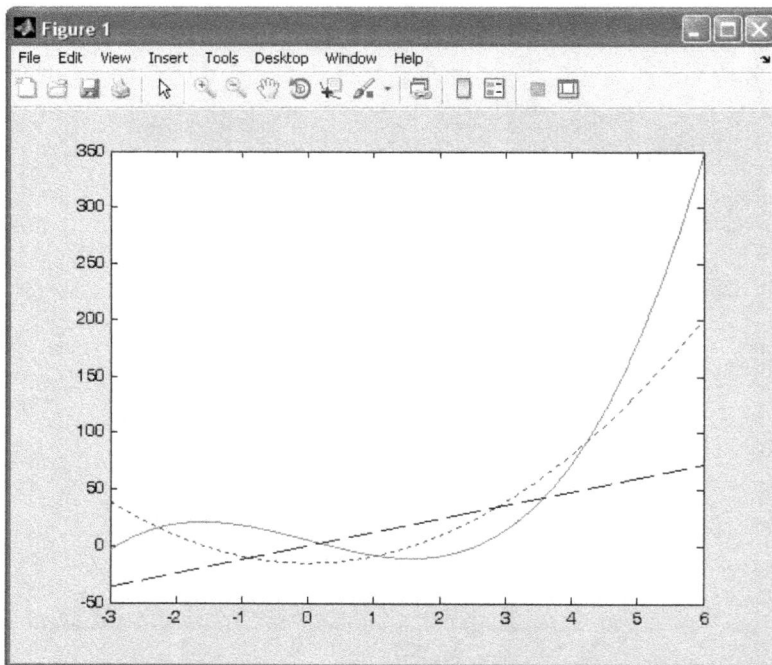

Plots in **MATLAB** can be formatted using commands that follow the *plot* commands, or by using the plot editor interactively in the Figure window. First, format the plot using commands as follows:

- Labels can be placed next to the axes with the *xlabel* ('*text as string* ') for the x-axis and *ylabel* ('*text as string* ') for the y-axis.
- The command *title*('*text as string* ') is a title command which can be added to the plot to place the title at the top of the figure as a text.
- There are two ways to place a text label in the plot. First, using *text* (*x,y,'text as string* ') command which is used to place the text in the figure such that the first character is positioned at the point with the coordinates x, y according to the

axes of the figure. Second, using *gtext ('text as string ')* command which is used to place the text at a position specified by the user's mouse in the figure window.
* The command *legend ('string1', 'string2',...,pos)* is used to place a legend on the plot. The legend command shows a sample of the line type of each graph that is plotted and, places a label specified by the user, beside the line sample. The *strings* in the command are the labels that are placed next to the line sample and their order corresponds to the order that the graphs were created. The *pos* in the command is an optional number that specifies where in the figure the legend is placed. Table B.10 shows the options that can be used for *pos*.

The command *axis* is used to change the range and the appearance of the axes of the plot, based on the minimum and maximum values of the elements of x and y. Table B.11 shows some common possible forms of *axis* command.

* The command *grid on* is used to add grid lines to the plot and the command *grid off is* used to remove grid lines from the plot.

**TABLE B.10** Options that can be used for *pos*

| Pos value | Description |
|---|---|
| -1 | Place the legend outside the axes boundaries on the right side |
| 0 | Place the legend inside the axes boundaries in a location that interferes the least with graph |
| 1 | Place the legend at the upper-right corner of the plot (this is the default) |
| 2 | Place the legend at the upper-left corner of the plot |
| 3 | Place the legend at the lower-left corner of the plot |
| 4 | Place the legend at the lower-right corner of the plot |

**TABLE B.11** Some common *axis* commands

| axis command | Description |
|---|---|
| axis ([xmin, xmax, ymin, ymax]) | Sets the limits of both the x and y axes (xmin, xmax, ymin, ymax are numbers) |
| axis equal | Sets the same scale for both axes |
| axis tight | Sets the axis limits to the range of the data |
| axis square | Sets the axes region to be square |

For example,

>> x=0:pi/30:2*pi;y1=exp(-2*x);y2=sin(x*3);

>> plot(x,y1,'-b',x,y2,'--r')

>> xlabel('x')

```
>> ylabel('y1 , y2')
>> title('y1=exp(-2*x), y2=sin(x*3)')
>> axis([0,11,-1, 1])
>> text(6,0.6,'Comparison between y1 and y2')
>> legend('y1','y2',0)
```

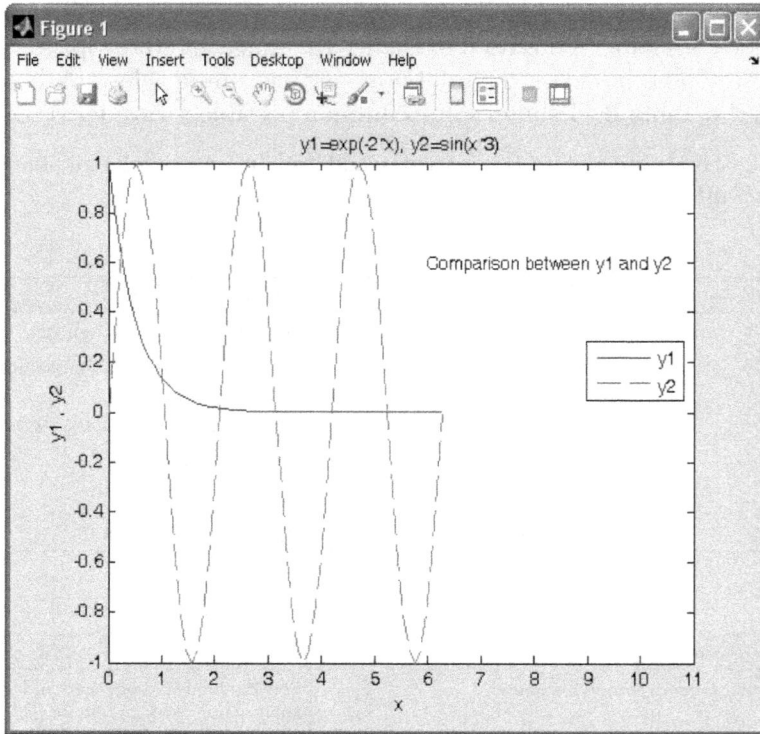

In MATLAB, users can use Greek characters in the text by typing \name of the letter within the string as in Table B.12.

**TABLE B.12** Some common Greek characters

| Greek characters in the string | Greek letter | Greek characters in the string | Greek letter |
|---|---|---|---|
| \alpha | $\alpha$ | \Phi | $\Phi$ |
| \beta | $\beta$ | \Delta | $\Delta$ |
| \gamma | $\gamma$ | \Gamma | $\Gamma$ |

**TABLE B.12** Some common Greek characters (Continued)

| Greek characters in the string | Greek letter | Greek characters in the string | Greek letter |
|---|---|---|---|
| \theta | $\theta$ | \Lambda | $\Lambda$ |
| \pi | $\pi$ | \Omega | $\Omega$ |
| \sigma | $\sigma$ | \Sigma | $\Sigma$ |

To get a lowercase Greek letter, the name of the letter must be typed in all lowercase. To get a capital Greek letter, the name of the letter must start with a capital letter.

Second, format the plot using the plot editor interactively in the Figure window. This can be done by clicking on the plot and/or using the menus as illustrated in the following figure.

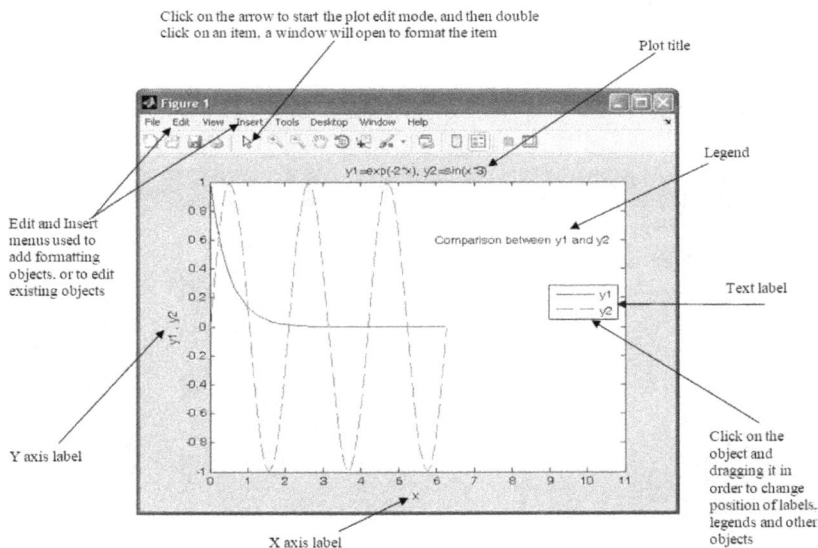

MATLAB can use logarithm scaling for a two-dimensional plot. Table B.13 shows MATLAB commands for logarithm scaling.

**TABLE B.13** Two-dimensional graphic for logarithm scaling

| Command | Description |
|---|---|
| Loglog | To plot log(y) versus log(x) |
| Semilogx | To plot y versus log(x) |
| Semilogy | To plot log(y) versus x |

Also, MATLAB can make plots with special graphics as in Table B.14.

**TABLE B.14** MATLAB plots with special graphics

| Command | Description |
|---------|-------------|
| *bar(x,y)* | Vertical bar plot |
| *barh(x,y)* | Horizontal bar plot |
| *stairs(x,y)* | Stairs plot |
| *stem(x,y)* | Stem plot |
| *pie(x)* | Pie plot |
| *hist(y)* | Histogram plot |
| *polar(x,y)* | Polar plot |

For example,

>> t=[0:pi/60:2*pi];

>> r=3+2*cos(t);

>> polar(t,r,'r.')

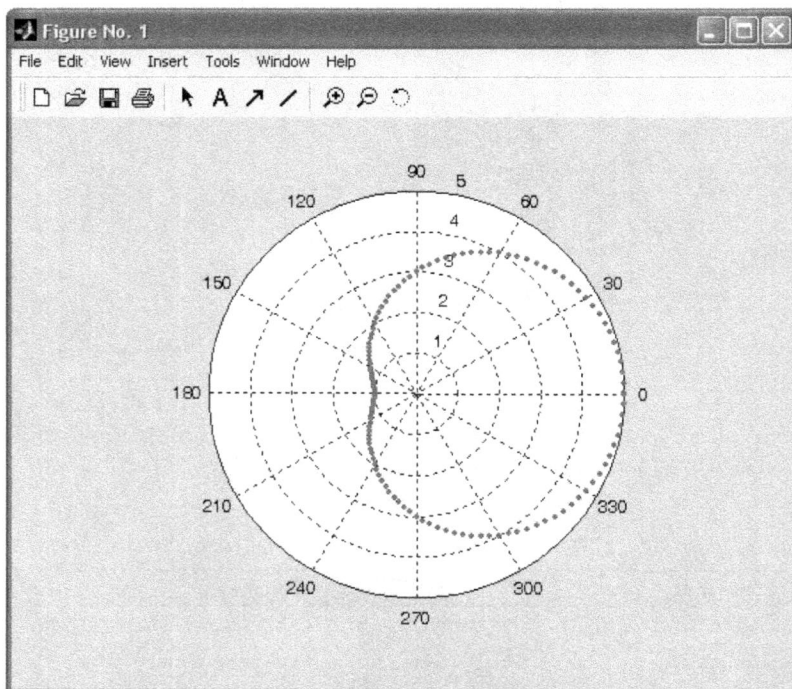

### B.2.2 Three-Dimensional Plotting

MATLAB has the capability to make a graph in three-dimensional plots using line, mesh, and surface plots.

The command *plot3(x,y,z)* is used in a three-dimensional line plot which is a line that is obtained by connecting points in three-dimensional space.

For example,

>> x=linspace(0,10*pi,100);

>> y=cos(x);z=sin(x);

>> plot3(x,y,z,'r');grid on

>> xlabel('x');ylabel('cos(x)');zlabel('sin(x)')

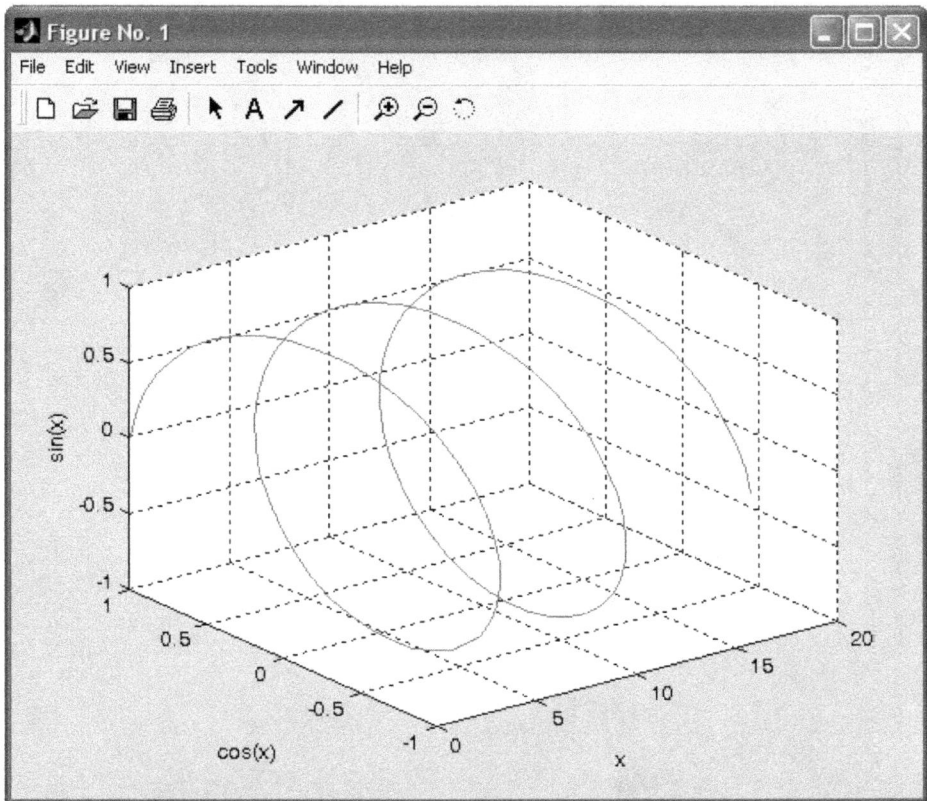

Another example,

```
> t = -5:0.1:5;
>> x = (3+t.^2).*sin(50*t);
>> y = (3+t.^2).*cos(50*t);
>> z = 5*t;
>> plot3(x,y,z,'g')
>> grid on
>> xlabel('x(t)'),ylabel('y(t)'),zlabel('z(t)')
>> title('plot3 example')
```

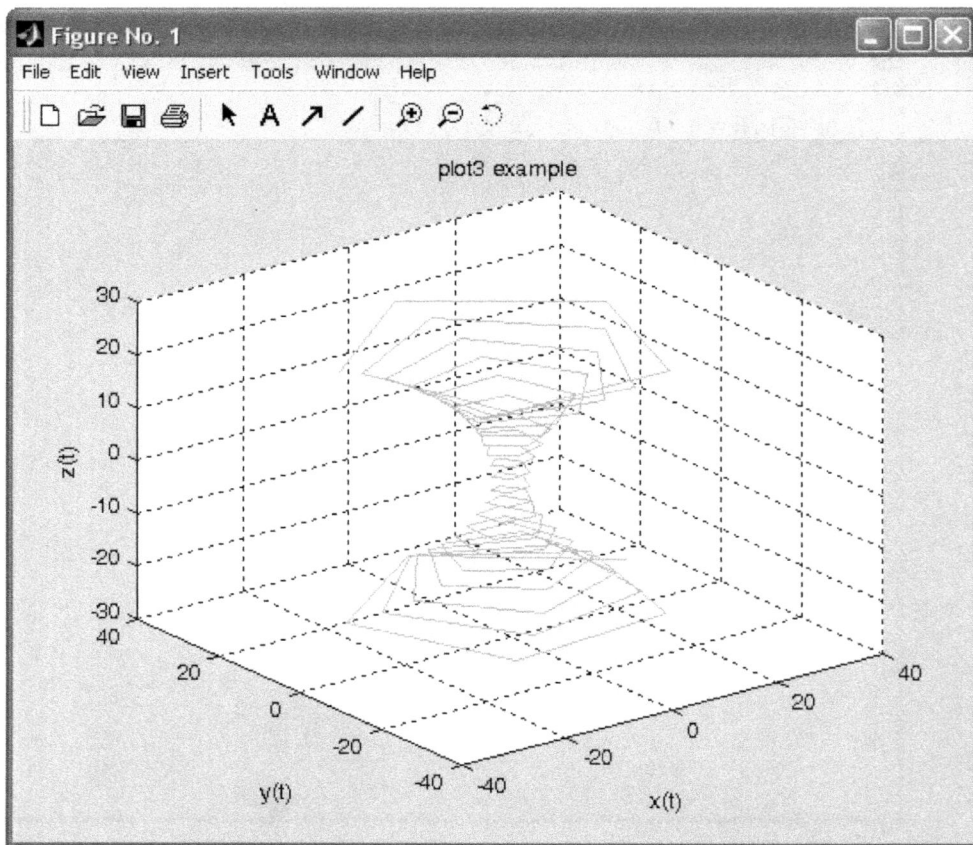

Also, The command *mesh(X,Y,Z)* is used in a three-dimensional plot that is applied to plotting functions z = f(x,y). This can be done by creating a grid in the x-y plane that covers the domain of the function, then calculating the value of z at each point of the grid, and then creating the plot.

For example,

>> x=(-6:0.1:6);y=(-6:0.1:6);[X,Y]=meshgrid(x,y);

>> Z=sin(X.^2+Y.^2).*exp(-0.6*(X.^2+Y.^2));

>> mesh(X,Y,Z)

>> zlabel('Z=sin(X.^2+Y.^2).*exp(-0.6*(X.^2+Y.^2))')

>> xlabel('X')

>> ylabel('Y')

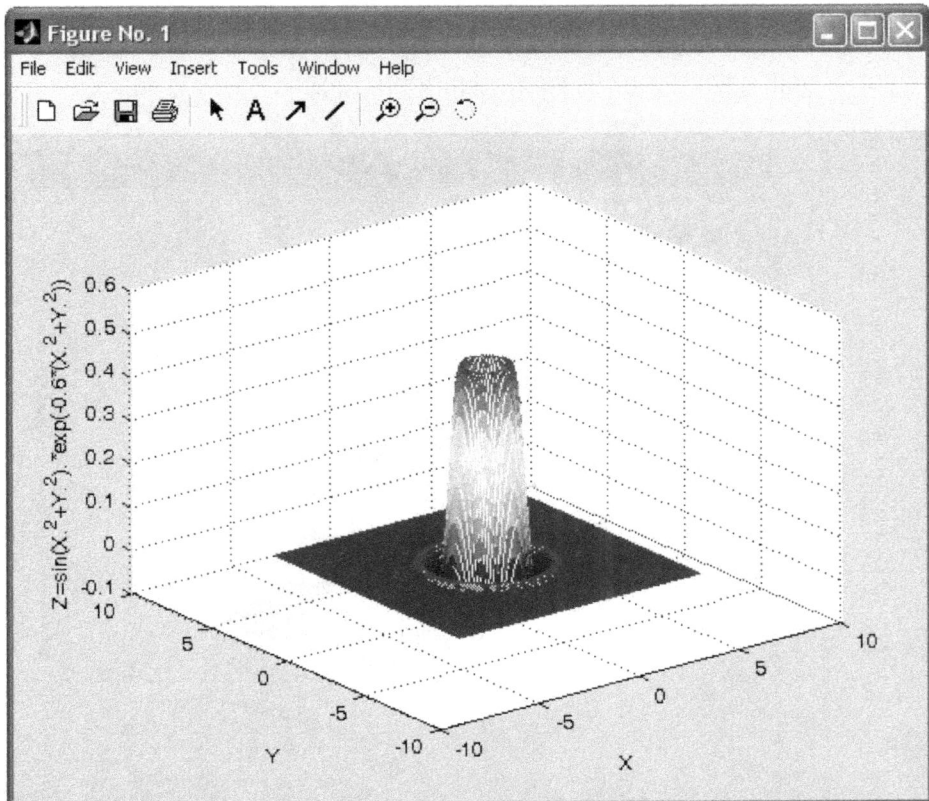

Table B.15 provides other common mesh plot types.

**TABLE B.15** Provides other common mesh plot types

| Mesh plot types | Description |
|---|---|
| *meshz(X,Y,Z)* | Mesh curtain plot which draws a curtain around the mesh |
| *meshc(X,Y,Z)* | Mesh and contour plot which draws a contour plot beneath the mesh |
| *waterfall(X,Y,Z)* | Draws a mesh in one direction only |

For example,

```
>> x=(-6:0.1:6);y=(-6:0.1:6);[X,Y]=meshgrid(x,y);
>> Z=sin(X.^2+Y.^2).*exp(-0.6*(X.^2+Y.^2));
>> meshz(X,Y,Z)
>> xlabel('X')
>> ylabel('Y')
>> zlabel('Z=sin(X.^2+Y.^2).*exp(-0.6*(X.^2+Y.^2))')
```

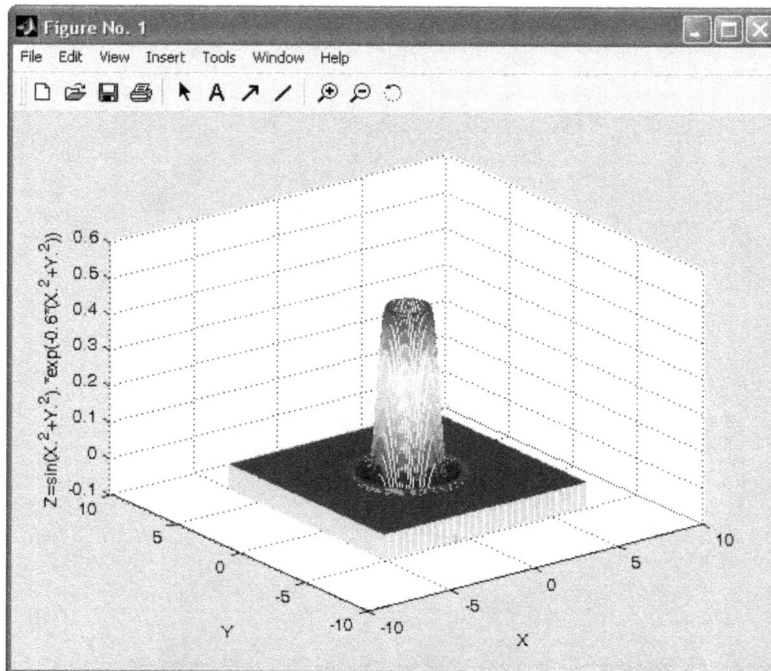

```
>> x=(-6:0.1:6);y=(-6:0.1:6);[X,Y]=meshgrid(x,y);
>> Z=sin(X.^2+Y.^2).*exp(-0.6*(X.^2+Y.^2));
>> meshc(X,Y,Z)
>> zlabel('Z=sin(X.^2+Y.^2).*exp(-0.6*(X.^2+Y.^2))')
>> ylabel('Y')
>> xlabel('X')
```

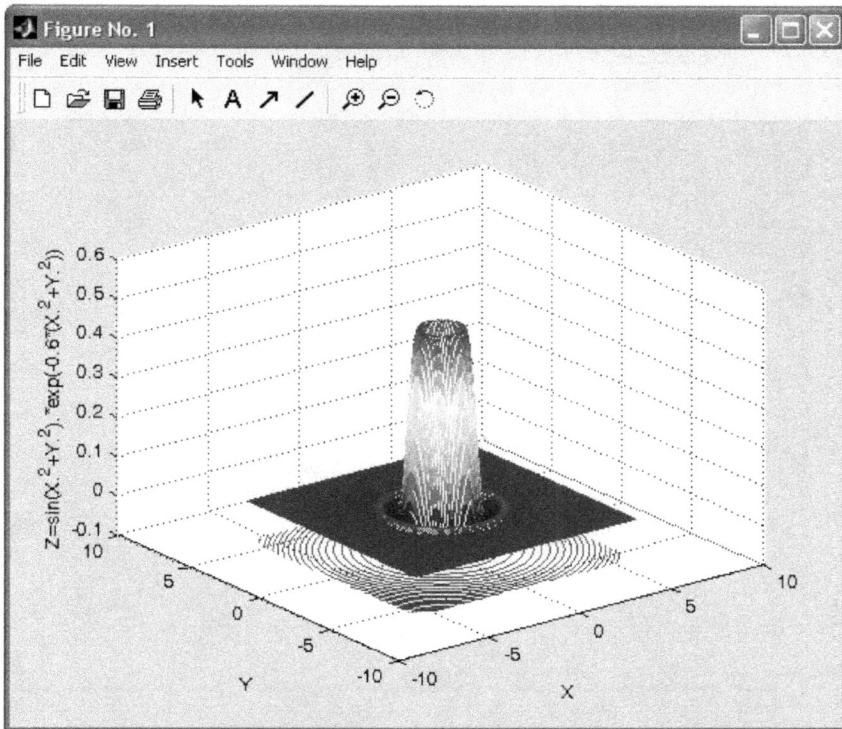

Another command *surf(X,Y,Z)* is used in a three-dimensional plot that is applied to plotting functions z = f(x,y) as in Mesh. This can be done by the same step for the mesh.

For example,

```
>> x=(-6:0.1:6);y=(-6:0.1:6);[X,Y]=meshgrid(x,y);
>> Z=sin(X.^2+Y.^2).*exp(-0.6*(X.^2+Y.^2));
```

```
>> surf(X,Y,Z)
>> xlabel('X')
>> ylabel('Y')
>> zlabel('Z=sin(X.^2+Y.^2).*exp(-0.4*(X.^2+Y.^2))')
```

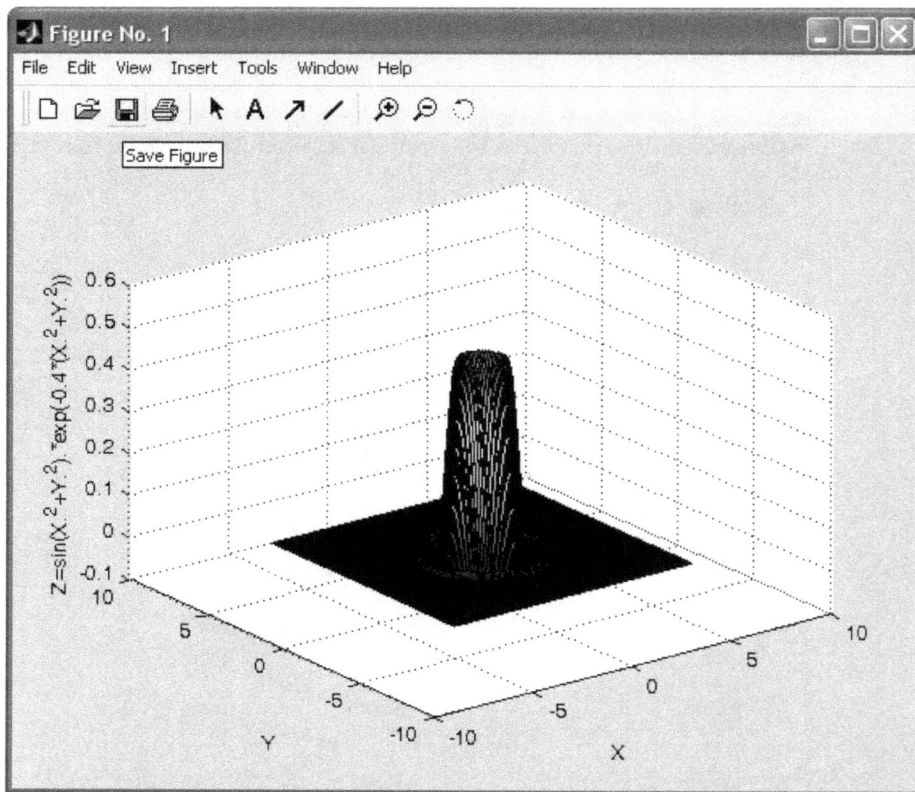

There are other common surface plot types as in Table B.16.

**TABLE B.16** Provides other common surface plot types

| Surface plot types | Description |
|---|---|
| *surfl(X,Y,Z)* | Surface plot with lighting |
| *surfc(X,Y,Z)* | Surface and contour plot which draws a contour plot beneath the mesh |

## B.3 PROGRAMMING IN MATLAB

So far we have used MATLAB commands and they were executed in the Command Window. This way is fine for a simple tasks, but for more complex ones, it becomes less convenient and difficult because the Command Window cannot be saved and executed again. Therefore, the commands and programs can be stored in a file. To begin, tell MATLAB to get its input from the file but this file must be created as an M-file. Do this by clicking on File/New/scripts to open a new file in the MATLAB Editor/Debugger or simple text editor, then type the program and save it by choosing save from the File menu. The file should be saved with an extension ".m".

For example, we created the program *nano1.m* using M-file as follows.

```
1   t=0:0.05*pi:40; %t vector, 0<=t<=30, increment is 0.05*pi
2   y1=exp(-0.05*t).*(2+sin(t))./(2-cos(0.25*t)); % calculate y1 values
3   y2=exp(-0.2*t).*(2+sin(t))./(2-cos(0.25*t)); % calculate y2 values
4   plot(t,y1,':r',t,y2,'-b') % create the plots for y1 and y2
5   xlabel('t(time)');  % label the x-axis
6   ylabel('functions y1(t)and y2(t)'); % label the y-axis
7   title('Functions y1(t)and y2(t)'); % create title
8   legend('y1','y2'); % create legend
```

We typed *nano1* in the Command Window, then hit enter to obtain the following figure.

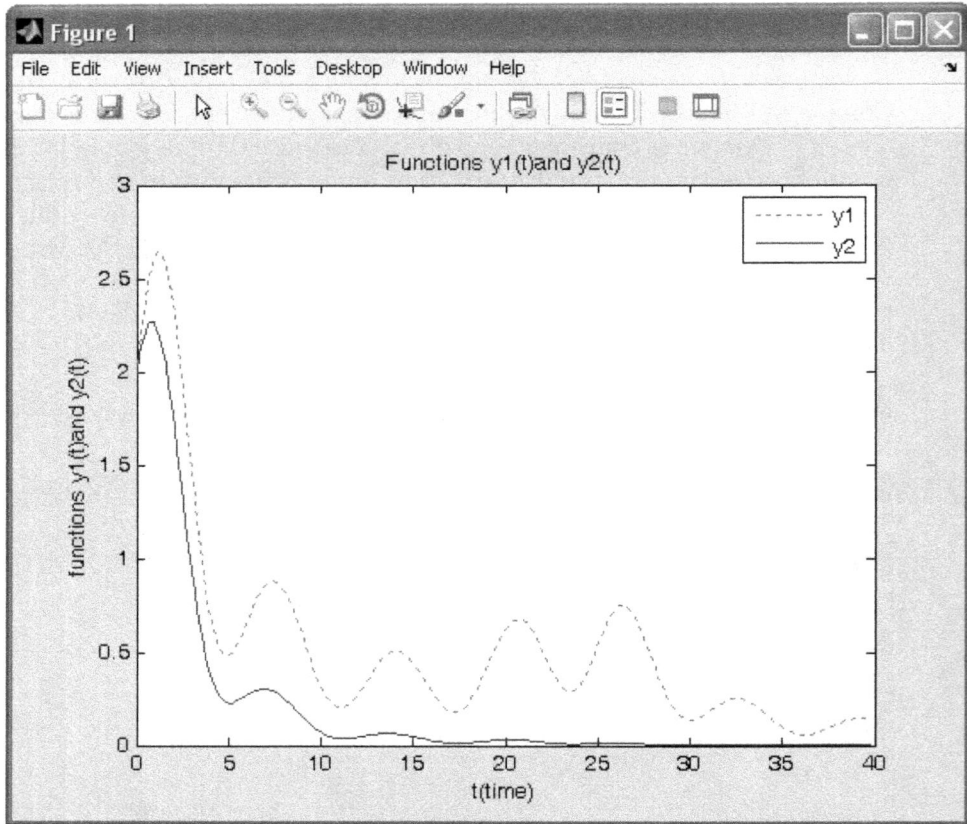

MATLAB uses flow control through its programs. To allow flow control in a program certain rational and logical operators are essential. These operators are shown in Tables B.17 and B.18.

**TABLE B.17** Rational operators

| Rational operators | Description |
| --- | --- |
| < | Less than |
| > | Greater than |
| <= | Less than or equal |
| > = | Greater than or equal |
| = = | Equal |
| ~ = | Not equal |

**TABLE B.18** Logical operators

| Logical operator | Description |
|---|---|
| ~ | NOT |
| & | AND |
| \| | OR |

There are four kind of statements used in MATLAB to control the flow through the user code. They are *for* loops, *while* loops, *if, else,* and *elseif* constructions, and *switch* constructions.

### B.3.1 For Loops

*for* loops allow a group of commands to be repeated a fixed number of times. The basic form of a *for* loop is:

> *for* index = start: increment: stop
>
> statements
>
> *end*

The increment can be omitted, but MATLAB will assume the increment is 1. Also, the increment can be positive or negative. For example,

```
>> for n=1:7
x(n)=sin(n*pi/10)
end
```

x = 0.3090   0.5878   0.8090   0.9511   1.0000   0.9511   0.8090

The general form of a *for* loop is:

> *for* x = array
>
> commands...
>
> *end*

For example,

>> m =[1 5 8; 10 17 22]

m =

   1   5   8

   10   17   22

>> for n = m

x=n(1)-n(2)

end

x =

  -9

x =

  -12

x =

  -14

## B.3.2 While Loops

*while* loop evaluates a group of statements an indefinite number of times in conjunction with a conditional statement. The general form of *while* loop is:

> *while* expression
>
>     commands…
>
> *end*

For example,

>> n=100;

>> x=[];

```
>> while (n>0)
n=n/2-1;
x=[x,n];
end
>> x

x =

  49.0000   23.5000   10.7500   4.3750   1.1875   -0.4063
```

### B.3.3  If, Else, and Elseif

The form of an *if* statement is:

> *if* expression
>
>     statements
>
> *end*

The expression can be either 1 (true) or 0 (false). The statements between the *if* and *end* statements are executed if the expression is true. If the expression is false the statements will be ignored and the execution will resume at the line after the *end* statement. An *end* keyword matches the *if* and terminates the last group of statements. For example,

```
>> x= 20;
>> if x>0
log(x)
end

ans =

  2.9957
```

The optional *elseif* and *else* keywords provide for the execution of alternate groups of statements.

The *if* and *else* can be presented as:

---

> *if* condition
>
>      statements
>
> *else*
>
>      statements
>
> *end*

---

For example, using just *if-else* statement as:

>> x= -24;

>> if x>0

log(x)

else

'x is negative number'

end

ans =

x is negative number

The *if, else,* and *elseif* can be presented as:

---

> *if*     condition 1
>
>      statements
>
> *elseif* condition 2
>
>      statements
>
> *else*     condition 3
>
>      statements
>
> *end*

---

For example using *if, else,* and *elseif* statements as:

```
>> x='28';
>> if ~isnumeric(x)
'x is not a number'
elseif isnumeric(x)&x<0
'x is a negative number'
else
log(x)
end

ans =

x is not a number
```

### 3.3.4   Switch

The *switch* statement executes groups of statements based on the value of a variable or expression. The basic form of a *switch* statement is:

```
switch expression
        case results 1
                statements
        case results 2
                statements
        .
        .
        .

                otherwise
                        statements
        end
```

The keywords *case* and *otherwise* delineate the groups of statements. These respective statements are executed if the value of the expression is equal to the respective results. If none of the *cases* are true, the *otherwise* statements are done. Only the first matching *case* is executed. The same statements can be done for different *cases* by enclosing several *results* in braces. For example,

>> x= 9;

>> switch x

case 1

disp('x is 1')

case {7 , 8, 9}

disp('x is 7, 8, and 9')

case 12

disp('x is 12')

otherwise

disp('x is not, 1, 7, 8, 9 and 12')

end

x is 7, 8, and 9

Considering the following tips can be helpful in working with MATLAB:

1. Variables and function names are case-sensitive.

2. Make comment in M-file by adding lines beginning with a % character.

3. Use a semicolon (;) at the end of each command to suppress output and the semicolon can be removed when debugging the file.

4. Retrieve previously executed commands by pressing the up (↑) and down (↓) arrow keys.

5. Use an ellipse (…) at the end of the line and continue on the next line, when an expression does not fit on one line.

## B.4  SYMBOLIC COMPUTATION

In previous sections, you learned that MATLAB can be a powerful programmable calculator. However, basic MATLAB uses numbers as in a calculator. Most calculators and basic MATLAB lack the ability to manipulate math expressions without using numbers. In this section, you see that MATLAB can manipulate and solve symbolic expressions that make you compute with math symbols rather than numbers. This process is called *symbolic math*. Table B.19 shows some common Symbolic commands. You can practice some symbolic expressions in the following section.

### B.4.1  Simplifying Symbolic Expressions

Symbolic simplification is not always straightforward; there is no universal simplification function because the meaning of a simplest representation of a symbolic expression cannot be defined clearly. MATLAB uses the *sym* or *syms* command to declare variables as symbolic variable. Then, the symbolic can be used in expressions and as arguments to many functions. For example, to rewrite a polynomial in a standard form, use the *expand* function:

>> syms x y; % creating a symbolic variables x and

>> x = sym('x'); y = sym('y'); % or equivalently

>> expand (sin(x+y))

ans =

$$\sin(x) *\cos(y) + \cos(x)* \sin(y)$$

You can use *subs* command to substitute a numeric value for a symbolic variable or replace one symbolic variable with another. For example,

>> syms x;

>> f=3*x^2-7*x+5;

>> subs(f,2)

ans =

3

>> simplify (sin(x)^2 + cos(x)^2) % Symbolic simplification

ans =

1

**TABLE B.19** Common symbolic commands

| Command | Description |
|---------|-------------|
| diff | Differentiate symbolic expression |
| int | Integrate symbolic expression |
| jacobian | Compute Jacobian matrix |
| limit | Compute limit of symbolic expression |
| symsum | Evaluate symbolic sum of series |
| taylor | Taylor series expansion |
| colspace | Return basis for column space of matrix |
| det | Compute determinant of symbolic matrix |
| diag | Create or extract diagonals of symbolic matrices |
| eig | Compute symbolic eigenvalues and eigenvectors |
| expm | Compute symbolic matrix exponential |
| inv | Compute symbolic matrix inverse |
| jordan | Compute Jordan canonical form of matrix |
| null | Form basis for null space of matrix |
| poly | Compute characteristic polynomial of matrix |
| rank | Compute rank of symbolic matrix |
| rref | Compute reduced row echelon form of matrix |
| svd | Compute singular value decomposition of symbolic matrix |
| tril | Return lower triangular part of symbolic matrix |
| triu | Return upper triangular part of symbolic matrix |
| coeffs | List coefficients of multivariate polynomial |
| collect | Collect coefficients |
| expand | Symbolic expansion of polynomials and elementary functions |
| factor | Factorization |
| horner | Horner nested polynomial representation |

**TABLE B.19** Common symbolic commands (Continued)

| Command | Description |
|---------|-------------|
| numden | Numerator and denominator |
| simple | Search for simplest form of symbolic expression |
| simplify | Symbolic simplification |
| subexpr | Rewrite symbolic expression in terms of common subexpressions |
| subs | Symbolic substitution in symbolic expression or matrix |
| compose | Functional composition |
| dsolve | Symbolic solution of ordinary differential equations |
| finverse | Functional inverse |
| solve | Symbolic solution of algebraic equations |
| cosint | Cosine integral |
| sinint | Sine integral |
| zeta | Compute Riemann zeta function |
| ceil | Round symbolic matrix toward positive infinity |
| conj | Symbolic complex conjugate |
| eq | Perform symbolic equality test |
| fix | Round toward zero |
| floor | Round symbolic matrix toward negative infinity |
| frac | Symbolic matrix elementwise fractional parts |
| imag | Imaginary part of complex number |
| log10 | Logarithm base 10 of entries of symbolic matrix |
| log2 | Logarithm base 2 of entries of symbolic matrix |
| mod | Symbolic matrix elementwise modulus |
| pretty | Pretty-print symbolic expressions |
| quorem | Symbolic matrix elementwise quotient and remainder |
| real | Real part of complex symbolic number |
| round | Symbolic matrix elementwise round |
| size | Symbolic matrix dimensions |
| sort | Sort symbolic vectors, matrices, or polynomials |
| sym | Define symbolic objects |

*(Continued)*

**TABLE B.19** Common symbolic commands (Continued)

| Command | Description |
|---------|-------------|
| syms | Shortcut for constructing symbolic objects |
| symvar | Find symbolic variables in symbolic expression or matrix |
| fourier | Fourier integral transform |
| ifourier | Inverse Fourier integral transform |
| ilaplace | Inverse Laplace transform |
| iztrans | Inverse $z$-transform |
| laplace | Laplace transform |
| ztrans | $z$-transform |

## B.4.2   Differentiating Symbolic Expressions

Use *diff* ( ) command for differentiation. For example,

\>> syms x;

\>> f =-cos(5*x)+2;

\>> diff(f)

ans =

5*sin(5*x)

\>> y=8*sin(x)*exp(x);

\>> diff(y)

ans =

8*cos(x)*exp(x)+8*sin(x)*exp(x)

\>> diff(diff(y))% second derivative of y

ans =

16*cos(x)*exp(x)

An example of a partial derivative is as follows:

\>> syms v u;

\>> f = cos(v*u);

\>> diff(f,u)% create partial derivative $\dfrac{\partial f}{\partial u}$

ans =

$$-\sin(u\ v)*\ v$$

\>> diff(f,v) % create partial derivative $\dfrac{\partial f}{\partial v}$

ans =

$$-\sin(u\ v)*\ u$$

\>> diff(f,u,2) % create second partial derivative $\dfrac{\partial^2 f}{\partial u^2}$

ans =

$$-\cos(u\ v)\wedge 2\ *v$$

## B.4.3  Integrating Symbolic Expressions

The *int(f)* function is used to integrate a symbolic expression f. For example,

\>> syms x;

\>> f=cos(x)^2;

\>> int(f)

ans =

$$1/2\ *\cos(x)\ *\sin(x)\ +\ 1/2*\ x$$

\>> int(1/(1+x^2))

ans =

$$\text{atan}(x)$$

## B.4.4  Limits Symbolic Expressions

The *limit(f)* command is used to calculate the limits of function f. For example,

`>> syms x y z;`

`>> limit((sin(x)/x), x, 0)` % $\lim\limits_{x \to 0} \dfrac{\sin x}{x} = 1$

ans =

$$1$$

`>> limit(1/x, x, 0, 'right')` % $\lim\limits_{x \to 0^+} \dfrac{1}{x} = \infty$

ans =

$$\text{infinity}$$

`>> limit(1/x, x, 0, 'left')` % $\lim\limits_{x \to 0^-} \dfrac{1}{x} = -\infty$

ans =

$$\text{-infinity}$$

## B.4.5  Taylor Series Symbolic Expressions

Use the *taylor( )* function to find the Taylor series of a function with respect to the variable given. For example,

`>> syms x; N =4;`

`>> taylor(exp(-x),N+1)` % $f(x) \cong \sum\limits_{n=0}^{N} \dfrac{1}{n!} f^n(0)$

ans =

$$1 - x + 1/2 *x^2 - 1/6 *x^3 + 1/24 *x^4$$

`>> f=exp(x);`

`>> taylor(f,3)`

ans =

$$1 + x + \tfrac{1}{2}* x^2$$

### B.4.6   Sums Symbolic Expressions

Use the *symsum( )* function to obtain the sum of a series. For example,

```
>> syms k n;
```

$$>> \text{symsum(k,0,n-1)} \quad \% \quad \sum_{k=0}^{n-1} k = 0 + 1 + 2 + \dots + n - 1 = \frac{1}{2}n^2 - \frac{1}{2}n$$

```
ans =
```

$$1/2 \text{ n}^2 - 1/2 \text{ n}$$

```
>> syms n N;
```

$$>> \text{symsum(1/n}^2\text{,1,inf)} \quad \% \quad \sum_{n=0}^{N} \frac{1}{n^2} = \frac{\pi^2}{6}$$

```
ans =
```

$$1/6 *\text{pi}^2$$

### B.4.7   Solving Equations as Symbolic Expressions

Many MATLAB commands and functions are used to manipulate the vectors or matrices consisting of symbolic expressions. For example,

```
>> syms a b c d;
>> M=[a b;c d];
>> det(M)

ans =
```

$$\text{a} * \text{d} - \text{b} * \text{c}$$

```
>> syms x y;
>> f=solve('5*x+4*y=3','x-6*y=2'); % solve the system 5x + 4y = 3, x − 6y = 2
>> x=f.x

x =
```

13/17

```
>> y=f.y

y =

                        -7/34

>> syms x;
>> solve (x^3-6*x^2+11*x-6)

ans =

                          1
                          2
                          3
```

Use *dsolve ( )* function to solve symbolic differential equations. For example,

```
>> syms x y t;

>> dsolve('Dy+3*y=8') % solve y' + 3y = 8

ans =

              8/3 -C1 +exp(-3 t) % C1 is undetermined constant
```

`>> dsolve('Dy=1+y^2','y(0)=1')` % solve $y' = 1 + y^2$ with initial condition $y(0) = 1$

```
ans =

                  tan(t + 1/4 pi)
```

`>> dsolve('D2y+9*y=0','y(0)=1','Dy(pi)=2')` % solve $y'' + 9y = 0$ with initial conditions                 % $y(0) = 1$, $y(\pi) = 2$

```
ans =

              - 2/3 *sin(3 t) + cos(3 t)
```

# ANSWERS TO ODD-NUMBERED EXERCISES

## CHAPTER 1

### 1.1    Points and Vectors on the Plane

**1.1.1**    **1.** Scalar.

**3.** Vector.

**5.** Scalar.

**7.** Vector.

**9.** Scalar.

**11.** Scalar.

**13.** Scalar.

**1.1.3**    $\vec{u} + \vec{v} = \begin{bmatrix} -3 \\ 1 \end{bmatrix}.$

$\vec{u} - \vec{v} = \begin{bmatrix} 1 \\ 9 \end{bmatrix}.$

$2\vec{u} = \begin{bmatrix} -2 \\ 10 \end{bmatrix}.$

**1.1.5** **1.**

**3.**

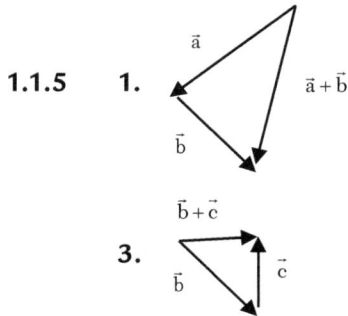

**1.1.7** $\sqrt{2a^2 - 2a = 1}$ .

**1.1.9** $a = \pm 1$ or $a = -8$.

**1.1.11** Since $ABCD$ is parallelogram, $\overrightarrow{AD} = \overrightarrow{BC}$, Hence,
$\overrightarrow{AC} + \overrightarrow{BD} = \overrightarrow{AD} + \overrightarrow{BC} = 2\overrightarrow{BC}$.

**1.1.13** Since

$$\overrightarrow{IA} + 3\overrightarrow{IA} = -3\overrightarrow{AB} \Leftrightarrow \overrightarrow{AI} = \frac{3}{4}\overrightarrow{AB} \ .$$

Similarly,

$$\overrightarrow{JA} = -\frac{1}{3}\overrightarrow{JB} \Leftrightarrow \overrightarrow{AJ} = \frac{1}{4}\overrightarrow{AB} \ .$$

Thus, we take I such that $\overrightarrow{AI} = \frac{3}{4}\overrightarrow{AB}$ , and J such that $\overrightarrow{AJ} = \frac{1}{4}\overrightarrow{AB}$ .

Now,

$$\overrightarrow{MA} + 3\overrightarrow{MB} = \overrightarrow{MI} + \overrightarrow{AI} + 3\overrightarrow{IB} = 4\overrightarrow{MI} \ , \text{ and}$$

$$3\overrightarrow{MA} + \overrightarrow{MB} = 3\overrightarrow{MJ} + 3\overrightarrow{JA} + \overrightarrow{MJ} + \overrightarrow{JB} = 4\overrightarrow{MJ} \ .$$

**1.1.15** **1.** $a = \frac{4}{7}, \beta = \frac{3}{7}$ .

**1.1.17** Since $\overrightarrow{BD} + \vec{u} = \vec{v}$, $\overrightarrow{BD} = \vec{v} - \vec{u}$. Then $\overrightarrow{BE} = x(\vec{v} - \vec{u})$.

Since $\overrightarrow{AC} = \vec{u} + \vec{v}$, $\overrightarrow{AE} = y(\vec{u} + \vec{v})$.

But $\overrightarrow{AB} = \overrightarrow{AE} + \overrightarrow{EB} = \overrightarrow{AE} - \overrightarrow{BE}$, i.e.
$\vec{u} = y(\vec{u} + \vec{v}) - x(\vec{v} - \vec{u}) = (x + y)\vec{u} + (y - x)\vec{v}$ .

Since $\vec{u}$ and $\vec{v}$ are non-collinear we have by Exercises 1.1.11,

$x + y = 1$ and $y - x = 0$, i.e., $x = y = \dfrac{1}{2}$ and E is the midpoint of both diagonals.

## 1.2 Scalar Product on the Plane

**1.2.1**    **1.**   $\vec{a} \cdot \vec{b} = 7$.

         **3.**   $\vec{a} \cdot \vec{b} = 28$.

**1.2.3**    **1.**   $\cos\left(\widehat{\vec{a}, \vec{b}}\right) = \dfrac{1}{2}$.

         Thus, $\left(\widehat{\vec{a}, \vec{b}}\right) = 60°$.

         **3.**   $\cos\left(\widehat{\vec{a}, \vec{b}}\right) = 0$.

         Thus, $\left(\widehat{\vec{a}, \vec{b}}\right) = 90°$.

**1.2.5**    $a = \dfrac{1}{2}$.

**1.2.7**    Let $\vec{a}$ and $\vec{b}$ be the vectors of two adjacent sides of the rhombus. Then $\|\vec{a}\| = \|\vec{b}\|$. Further, the diagonals are the vectors $\vec{a} + \vec{b}$ and $\vec{a} - \vec{b}$. Now compute $\left(\vec{a} + \vec{b}\right) \cdot \left(\vec{a} - \vec{b}\right) = a^2 - b^2 = 0$, since $a = b$.

**1.2.9**    $\vec{a} \cdot \vec{b} = -\sqrt{675} + \sqrt{675} = 0$.

**1.2.11**    $\left\|\vec{a} + \vec{b}\right\| = \left\|\vec{a} - \vec{b}\right\| \Leftrightarrow \left\|\vec{a} + \vec{b}\right\|^2 = \left\|\vec{a} - \vec{b}\right\|^2$.

                     $\Leftrightarrow 4\vec{a} \cdot \vec{b} = 0$.

                     $\Leftrightarrow \vec{a} \cdot \vec{b} = 0$.

This proves what we want.

**1.2.13** $\vec{c} \cdot \vec{b} = \left( \vec{a} \cdot \vec{b} \right) - \dfrac{\left( \vec{a} \cdot \vec{b} \right) \left\| \vec{b} \right\|^2}{\left\| \vec{b} \right\|^2} = \left( \vec{a} \cdot \vec{b} \right) - \left( \vec{a} \cdot \vec{b} \right) = 0.$

**1.2.15** Since,

$\left( \vec{a} \cdot \vec{b} \right) = 0$, we have

$\left\| \vec{a} + \vec{b} \right\|^2 = \left( \vec{a} + \vec{b} \right) \cdot \left( \vec{a} + \vec{b} \right) = \vec{a} \cdot \vec{a} + 0 + \vec{b} \cdot \vec{b} = \left\| \vec{a} \right\|^2 + \left\| \vec{b} \right\|^2.$

**1.2.17** $\left( \vec{a} - \vec{b} \right) \cdot \left( \vec{a} - \vec{b} \right) = \left\| \vec{a} - \vec{b} \right\|^2 = 0$, but the norm of a vector is 0 if and only if the vector is the $\vec{0}$ vector.

**1.2.19** $\begin{bmatrix} 1 \\ m_1 \end{bmatrix} \cdot \begin{bmatrix} 1 \\ m_2 \end{bmatrix} = 0 \Leftrightarrow 1 + m_1 m_2 = 0 \Leftrightarrow m_1 m_2 = -1.$

**1.2.21** $\left( \vec{v} - \dfrac{\vec{v} \cdot \vec{w}}{\left\| \vec{w} \right\|^2} \vec{w} \right) \cdot \vec{w} = \vec{v} \cdot \vec{w} - \dfrac{\vec{v} \cdot \vec{w}}{\left\| \vec{w} \right\|^2} \left\| \vec{w} \right\|^2 = 0.$

## 1.3    Linear Independence

**1.3.1** We must show that there exist scalars $a$ and $b$ such that

$a \begin{bmatrix} 1 \\ 0 \end{bmatrix} + b \begin{bmatrix} 3 \\ 1 \end{bmatrix} = \begin{bmatrix} 7 \\ 3 \end{bmatrix}$, we get the vector equation $\begin{bmatrix} 3b + a \\ b \end{bmatrix} = \begin{bmatrix} 7 \\ 3 \end{bmatrix}$. The

solution of this system is given by $a = -2$ and $b = 3$. Thus, we can

write $\begin{bmatrix} 7 \\ 3 \end{bmatrix} = (-2) \begin{bmatrix} 1 \\ 0 \end{bmatrix} + 3 \begin{bmatrix} 3 \\ 1 \end{bmatrix}.$

**1.3.3** $\begin{bmatrix} a \\ b \end{bmatrix} = \dfrac{b - a}{2} \begin{bmatrix} -1 \\ 1 \end{bmatrix} + \dfrac{a + b}{2} \begin{bmatrix} 1 \\ 1 \end{bmatrix}.$

**1.3.5** $a \begin{bmatrix} 1 \\ 0 \end{bmatrix} + b \begin{bmatrix} 0 \\ 1 \end{bmatrix} = \begin{bmatrix} 0 \\ 0 \end{bmatrix}$, then $a = 0$ and $b = 0$. Thus the vectors

$\vec{x} = \begin{bmatrix} 1 \\ 0 \end{bmatrix}$ and $\vec{y} = \begin{bmatrix} 0 \\ 1 \end{bmatrix}$ are linearly independent.

**1.3.7**   $2a + b + 8c = 0$

$a + 2b + 7c = 0,$

By solving this homogenous system of linear equations with more unknowns than equations, we get $a = -3c$, $b = -2c$, thus for $c = -1$, we get $a = 3$ and $b = 2$. Therefore, we have shown that the vectors $\begin{bmatrix} 2 \\ 1 \end{bmatrix}, \begin{bmatrix} 1 \\ 2 \end{bmatrix}$, and $\begin{bmatrix} 8 \\ 7 \end{bmatrix}$ in $\mathbb{R}^2$ linearly dependent because in addition to solution $a = b = c = 0$ for the vector equation

$a \begin{bmatrix} 2 \\ 1 \end{bmatrix} + b \begin{bmatrix} 1 \\ 2 \end{bmatrix} + c \begin{bmatrix} 8 \\ 7 \end{bmatrix} = \begin{bmatrix} 0 \\ 0 \end{bmatrix}$, we also have the solution

$a = 3, b = 2,$ and $c = -1$.

**1.3.9**   Suppose that vectors $\vec{x}$ and $\vec{y}$ are linearly dependent, that is, $a\vec{x} + b\vec{y} = \vec{0}$, where $a$ and $b$ are not both zero. If $a \neq 0$, then $\vec{x} = -\dfrac{b}{a}\vec{y}$, and if $b \neq 0$, then $\vec{y} = -\dfrac{a}{b}\vec{x}$; in either case, the vectors are parallel. Conversely, if $\vec{x}$ and $\vec{y}$ are parallel, then either $\vec{x} = c\vec{y}$ or $\vec{y} = c\vec{x}$. In the first case, $1 \cdot \vec{x} + (-c)\vec{y} = \vec{0}$; in the second case, $c\vec{x} + (-1)\vec{y} = \vec{0}$; in either case, the pair is linearly dependent.

**1.3.11**   **1.**   $-7$.

**1.3.13**   **1.**   By Chasles' Rule

$$\overrightarrow{AE} = \frac{1}{4}\overrightarrow{AC} \Leftrightarrow \overrightarrow{AB} + \overrightarrow{BE} = \frac{1}{4}\overrightarrow{AC}$$

and

$$\overrightarrow{AF} = \frac{3}{4}\overrightarrow{AC} \Leftrightarrow \overrightarrow{AD} + \overrightarrow{DF} = \frac{3}{4}\overrightarrow{AC}$$

$$\overrightarrow{AB} + \overrightarrow{BE} + \overrightarrow{AD} + \overrightarrow{DF} = \overrightarrow{AC} \Leftrightarrow \overrightarrow{BE} + \overrightarrow{DF} = \overrightarrow{AD} + \overrightarrow{DC} - \overrightarrow{AB} - \overrightarrow{AD}$$

$$\Leftrightarrow \overrightarrow{BE} = -\overrightarrow{DF}$$

## 1.4    Geometric Transformations in Two Dimensions

**1.4.1**    $a = -3, b = -\dfrac{1}{2}.$

**1.4.3**    $\begin{bmatrix} a & b \\ c & -a \end{bmatrix}, bc = -a^2.$

**1.4.5**    From $L\begin{pmatrix} x \\ y \end{pmatrix} = \begin{pmatrix} 1 \\ 2 \end{pmatrix}$, it follows that $L\begin{pmatrix} 1 \\ 1 \end{pmatrix} = \begin{pmatrix} 1 \\ 2 \end{pmatrix}, L\begin{pmatrix} 2 \\ 2 \end{pmatrix} = \begin{pmatrix} 1 \\ 2 \end{pmatrix}$ and also

$L\begin{pmatrix} 3 \\ 3 \end{pmatrix} = \begin{pmatrix} 1 \\ 2 \end{pmatrix}$. But, $L\left( \begin{pmatrix} 1 \\ 1 \end{pmatrix} + \begin{pmatrix} 2 \\ 2 \end{pmatrix} \right) = L\begin{pmatrix} 3 \\ 3 \end{pmatrix} = \begin{pmatrix} 1 \\ 2 \end{pmatrix}$ and

$L\begin{pmatrix} 1 \\ 1 \end{pmatrix} + L\begin{pmatrix} 2 \\ 2 \end{pmatrix} = \begin{pmatrix} 1 \\ 2 \end{pmatrix} + \begin{pmatrix} 1 \\ 2 \end{pmatrix} = \begin{pmatrix} 2 \\ 4 \end{pmatrix}$. Thus, $L$ is not linear.

**1.4.7**    We show that $L$ preserves addition. Let $\begin{pmatrix} x_1 \\ y_1 \end{pmatrix}$ and $\begin{pmatrix} x_2 \\ y_2 \end{pmatrix}$ be elements of $\mathbb{R}^2$. Then

$$L\left( \begin{pmatrix} x_1 \\ y_1 \end{pmatrix} + \begin{pmatrix} x_2 \\ y_2 \end{pmatrix} \right) = L\begin{pmatrix} x_1 + x_2 \\ y_1 + y_2 \end{pmatrix}$$

$$= \begin{pmatrix} x_1 + x_2 - y_1 - y_2 \\ 3x_1 + 3x_2 \end{pmatrix}$$

$$= \begin{pmatrix} x_1 - y_1 \\ 3x_1 \end{pmatrix} + \begin{pmatrix} x_2 - y_2 \\ 3x_2 \end{pmatrix}$$

$$= L\begin{pmatrix} x_1 \\ y_1 \end{pmatrix} + L\begin{pmatrix} x_2 \\ y_2 \end{pmatrix}.$$

Thus $L$ preserves vector addition.

Now, we show that $L$ preserves scalar multiplication. Let $k$ be a scalar.

$$L\left( k\begin{pmatrix} x \\ y \end{pmatrix} \right) = L\begin{pmatrix} kx \\ ky \end{pmatrix}$$

$$= \begin{pmatrix} kx - ky \\ 3kx \end{pmatrix}$$

$$= k \begin{pmatrix} x - y \\ 3x \end{pmatrix}$$

$$= kL \begin{pmatrix} x \\ y \end{pmatrix}.$$

Thus $L$ preserves scalar multiplication. Thus, $L$ is linear.

**1.4.9** The desired transformations are in Figures C.3 through C.5.

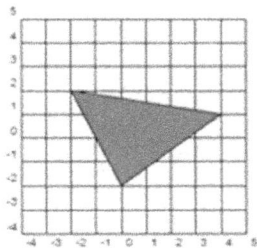

Figure C.3: Horizontal Stretch.

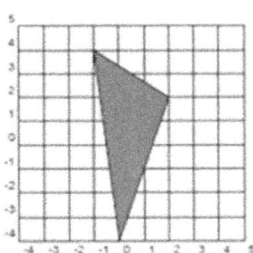

Figure C.4: Vertical Stretch.

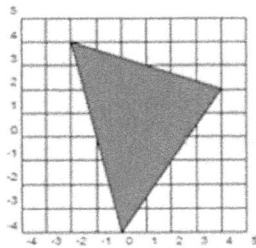

Figure C.5: Horizontal and Vertical Stretch.

**1.4.11** The desired transformations are in Figures C.6 through C.9.

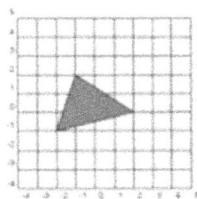

Figure C.6: Levogyrate rotation $\frac{\pi}{2}$ radians.

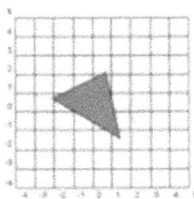

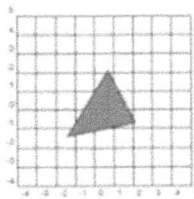

Figure C.7: Levogyrate rotation $\frac{\pi}{4}$ radians.

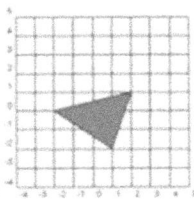

Figure C.8: Dextrogyrate rotation $\frac{\pi}{2}$ radians.

Figure C.9: Dextrogyrate rotation $\frac{\pi}{4}$ radians.

**1.4.13** $R_H = \begin{bmatrix} -1 & 0 \\ 0 & 1 \end{bmatrix}, \begin{bmatrix} -3 \\ 2 \end{bmatrix}.$

**1.4.15** The rotation through $\pi/2$ about the point $(5,1)$ is $\begin{bmatrix} 0 & -1 & 0 \\ 1 & 0 & 0 \\ 0 & 0 & 1 \end{bmatrix}.$

The image of the unit square under this rotation is $(6, -3)$, $(5, -3)$, $(5, -4)$, and $(6, -4)$.

**1.4.17** 1. $L_2 \circ L_1 = \begin{bmatrix} 27 \\ -11 \end{bmatrix}.$

## 1.5    Determinants in Two Dimensions

**1.5.1**    **1.** $-9$.

**3.** $22$.

**5.** $x^2 - 5x + 21$.

**1.5.3**    $1$.

**1.5.5**    $\det(kA) = k^2 \det(A)$.

**1.5.7**    $\det(AB) = \det(A)\det(B) = \det(B)\det(A) = \det(BA)$.

**1.5.9**    $\vec{r_3} = \begin{bmatrix} -b \\ a \end{bmatrix} \cdot \begin{bmatrix} c \\ d \end{bmatrix} = -bc + ad = \det \begin{bmatrix} a & c \\ b & d \end{bmatrix}$.

Since the dot product of two vectors is positive if the angle between them is less than $\pi / 2$, the determinant is positive if the angle between $\vec{r_2}$ and $\vec{r_3}$ is less than $\pi / 2$. Thus $\vec{r_2}$ lies counterclockwise from $\vec{r_1}$.

## 1.6    Parametric Curves on the Plane

**1.6.1**    $y - x = 4t \Rightarrow t = \dfrac{y - x}{4}$ and $x = \left( \dfrac{y - x}{4} \right)^3 - 2\left( \dfrac{y - x}{4} \right)$.

**1.6.3**    **1.** $ay - cx = ad - bc$, this is a straight line with positive slope.

**3.** $\dfrac{x^2}{a^2} - \dfrac{y^2}{b^2} = 1, x > 0$. This is one branch of a hyperbola.

**1.6.5**    Observe that $y = 2x + 1$, so the trace is part of this line. Since in the interval $[0, 4\pi]$, $-1 \leq \sin t \leq 1$, we want the portion of the line $y = 2x + 1$ with $-1 \leq x \leq 1$ and thus $-1 \leq y \leq 3$. The curve starts at the middle point $(0,1)$ at $t = 0$, reaches the high point $(1,3)$ at $t = \dfrac{\pi}{2}$, reaches its low point $(-1,1)$ at $t = \dfrac{7\pi}{2}$, and finishes in the middle point $(0,1)$ when $t = 4\pi$.

**1.6.7**    $\dfrac{2}{5}$.

**1.6.9**    $1$.

**1.6.11**    $\dfrac{3}{2}$.

**1.6.13**    $8\rho$.

**1.6.15**    $\left(gh - V^2\right)^2 \geq g^2\left(h^2 + u^2\right) \Rightarrow g^2 u^2 \leq V^2\left(V^2 - 2gh\right)$.

## 1.7    Vectors in Space

**1.7.1**    $22$.

**1.7.3**    $-2$.

**1.7.5**    $(x-1) - 2(y-1) - (z-1) = 0$.

**1.7.7**    $abc + 2(ab + bc + ca) + \pi(a + b + c) + \dfrac{4\pi}{3}$.

**1.7.9**    $\begin{bmatrix} x \\ y \\ z \end{bmatrix} = \begin{bmatrix} 0 \\ 0 \\ 1 \end{bmatrix} + t \begin{bmatrix} a \\ a^2 \\ a^2 \end{bmatrix}, t \in \mathbb{R}$.

**1.7.11**    $\dfrac{\sqrt{3}}{3}$.

**1.7.13**

$$\left| x_1 y_1 + x_2 y_2 + x_3 y_3 \right| \leq \sqrt{x_1^{\,2} + x_2^{\,2} + x_3^{\,2}} \Rightarrow \left(a^2 + b^2 + c^2\right)^2 \leq \left(a^4 + b^4 + c^4\right)(3).$$

**1.7.15**    $x + y + z = 1$.

$\dfrac{1}{6}$.

**1.7.17** $\left\| proj_n^{\vec{r}_0 - \vec{b}} \right\| = \left\| \dfrac{\left(\vec{r}_0 - \vec{b}\right)\bullet\vec{n}}{\|\vec{n}\|} \, \vec{n} \right\| = \dfrac{\left|\left(\vec{r}_0 - \vec{b}\right)\bullet\vec{n}\right|}{\|\vec{n}\|}$

$\left\| proj_n^{\vec{r}_0 - \vec{b}} \right\| = \dfrac{\left|\vec{r}_0\bullet\vec{n} - \vec{b}\bullet\vec{n}\right|}{\|\vec{n}\|} = \dfrac{\left|\vec{a}\bullet\vec{n} - \vec{b}\bullet\vec{n}\right|}{\|\vec{n}\|} = \dfrac{\left|\left(\vec{a}_0 - \vec{b}\right)\bullet\vec{n}\right|}{\|\vec{n}\|}.$

## 1.8    Cross Product

**1.8.1**

$$\left(\vec{a} - \vec{b}\right)\times\left(\vec{a} + \vec{b}\right) = \vec{a}\times\vec{a} + \vec{a}\times\vec{b} - \vec{b}\times\vec{a} + \vec{b}\times\vec{b} = \vec{0} + \vec{a}\times\vec{b} + \vec{a}\times\vec{b} + \vec{0} = 2\vec{a}\times\vec{b}.$$

**1.8.3**    $\sqrt{3}$.

**1.8.5**    It is not associative, since $\vec{i}\times\left(\vec{i}\times\vec{j}\right) = \left(\vec{i}\times\vec{k}\right) = -\vec{j}$, but

$\left(\vec{i}\times\vec{i}\right)\times\vec{j} = \vec{0}\times\vec{j} = \vec{0}$.

**1.8.7**    $-\vec{i} - \vec{k}$.

**1.8.9**    $2ax + 3a^2y - az = a^2$.

**1.8.11**    $\vec{a}\times\left(\vec{b}\times\vec{c}\right) = (\vec{a}\bullet\vec{c})\vec{b} - \left(\vec{a}\bullet\vec{b}\right)\vec{c}$, $\vec{b}\times(\vec{c}\times\vec{a}) = \left(\vec{b}\bullet\vec{a}\right)\vec{c} - \left(\vec{b}\bullet\vec{c}\right)\vec{a}$,

$\vec{c}\times\left(\vec{a}\times\vec{b}\right) = \left(\vec{c}\bullet\vec{b}\right)\vec{a} - \left(\vec{c}\bullet\vec{a}\right)\vec{b}$, and adding yields the result.

**1.8.13**    $\left\{\vec{x} : \vec{x} \in \mathbb{R}\vec{a}\times\vec{b}\right\}$.

**1.8.15**    $\vec{a}\bullet\left(\vec{b}\times\vec{c}\right) = \det\begin{bmatrix} a_1 & a_2 & a_3 \\ b_1 & b_2 & b_3 \\ c_1 & c_2 & c_3 \end{bmatrix} = 0$ if and only if the row vectors

$\vec{a}, \vec{b},$ and $\vec{c}$ are linearly independent, i.e., if and only if one vector lies in the plane of the other two vectors.

**1.8.17** Observe that, $\vec{x}\bullet(\vec{y}\times\vec{z})=(\vec{x}\times\vec{y})\bullet\vec{z}$, now putting $\vec{x}=\vec{a}\times\vec{b}$, $\vec{y}=\vec{c}$, and $\vec{z}=\vec{d}$, this gives, $\left(\left(\vec{a}\times\vec{b}\right)\times\vec{c}\right)\bullet\vec{d}=\left(\left(\vec{c}\bullet\vec{a}\right)\vec{b}-\left(\vec{c}\bullet\vec{b}\right)\vec{a}\right)\bullet\vec{d}=\left(\vec{c}\bullet\vec{a}\right)\left(\vec{b}\bullet\vec{d}\right)$ $-\left(\vec{c}\bullet\vec{b}\right)\left(\vec{a}\bullet\vec{d}\right)$.

**1.8.19**   **1.**  $\begin{bmatrix} 12 \\ 18 \\ 24 \end{bmatrix}$.

  **3.**  $x=6t, y=0, z=3-3t$.

  **5.**  $\left(3,0,\dfrac{3}{2}\right)$.

  **7.**  $\dfrac{33}{16}\sqrt{29}$.

## 1.9   Matrices in Three Dimensions

**1.9.1**   **1.**  $\begin{bmatrix} 1 & 1 & 5 \\ -1 & 3 & 2 \\ 1 & 9 & 1 \end{bmatrix}$.

  **3.**  $\begin{bmatrix} 0 & 7 & -9 \\ -2 & 4 & 0 \\ -1 & 16 & -7 \end{bmatrix}$.

**1.9.3**  $\begin{bmatrix} a \\ 2a-3b \\ 5b \\ a+2b \\ a \end{bmatrix} = a\begin{bmatrix} 1 \\ 2 \\ 0 \\ 1 \\ 1 \end{bmatrix} + b\begin{bmatrix} 0 \\ -3 \\ 5 \\ 2 \\ 0 \end{bmatrix}$, so clearly the family,

$$\left\{ \begin{bmatrix} 1 \\ 2 \\ 0 \\ 1 \\ 1 \end{bmatrix}, \begin{bmatrix} 0 \\ -3 \\ 5 \\ 2 \\ 0 \end{bmatrix} \right\}$$

spans the subspace. To show that this is a linearly independent family, assume that

$$a \begin{bmatrix} 1 \\ 2 \\ 0 \\ 1 \\ 1 \end{bmatrix} + b \begin{bmatrix} 0 \\ -3 \\ 5 \\ 2 \\ 0 \end{bmatrix} = \begin{bmatrix} 0 \\ 0 \\ 0 \\ 0 \\ 0 \end{bmatrix}.$$

Then it clearly follows that $a = b = 0$, and so this is a linearly independent family.

**1.9.5**
$$A^2 = \begin{bmatrix} 1 & 2 & 3 & 4 & \cdots & n-1 & n \\ 0 & 1 & 2 & 3 & \cdots & n-2 & n-1 \\ 0 & 0 & 1 & 2 & \cdots & n-3 & n-2 \\ \vdots & \vdots & \vdots & \vdots & \vdots & \vdots & \vdots \\ 0 & 0 & 0 & 0 & \cdots & 0 & 1 \end{bmatrix}.$$

$$A^3 = \begin{bmatrix} 1 & 3 & 6 & 10 & \cdots & \dfrac{(n-1)}{2} & \dfrac{n(n+1)}{2} \\ 0 & 1 & 3 & 6 & \cdots & \dfrac{(n-2)(n-1)}{2} & \dfrac{(n-1)n}{2} \\ 0 & 0 & 1 & 3 & \cdots & \dfrac{(n-3)(n-2)}{2} & \dfrac{(n-2)(n-1)}{2} \\ \vdots & \vdots & \vdots & \vdots & \vdots & \vdots & \vdots \\ 0 & 0 & 0 & 0 & \cdots & 0 & 1 \end{bmatrix}.$$

**1.9.7** $\begin{bmatrix} 1 & -1 & 0 \\ 1 & 0 & 1 \\ 0 & 1 & -1 \end{bmatrix}.$

**1.9.9** $\begin{bmatrix} 0 & 0 & 0 \\ 0 & 1 & 0 \\ 0 & 0 & 0 \end{bmatrix}, \begin{pmatrix} 0 \\ 4 \\ 0 \end{pmatrix}.$

**1.9.11** Not linear, because

$$L\left(a\begin{pmatrix} x_1 \\ x_2 \\ x_3 \end{pmatrix}\right) = L\begin{pmatrix} a\,x_1 \\ a\,x_2 \\ a\,x_3 \end{pmatrix} = \begin{pmatrix} 1 \\ a\,x_2 \\ a\,x_3 \end{pmatrix}, \text{but } a\,L\begin{pmatrix} x_1 \\ x_2 \\ x_3 \end{pmatrix} = a\begin{pmatrix} 1 \\ x_2 \\ x_3 \end{pmatrix} = \begin{pmatrix} a \\ a\,x_2 \\ a\,x_3 \end{pmatrix}.$$

**1.9.13** $\begin{bmatrix} 1 & 0 & 0 \\ 0 & -1 & 0 \\ 0 & 0 & 1 \end{bmatrix}.$

## 1.10  Determinants in Three Dimensions

**1.10.1** $-10$.

**1.10.3** $aef$.

**1.10.5** $2$.

**1.10.7** $-t^4 - t^3 + 18t^2 + 9t - 21$.

**1.10.9** $c = -5$.

**1.10.11** $\det\begin{bmatrix} a & b & c \\ b & c & a \\ c & a & b \end{bmatrix} = a\begin{bmatrix} c & a \\ a & b \end{bmatrix} - b\begin{bmatrix} b & a \\ c & b \end{bmatrix} + c\begin{bmatrix} b & c \\ c & a \end{bmatrix}$

$$= acb - a^3 - b^3 + abc + abc - c^3 = 3abc - a^3 - b^3 - c^3$$

**1.10.13** $\lambda = -1$ and $1$.

**1.10.15** $37$.

## 1.11    Some Solid Geometry

**1.11.1**    $\dfrac{a}{\sqrt{2}}, \dfrac{\pi}{4}$.

## 1.12    Cavalieri and the Pappus-Guldin Rules

**1.12.1**    The lateral area is thus $\pi r\sqrt{r^2 + h^2}$.

The volume of the cone is $\dfrac{2}{3}\pi r \times \dfrac{rh}{2} = \dfrac{\pi}{3}r^2 h$.

**1.12.3**    Area $= \dfrac{1}{2}\pi r^2$.

Volume $= \dfrac{4}{3}\pi r^3$.

**1.12.5**    $A = 11309.7 \ m^2$
$V = 113097.3 \ m^3$.

## 1.13    Dihedral Angles and Platonic Solids

**1.13.1**    **1.**    Tetrahedron.
**2.**    Cube or hexahedron.
**3.**    Octahedron.
**4.**    Dodecahedron.
**5.**    Icosahedron.

**1.13.3**    $\sin\dfrac{\theta}{2} = \dfrac{\cos(\pi/n)}{\sin(\pi/m)}$.

## 1.14    Spherical Trigonometry

**1.14.1**    $\cos\theta = \dfrac{1}{3}$, where $\theta$ is the angle between $\vec{a}$ and $\vec{b}$.

**1.14.3**    $V = \dfrac{1}{3} \times \left(6a^2\right) \times \left(\dfrac{a}{2}\right) = a^3$.

$$V = \frac{a^3}{4}\left(\sqrt{25 + 10\sqrt{5}}\right) \times \left(\sqrt{10 + 22\sqrt{\frac{1}{5}}}\right).$$

**1.14.5**

$$V = \frac{a^3}{4}\left(15 + 7\sqrt{5}\right)$$

## 1.15    Canonical Surfaces

**1.15.1**    $4x^2 + 4y^2 + z^2 = 1$.

**1.15.3**    A spiral staircase.

**1.15.5**    The *Maple*$^{TM}$ commands to graph this surface are:

```
>  with(plots) :
>   implicitplot3d(3 x² + 5 y² = 1, x = -1 ..1, y = -1 ..1, z = -10 ..10);
```

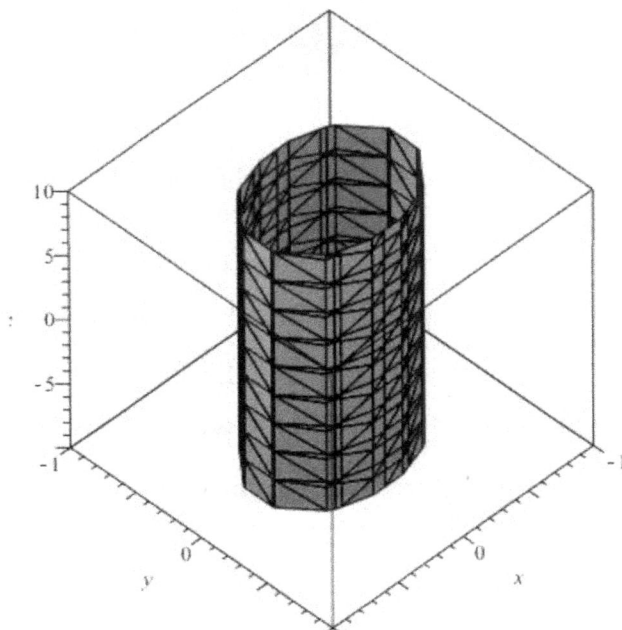

**FIGURE C1.15.1 (a)** Exercise 1.15.5.

> *plot3d*( [ cos(*s*), sin(*s*), *t* ], *s* =− 10 ..10, *t* =− 10 ..10, *numpoints* = 5001 );

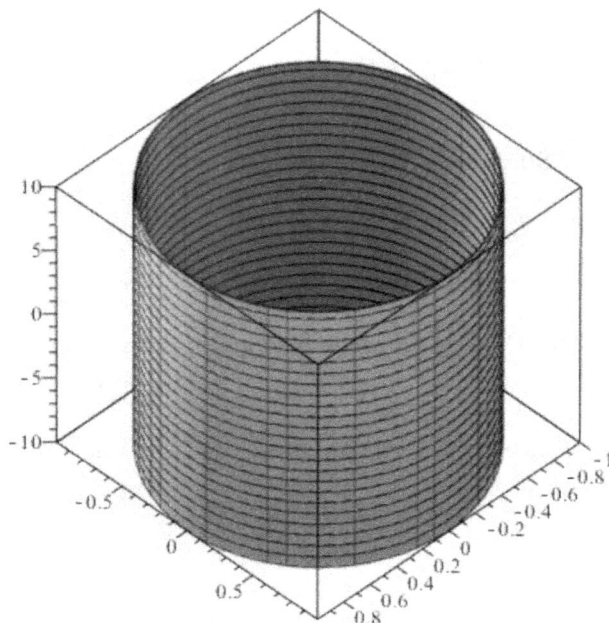

**FIGURE C1.15.1 (b)** Exercise 1.15.5.

**1.15.7** We may take $A : x + y + z = 0$, $\Sigma : x^2 + y^2 + z^2 = 0$, showing that the surface is of revolution. Its axis is the line in the direction $\vec{i} + \vec{j} + \vec{k}$.

**1.15.9** We may take $A : x + y + z = 0$, $\Sigma : x^2 + y^2 + z^2 = 0$ as our plane and sphere. The axis of revolution is then in the direction of $\vec{i} + \vec{j} + \vec{k}$.

**1.15.11** Every plane through the origin, which makes a circular cross section, must intersect the $yz$ -plane, and the diameter of any such cross section must be a diameter of the ellipse $x = 0$, $\dfrac{y^2}{b^2} + \dfrac{z^2}{c^2} = 1$.

Therefore, the radius of the circle is at most $b$. Arguing similarly on the $xy$-plane shows that the radius of the circle is at least $b$. To show that circular cross section of radius $b$ actually exist, one may verify that the two planes given by $a^2 \left( b^2 - c^2 \right) z^2 = c^2 \left( a^2 - b^2 \right) x^2$ give circular cross sections of radius $b$.

## 1.16 Parametric Curves in Space

**1.16.1**   42.

**1.16.3**

$$t_1 t_2 t_3 + t_1 t_2 t_4 + t_1 t_3 t_4 + t_2 t_3 t_4 = 0 \Rightarrow \frac{t_1 t_2 t_3 + t_1 t_2 t_4 + t_1 t_3 t_4 + t_2 t_3 t_4}{t_1 t_2 t_3 t_4} = 0$$

$$\Rightarrow \frac{1}{t_1} + \frac{1}{t_2} + \frac{1}{t_3} + \frac{1}{t_4} = 0,$$

as required.

**1.16.5**   You can parameterize the cylinder $y^2 + z^2 = 16$ by $y = 4\cos t$, and $z = 4\sin t$. From the equation $x = 8 - y^2 - z$, you obtain that $x = 8 - 16\cos^2 t - 4\sin t$.

## 1.17 Multidimensional Vectors

**1.17.1**   **1.**   If $f'(x) = 0$, then $e^{x-1} = 1$ implying that $x = 1$. Thus $f$ has a single minimum point at $x = 1$. Thus for all real numbers $x$, $0 = f(1) \geq f(x) = e^{x-1} - x$, which gives the desired result.

**3.**   Easy Algebra!

**1.17.3**   The maximum value $e = \dfrac{16}{5}$ is reached when $a = b = c = d = \dfrac{6}{5}$.

**1.17.5**   Applying the AM-GM inequality, for $1, 2, \ldots, n$:

$$n!^{1/n} = \left( 1 \times 2 \times \cdots \times n \right)^{1/n} < \frac{1 + 2 + \cdots + n}{n} = \frac{n+1}{2}, \text{ with strict inequality}$$

for $n > 1$.

**1.17.7**   Observe that $\sum_{k=1}^{n} 1 = n$. Then we have

$$n^2 = \left(\sum_{k=1}^{n} 1\right)^2 = \left(\sum_{k=1}^{n} (a_k)\frac{1}{a_k}\right)^2 \le \left(\sum_{k=1}^{n} a_k^2\right)\left(\sum_{k=1}^{n} \frac{1}{a_k^2}\right).$$

**1.17.9**   The only linear combination giving the zero vector is the trivial linear combination, which that the vectors are linearly independent.

# CHAPTER 2

**2.1.1**   **1.** Closed in $\mathbb{R}^2$.

**3.** Open in $\mathbb{R}^2$.

**5.** Open in $\mathbb{R}^2$.

**7.** Open in $\mathbb{R}^2$.

**9.** Open in $\mathbb{R}^2$.

**11.** Open in $\mathbb{R}^2$.

**13.** Closed in $\mathbb{R}^2$.

**15.** Open in $\mathbb{R}^2$.

**17.** $\mathbb{R}^2$.

**19.** Closed in $\mathbb{R}^2$.

**21.** Closed in $\mathbb{R}^2$.

**2.1.3**   **1.** Open in $\mathbb{R}^3$.

**3.** Neither open nor closed in $\mathbb{R}^3$.

**5.** Closed ball in $\mathbb{R}^3$.

**7.** Closed in $\mathbb{R}^3$.

**9.** Closed in $\mathbb{R}^3$.

**11.** Closed in $\mathbb{R}^3$.

**13.** Open in $\mathbb{R}^3$.

**15.** Open in $\mathbb{R}^3$.

**2.1.5**   Suppose $S_1$ and $S_2$ are open sets and $P$ is any point in $S_1 \cup S_2$. To prove that $S_1 \cup S_2$ is an open set, we must show that $P$ is an interior point of $S_1 \cup S_2$. From the defining property of $P$, the point is either in $S_1$ or $S_2$. Suppose that $P \in S_1$ as the proof is analogous when $P \in S_2$. Since $P \in S_1$, which is an open set, $P$ is an interior point of $S_1$ and there is some open ball with center $P$, which contains only points in $S_1$. Hence, $P$ is also an interior of $S_1 \cup S_2$ and thus this set is an open set.

**2.1.7**   **1.**   $V$ is closed set.

**3.**   $D$ is neither open nor closed set.

**5.**   $B$ is neither open nor closed set.

**7.**   $F$ is closed set.

**9.**   $K$ is closed set.

## 2.2   Multivariable Functions

**2.2.1**   **1.**         $>$  *with( plots )* :
$>$

*contourplot*$(x + y, x = -2 ..2, y = 0 ..2, color = blue)$

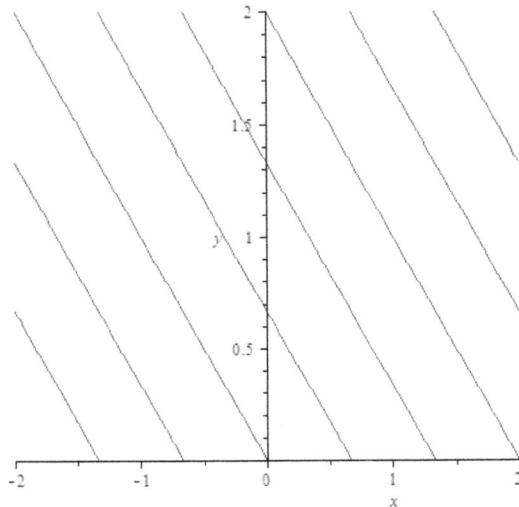

> $plot3d(x + y, x = 0\,..2, y = 0\,..2, axes = frame, style = contour, color = blue)$

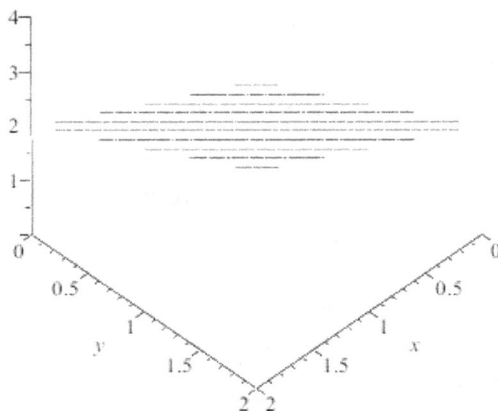

3.  > $with(plots)$ :

> $contourplot(x^3 - y, x = -2\,..2, y = -2\,..2, color = blue)$

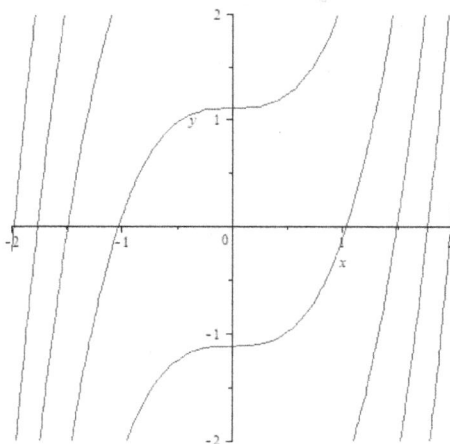

> $plot3d(x^3 - y, x = -2\,..2, y = -2\,..2, axes = frame, style = contour, color = blue)$

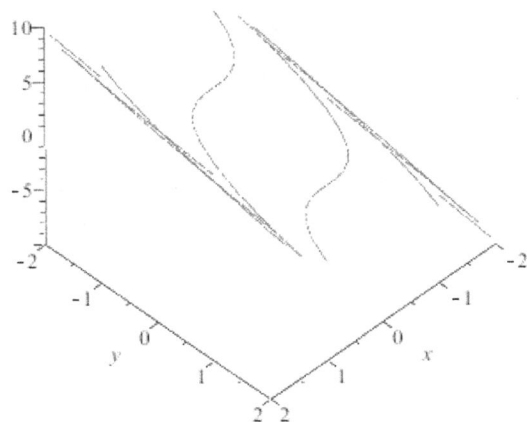

**5.** > *with*(*plots*) :

> *contourplot*$(x^2 + 4y^2, x = -2 ..2, y = -2 ..2, color = blue)$

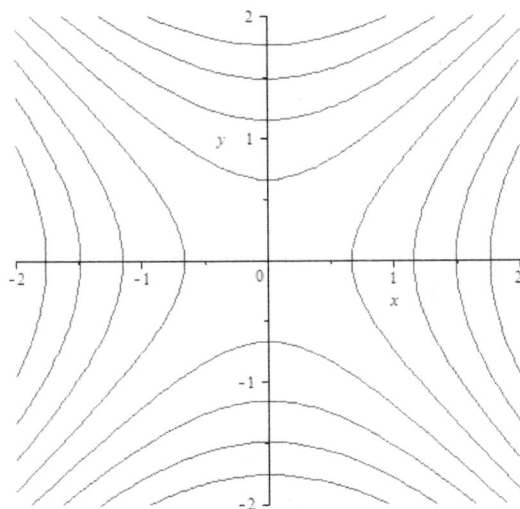

> *plot3d*$(y^2 - x^2, x = -2 ..2, y = -2 ..2, axes = frame, style = contour, color = blue)$

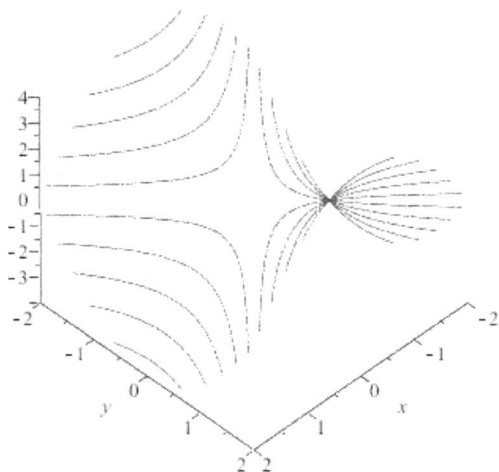

**7.** > *with*(*plots*) :

> *contourplot*$\left(\sin\left(x^2 + y^2\right), x = -3\ ..3, y = -3\ ..3, color = blue, contours = 3\right)$

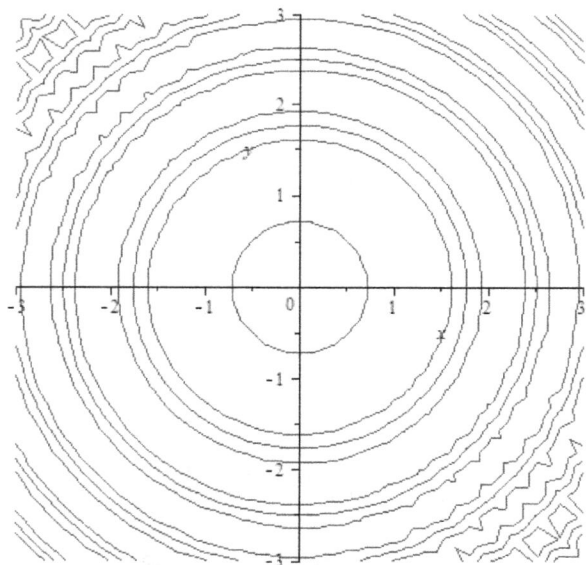

> *plot3d*$\left(\sin\left(x^2 + y^2\right), x = -3\ ..3, y = -3\ ..3, axes = box, contours = 3\right)$

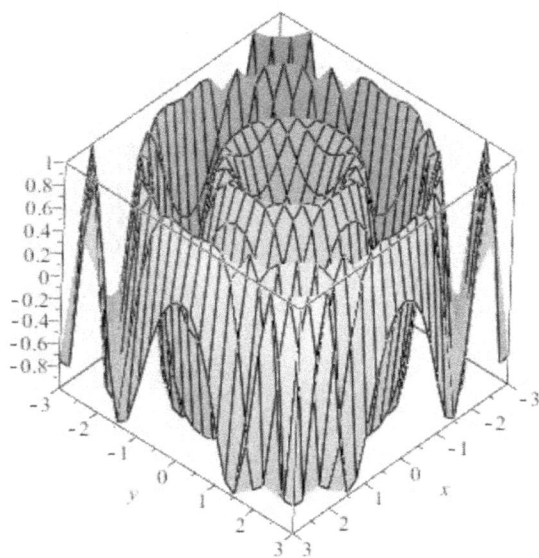

**9.**

> $with(plots)$ :

> $contourplot(5 - x^2 - y^2, x = -3 ..3, y = -3 ..3, color = blue, contours = 3)$

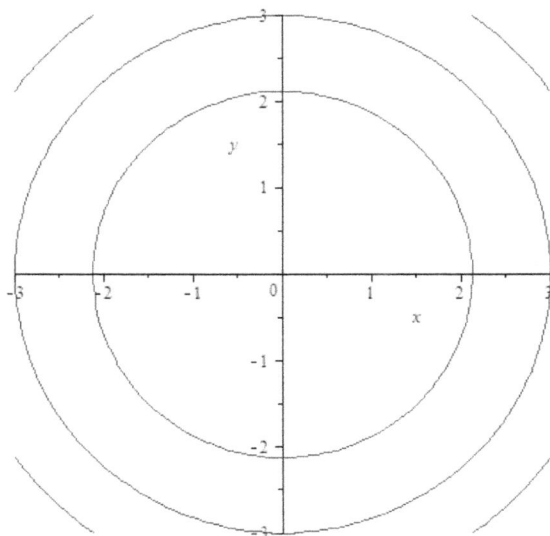

> $plot3d(5 - x^2 - y^2, x = -3 ..3, y = -3 ..3, axes = box, contours = 3)$

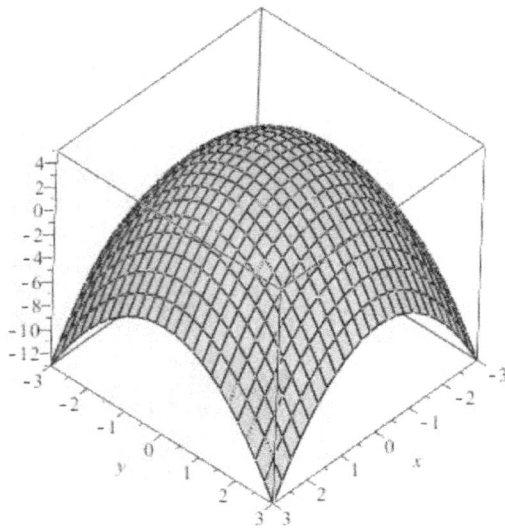

**11.** > *with*(*plots*) :

> *contourplot*( sin(*x*) · sin(*y*), *x* = −2 ..2, *y* = −2 ..2, *scaling* = *constrained*)

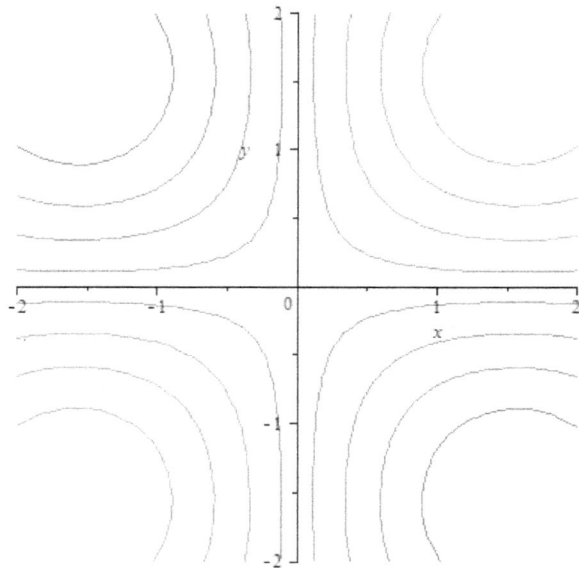

Or

>

$contourplot(g(x, y), x = -2 .. 2, y = -2 .. 2, contours = [-3.5, -3, -2.5, -2, -1.5, -1, -0.5, 0, 0.5, 1, 1.5, 2, 2.5, 3, 3.5], filled = true, coloring = [white, red]);$

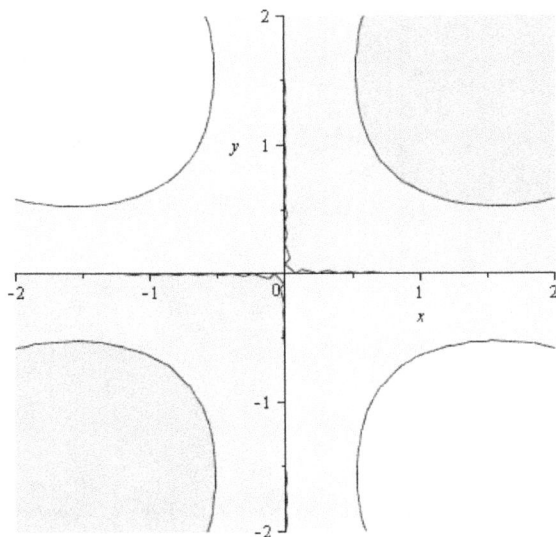

In 3D

> $g := (x, y) \rightarrow \sin(x) \cdot \sin(y);$

$$g := (x, y) \rightarrow \sin(x) \sin(y)$$

> $plot3d(g(x, y), x = -2 .. 2, y = -2 .. 2, axes = framed, orientation = [150, 70]);$

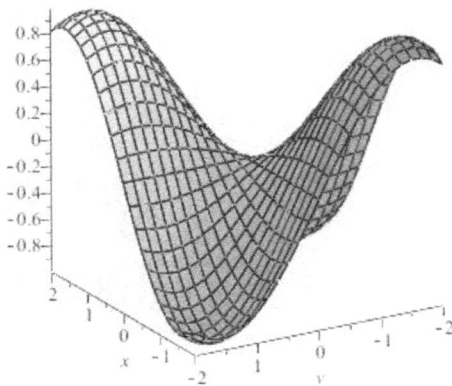

Or

> 

*plot3d*(*g*(*x*, *y*), *x* = −2 ..2, *y* = −2 ..2, *axes* = *framed*, *style* = *patchcontour*,
  *orientation* = [150, 70]);

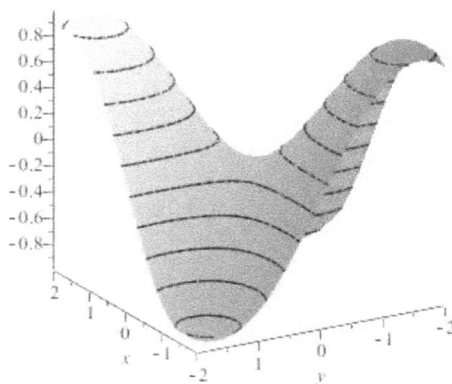

Or

> *p1* := *plot3d*(*g*(*x*, *y*), *x* = −2 ..2, *y* = −2 ..2, *color* = *black*) :
> *p2* := *plot3d*(−2, *x* = −2 ..2, *y* = −2 ..2, *color* = *blue*) :
> *p3* := *plot3d*(0, *x* = −2 ..2, *y* = −2 ..2, *color* = *blue*) :
> *p4* := *plot3d*(2, *x* = −2 ..2, *y* = −2 ..2, *color* = *blue*) :
> *display*([*p1*, *p2*, *p3*, *p4*], *axes* = *framed*, *style* = *hidden*, *orientation*
  = [150, 70]);

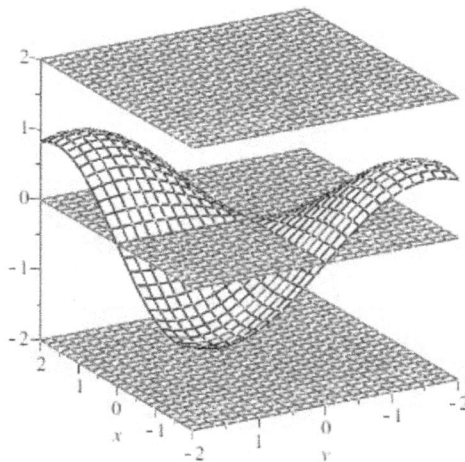

**13.**

> $g := (x, y) \rightarrow \tan^{-1}\left(\dfrac{y}{x+1}\right);$

$$g := (x, y) \rightarrow \arctan\left(\dfrac{y}{x+1}\right)$$

> $contourplot(g(x, y), x = -2 ..2, y = -2 ..2);$

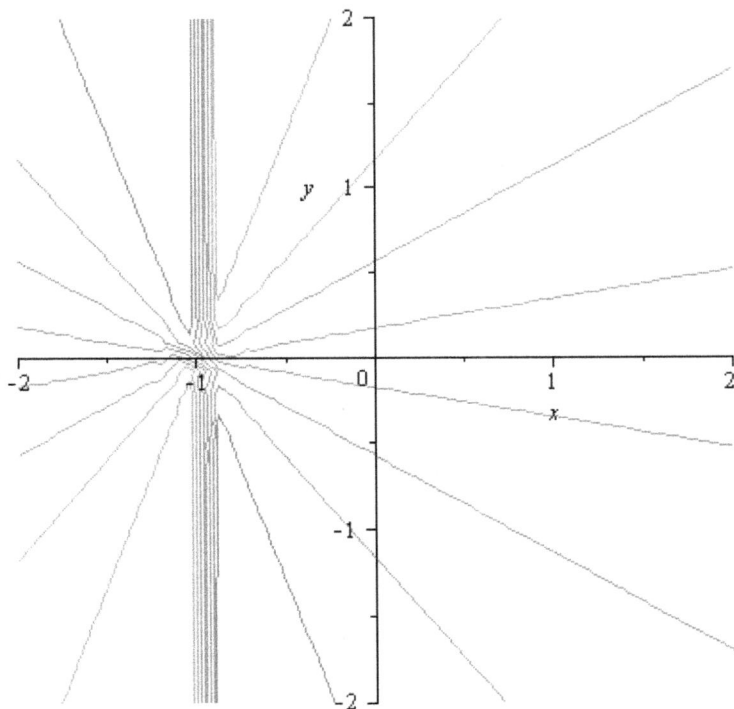

In 3D

>

$plot3d(g(x, y), x = -2 ..2, y = -2 ..2, axes = framed, style = patchcontour,$
$\quad orientation = [150, 70]);$

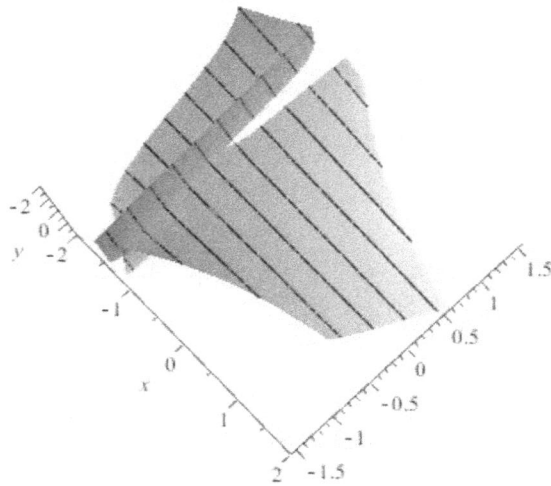

**15.**

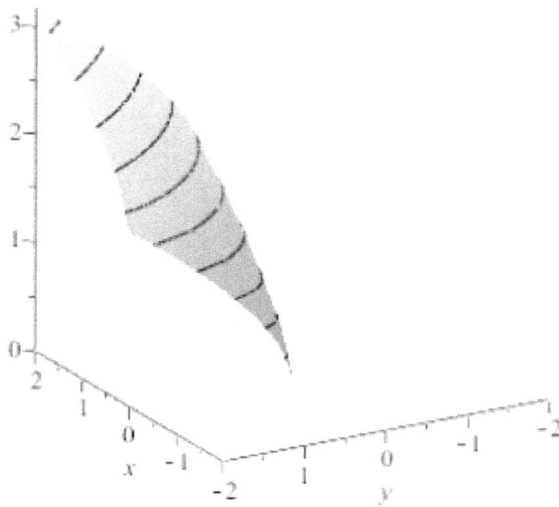

```
>  g := (x, y) → (x + 1)² + y²;
```

$$g := (x, y) \rightarrow (x + 1)^2 + y^2$$

```
>  contourplot(g(x, y), x = -2 ..2, y = -2 ..2);
```

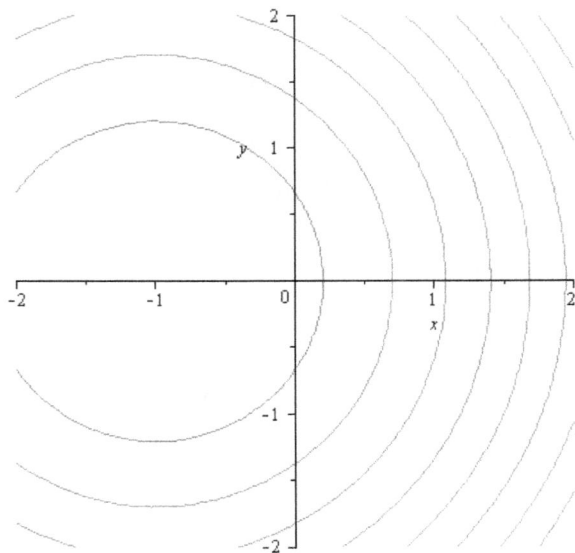

## In 3D

> $plot3d(g(x, y), x = -2\,..2, y = -2\,..2, axes = framed, style = patchcontour,\\ orientation = [\,150, 70\,]);$

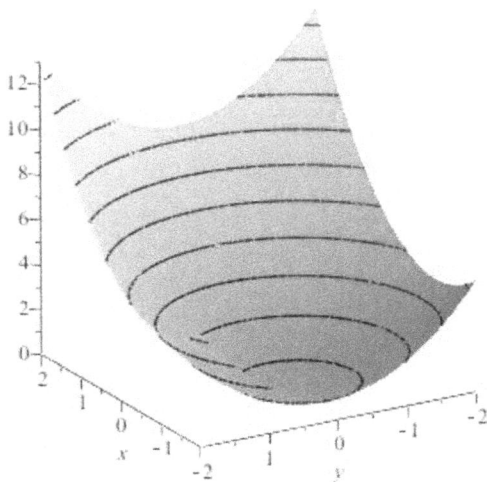

**2.2.3**   **1.** Shift $g(x,y)$ upward 2 units.

   **3.** Reflect $g(x,y)$ about the $xy$-plane.

   **5.** Reflect $g(x,y)$ in the plane $x = 0$.

   **7.** Reflect $g(x,y)$ in the origin.

## 2.3   Limits.

**2.3.1**   $>$  *with(plots)* :
       $>$

   $plot3d\left(\sqrt{4 - x^4 - y^2}, x = -10 .. 10, y = -10 .. 10, axes = boxed, style = surface\right)$

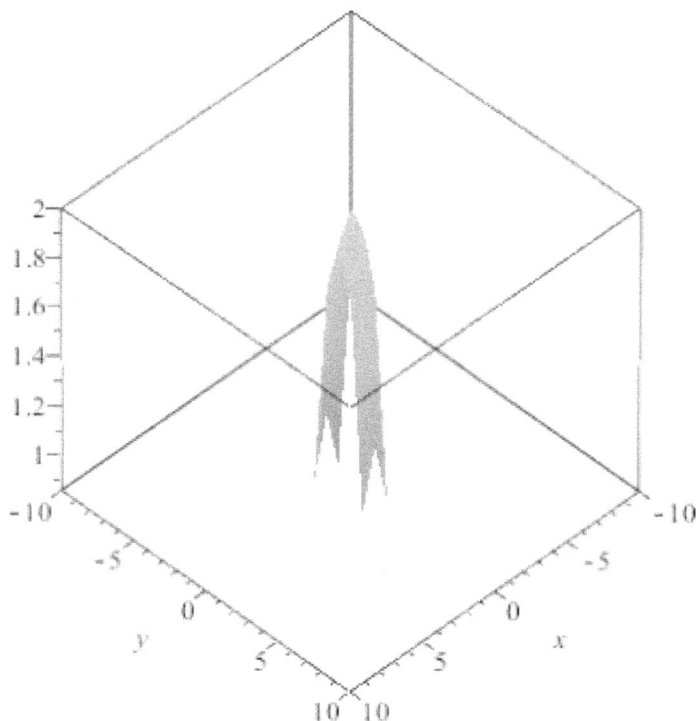

**FIGURE 2.3.8** Exercise 2.3.1 for $(x, y) \mapsto \sqrt{4 - x^2 - y^2}$.

**2.3.3** &gt; *with*(*plots*) :

&gt;

$$plot3d\left(\dfrac{1}{x^2 + y^2}, x = -10..10, y = -10..10, axes = boxed, style \right.$$

$$\left. = surface \right)$$

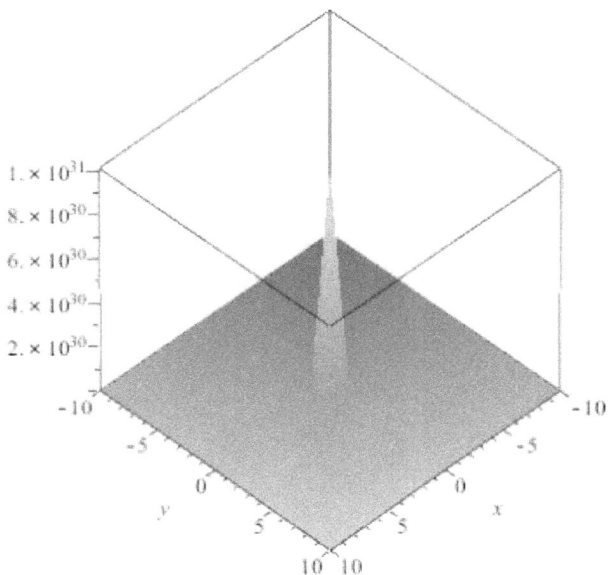

**FIGURE 2.3.10** Exercise 2.3.2 for $(x, y) \mapsto \dfrac{1}{x^2 + y^2}$ .

**2.3.5**   0.

**2.3.7**   0.

**2.3.9**   $\dfrac{1}{3}$ .

**2.3.11**   Does not exist.

**2.3.13**   Show that $\left| \dfrac{x^3 y}{\left(x^2 + y^2\right)} \right| \leq |xy|$ for $(x, y) \neq (0, 0)$. So the limit is 0.

**2.3.15** $\lim\limits_{x \to 0} f(x,x) = \lim\limits_{x \to 0} \left( \dfrac{x^2}{x^2 + x^2} \right) = \dfrac{1}{2}$; $\lim\limits_{x \to 0} f(x,0) = \lim\limits_{x \to 0} \left( \dfrac{0}{x^2 + 0} \right) = 0$.

**2.3.17** Since $\lim\limits_{x \to 0} \left( \lim\limits_{y \to 0} \dfrac{x^2 - y^2}{x^2 + y^2} \right) = 1$, and $\lim\limits_{y \to 0} \left( \lim\limits_{x \to 0} \dfrac{x^2 - y^2}{x^2 + y^2} \right) = -1$.

Thus, the iterated limits are not equal and therefore, $\lim\limits_{(x,y) \to (0,0)} g(x,y)$,

does not exist. Hence, $g(x,y)$ is discontinuous at $(0,0)$.

**2.3.19** $c = 0$.

## 2.4 Definition of the Derivative

**2.4.1** By the Cauchy-Bunyakovsky-Schwarz inequality,

$$\left\| L\left( \vec{h} \right) \right\| \le \left\| \vec{h} \right\| \left( \sum_{k=1}^{n} \left\| L\left( \vec{e}_k \right) \right\|^2 \right)^{1/2}$$

and

$$\left\| \vec{h} \times L\left( \vec{h} \right) \right\| \le \left\| \vec{h} \right\| \left\| L\left( \vec{h} \right) \right\| \le \left\| \vec{h} \right\|^2 \left( \left\| L\left( \vec{e}_k \right) \right\|^2 \right)^{1/2} = o\left( \left\| \vec{h} \right\| \right),$$

as it was to be shown.

## 2.5 The Jacobi Matrix

**2.5.1** **1.** $f_x(x,y) = 3\left( x^3 - y^2 \right)\left( 3x^2 \right)$

$f_y(x,y) = 3\left( x^3 - y^2 \right)\left( -2y \right)$.

$f_z(x,y,z) = 0$

**3.** $f_x(x,y) = \dfrac{3y\left( y^2 - x^2 \right)}{\left( x^2 + y^2 \right)^2}$

$f_y(x,y) = \dfrac{3x\left( x^2 - y^2 \right)}{\left( x^2 + y^2 \right)^2}$.

$f_z(x,y,z) = 0$

**2.5.3**  $\dfrac{\partial f}{\partial x} = 2x \log\left(x^2 y^2 + 1\right) + \dfrac{2\left(z^2 + x^2\right)xy^2}{x^2 y^2 + 1}$

$\dfrac{\partial f}{\partial y} = \dfrac{2\left(z^2 + x^2\right)x^2 y}{x^2 y^2 + 1}$

$\dfrac{\partial f}{\partial z} = 2z \log\left(x^2 y^2 + 1\right)$

**2.5.5**  $\dfrac{\partial}{\partial x} f(x,y) = \begin{cases} 1 & \text{if} \quad x < y^2 \\ 0 & \text{if} \quad x > y^2 \end{cases}$

$\dfrac{\partial}{\partial y} f(x,y) = \begin{cases} 0 & \text{if} \quad x < y^2 \\ 2y & \text{if} \quad x > y^2 \end{cases}$

**2.5.7**  $\dfrac{\partial w}{\partial x} = \left(3u^2 + 2uv\right)\left(y\cos xy\right) + \left(u^2 - 3\right)\left(\dfrac{y}{x}\right)$

$\dfrac{\partial w}{\partial y} = \left(3u^2 + 2uv\right)\left(x\cos xy\right) + \left(u^2 - 3\right)\ln x$

**2.5.9**  $\begin{bmatrix} 0 & -1 \\ 0 & 0 \\ 2 & 1 \end{bmatrix}.$

**2.5.11**  $f'(r,\theta) = \begin{bmatrix} \cos\theta & -r\sin\theta \\ \sin\theta & r\cos\theta \end{bmatrix}.$

$J(r,\theta) = r.$

**2.5.13**  $f'(u,v) = \begin{bmatrix} \dfrac{v^2 - u^2}{\left(v^2 + u^2\right)^2} & \dfrac{-2uv}{\left(v^2 + u^2\right)^2} \\ \dfrac{-2uv}{\left(v^2 + u^2\right)^2} & \dfrac{u^2 - v^2}{\left(u^2 - v^2\right)^2} \end{bmatrix}.$

$J(u,v) = \dfrac{-\left(u^2 + v^2\right)^2}{\left(u^2 + v^2\right)^4} = \dfrac{-1}{\left(u^2 + v^2\right)^2}.$

**2.5.15** 1.

**2.5.17** $6t^2 + 6t^5$.

**2.5.19** $\dfrac{1}{2a^3}\arctan\dfrac{b}{a} + \dfrac{b}{2a^2\left(a^2+b^2\right)}$.

## 2.6 Gradients and Directional Derivatives

**2.6.1** $\begin{bmatrix} \dfrac{1}{x}e^y \\ \ln xe^y \end{bmatrix}$.

**2.6.3** $\begin{bmatrix} 2 \\ 1 \end{bmatrix}$.

**2.6.5** $\begin{bmatrix} 2xy\sin(yz) \\ x^2\left(y\cos(yz)z + \sin(yz)\right) \\ x^2y^2\cos(yz) \end{bmatrix}$.

**2.6.7** $\nabla f(x,y,z) = \begin{bmatrix} 2xe^y \\ x^2e^y \\ 0 \end{bmatrix}, \nabla k(x,y,z) = \begin{bmatrix} zy^2e^{xz} \\ 2ye^{xz} \\ xy^2e^{xz} \end{bmatrix}, fk = x^2y^2e^{y+xz}$

$\nabla(fk) = \begin{bmatrix} 2xy^2e^{y+xz} + x^2y^2ze^{y+xz} \\ 2x^2ye^{y+xz} + x^2y^2e^{y+xz} \\ x^3y^2e^{y+xz} \end{bmatrix}$

$f\nabla k + k\nabla f = x^2e^y\begin{bmatrix} zy^2e^{xz} \\ 2ye^{xz} \\ xy^2e^{xz} \end{bmatrix} + y^2e^{xz}\begin{bmatrix} 2xe^y \\ x^2e^y \\ 0 \end{bmatrix}$

Therefore,

$\nabla(fk) = f\nabla k + k\nabla f$.

**2.6.9** 1.1.

**2.6.11** Consider an arbitrary unit vector $\vec{u}$. Then the directional derivative satisfies

$$\frac{\partial f}{\partial \vec{u}} = \nabla f \cdot \vec{u} = \|\nabla f\| \|\vec{u}\| \cos\theta = \|\nabla f\| \cos\theta,$$

Where $\theta$ is the angle between $\nabla f$ and $\vec{u}$. Consequently,

$$-\|\nabla f\| \le \frac{\partial f}{\partial \vec{u}} \le \|\nabla f\|.$$

Thus $\frac{\partial f}{\partial \vec{u}}$ is the largest, when $\theta = 0$ (i.e, same direction of $\nabla f$ ), and the smallest when $\theta = \pi$ (i.e, opposite direction of $\nabla f$ ).

**2.6.13** $\nabla f(x,y) = 2y(x+y)^{-2}\,\vec{i} - 2x(x+y)^{-2}\,\vec{j};$
$\nabla f(2,-1) = -2\vec{i} - 4\vec{j};$
$\vec{u} = (1/5)(3\vec{i} + 4\vec{j});$
$D_{\vec{v}} f(-2,1) = -22/5.$

**2.6.15** $f(x,y,z) = \begin{bmatrix} \dfrac{1}{y} \\ \left(\dfrac{-x}{y^2}\right) - \left(\dfrac{1}{z}\right) \\ \dfrac{y}{z^2} \end{bmatrix}; \ \nabla f(0,-1,2) = -\vec{i} - \left(\dfrac{1}{2}\right)\vec{j} - \left(\dfrac{1}{4}\right)\vec{k};$

$\overrightarrow{P_1 P_2} = \begin{bmatrix} 3 \\ 2 \\ -6 \end{bmatrix}, \ \vec{u} = \dfrac{\overrightarrow{P_1 P_2}}{7}. \ D_{\overrightarrow{P_1 P_2}} f(0,-1,2) = \dfrac{(-6-2+3)}{14} = \dfrac{-5}{14}.$ The

maximal direction is $\nabla f(0,-1,2)$; maximum rate is

$$\left|\nabla f(0,-1,2)\right| = \frac{\sqrt{21}}{4}.$$

**2.6.17** **1.** $\nabla \bullet \left( \phi \vec{U} \right) = \dfrac{\partial}{\partial x} \left( \phi U_1 \right) + \dfrac{\partial}{\partial y} \left( \phi U_2 \right) + \dfrac{\partial}{\partial z} \left( \phi U_3 \right)$

$$= \dfrac{\partial \phi}{\partial x} U_1 + \dfrac{\partial \phi}{\partial y} U_2 + \dfrac{\partial \phi}{\partial z} U_3 + \phi \dfrac{\partial U_1}{\partial x} + \phi \dfrac{\partial U_2}{\partial y} + \phi \dfrac{\partial U_3}{\partial z}.$$

$$= \nabla \phi \bullet \vec{U} + \phi \nabla \bullet \vec{U}$$

**3.**

$$\nabla \bullet \left( \vec{U} \times \vec{V} \right) = \dfrac{\partial}{\partial x} \left( U_2 V_3 - U_3 V_2 \right) + \dfrac{\partial}{\partial y} \left( U_3 V_1 - U_1 V_3 \right) + \dfrac{\partial}{\partial z} \left( U_1 V_2 - U_2 V_1 \right)$$

$$= V_1 \left( \dfrac{\partial U_3}{\partial y} - \dfrac{\partial U_2}{\partial z} \right) + V_2 \left( \dfrac{\partial U_1}{\partial z} - \dfrac{\partial U_3}{\partial x} \right) + V_3 \left( \dfrac{\partial U_2}{\partial x} - \dfrac{\partial U_1}{\partial y} \right)$$

$$- U_1 \left( \dfrac{\partial V_3}{\partial y} - \dfrac{\partial V_2}{\partial z} \right) + U_2 \left( \dfrac{\partial V_1}{\partial z} - \dfrac{\partial V_3}{\partial x} \right) + U_3 \left( \dfrac{\partial V_2}{\partial x} - \dfrac{\partial V_1}{\partial y} \right)$$

$$= \vec{V} \bullet \nabla \times \vec{U} - \vec{U} \bullet \nabla \times \vec{V}.$$

**2.6.19**  Parallel planes have proportional gradients. Therefore, if
$F \left( x, y, z \right) = x^2 - 2y^2 - 4z^2 - 16$ and $G \left( x, y, z \right) = 4x - 2y + 4z - 5$, then

$$\nabla F \left( x, y, z \right) = \begin{bmatrix} 2x \\ -4y \\ -8z \end{bmatrix} \text{ and } G \left( x, y, z \right) = \begin{bmatrix} 4 \\ -2 \\ 4 \end{bmatrix} \text{ must}$$

be proportional, i.e.,

$$2x = 4k, -4y = -2k, -8z = 4k, \Rightarrow x = 2k, y = \dfrac{k}{2}, z = \dfrac{-k}{2}$$

$$\Rightarrow 4k^2 - 2 \left( \dfrac{k^2}{4} \right) - 4 \left( \dfrac{k^2}{4} \right) = 16 \Rightarrow k = \pm \dfrac{4\sqrt{2}}{\sqrt{5}}.$$

Thus, the points are $\left( x, y, z \right) = \left( \pm \dfrac{8\sqrt{2}}{\sqrt{5}}, \pm \dfrac{2\sqrt{2}}{\sqrt{5}}, \mp \dfrac{2\sqrt{2}}{\sqrt{5}}, \right).$

**2.6.21** $\nabla f(x,y) = \begin{bmatrix} -2x \\ -2y \end{bmatrix}; \nabla f(-1,2) = \begin{bmatrix} 2 \\ -4 \end{bmatrix}; -|\nabla f(-1,2)| = -2\sqrt{5}$

Thus, the $\vec{v}$ direction is $\begin{bmatrix} \dfrac{-2}{2\sqrt{5}} \\ \dfrac{-4}{-2\sqrt{5}} \end{bmatrix} = \begin{bmatrix} \dfrac{-1}{\sqrt{5}} \\ \dfrac{2}{\sqrt{5}} \end{bmatrix}$.

**2.6.23** $\nabla \times \vec{f} = \begin{bmatrix} y \\ 6xz - 1 \\ 0 \end{bmatrix}$

$\nabla \times (\nabla \times \vec{f}) = \begin{bmatrix} -6x \\ 0 \\ 6z - 1 \end{bmatrix}$

$\nabla \bullet \vec{f} = 3z^2 - z + 2$ .

$\nabla (\nabla \bullet \vec{f}) = \begin{bmatrix} 0 \\ 0 \\ 6z - 1 \end{bmatrix}$

$\nabla^2 \vec{f} = \begin{bmatrix} 6x \\ 0 \\ 0 \end{bmatrix}$

Thus

$\nabla (\nabla \bullet \vec{f}) - \nabla^2 \vec{f} = \begin{bmatrix} -6x \\ 0 \\ 6z - 1 \end{bmatrix} = \nabla \times (\nabla \times \vec{f})$.

## 2.7    Levi-Civitta and Einstein

**2.7.1**  $\varepsilon_{ijk} x_j (\varepsilon_{klm} y_l z_m) = x_j y_i z_j - x_j y_j z_i = (x_j z_j) y_i - (x_j y_j) z_i$.

**2.7.3**  $\nabla \bullet (\nabla \times \vec{u}) = -\varepsilon_{ijk} \partial_i \partial_j h_k = 0$.

**2.7.5**  $v_a \partial_a v_b = v_a \partial_b v_a - v_a (\partial_b v_a) + v_a (\partial_a v_b) = v_a \partial_a v_b$.

## 2.8    Extrema

**2.8.1**    There is one critical point $(1,1)$.

**2.8.3**    The critical point is thus a saddle point.

**2.8.5**    $\left(1,-1,\dfrac{1}{2}\right)$, and $\left(-1,1,-\dfrac{1}{2}\right)$ are the critical points.

$\left(1,-1,\dfrac{1}{2}\right)$ is a saddle point. $\left(-1,1,-\dfrac{1}{2}\right)$ is also a saddle point.

**2.8.7**    we have a local maximum at $(1,1,1)$. In some neighborhood of $(0,0,0)$ is a saddle point.

**2.8.9**    By the Fundamental Theorem of Calculus, there exists a continuously differentiable function $G$ such that

$$f(x,y) = \int_{y^2-x}^{x^2+y} g(t)\,dt = G(x^2+y) - G(y^2-x).$$

This gives, $\dfrac{\partial f}{\partial x}(0,0) = \dfrac{\partial f}{\partial y}(0,0) = g(0)$,

So $(0,0)$ is a critical point and $H_f(0,0) = \begin{bmatrix} -g'(0) & 0 \\ 0 & g'(0) \end{bmatrix}$.

Regardless of the sign of, $g'(0)$ the determinant of this last matrix is $-(g'(0))^2 < 0$, and so $(0,0)$ is a saddle point.

**2.8.11**    $\nabla f(x,y) \Rightarrow f(18/13, -14/13)$ is not extremum.

**2.8.13**    $\nabla f(x,y) \Rightarrow f(1,3/2) = 37/4$ is a local maximum.

**2.8.15**    The graph of $g$ is $z - g(x,y) = 0$; with $G(x,y,z) = z - g(x,y)$, the general normal is $\left(-g_x, -g_y, 1\right)$ so that the normal above a critical point for $g$ is $(0,0,1)$, which is a vertical vector. Thus, the tangent plane is horizontal at any critical point.

**2.8.17** Let $g(x,y,z) = x^2 + y^2 - 1, h(x,y,z) = x - yz$.

Hence $y = -1$ or $y = 1/2$, $y = -1 \Rightarrow x = 0$ and $z = 0$. Solution

$(0,-1,0)$ and $f(0,-1,0) = 0$. If $y = 1/2$, $x = \pm\dfrac{\sqrt{3}}{2}, z = \pm\sqrt{3}$.

Maximum $= \dfrac{3\sqrt{3}}{4}$, minimum $= -\dfrac{3\sqrt{3}}{4}$.

**2.8.19** If $x, y$, and $z$ are perpendicular distances to the sides of length

$L_1, L_2$, and $L_3$, then it is necessary to maximize $d$, where

$d^2 = x^2 + y^2 + z^2$, subject constraint area, $A$ where

$A = \dfrac{L_1 x + L_2 y + L_3 z}{2}$. Thus, minimum $= \dfrac{4A^2}{L_1 + L_2 + L_3}$.

## 2.9    Lagrange Multipliers

**2.9.1** We have $a = \dfrac{\sqrt{S}}{6}$, and the maximum volume is $abc = \dfrac{\left(\sqrt{S}\right)^3}{\left(\sqrt{6}\right)^3}$.

**2.9.3** Using CBS, $x + 3y \le 2^{5/4}\sqrt{3}$.

**2.9.5** The desired maximum is thus $f\left(-\sqrt{2}, \sqrt{2}\right) = f\left(\sqrt{2}, -\sqrt{2}\right)$, and the

minimum is

$f\left(1/\sqrt{2}, 1/\sqrt{2}\right) = f\left(-1/\sqrt{2}, -1/\sqrt{2}\right) = 1$.

**2.9.7** The first point gives an absolute maximum of $18 + \dfrac{12\sqrt{14}}{7}$ and the

second an absolute minimum of $18 - \dfrac{12\sqrt{14}}{7}$.

**2.9.9** $(0, \sqrt{2}, 1)$ yields a maximum and that $(0, -\sqrt{2}, 1)$ yields a minimum.

**2.9.11** $f(x,y) = x^a y^b e^{-1} \le \left(\dfrac{a}{a+b}\right)^a \left(\dfrac{b}{a+b}\right)^b e^{-1}$.

**2.9.13** The maximum $= \dfrac{k_1 k_2 k_3}{27}$.

# CHAPTER 3

## 3.1    Differential Forms

**3.1.1**    **1.**  0-forms → C. Functions forms.

**3.**  2-forms → A. Surface elements.

**3.1.3**    $dw = (-x - y)dx \wedge dy + y^2 z^3 dx \wedge dz + 2xyz^3 dy \wedge dz$.

**3.1.5**    $dx \wedge dy = r dr + d\theta$.

**3.1.7**    $dw = \left( \dfrac{\partial h}{\partial y} - \dfrac{\partial g}{\partial z} \right) dy \wedge dz - \left( \dfrac{\partial f}{\partial z} - \dfrac{\partial h}{\partial x} \right) dx \wedge dz + \left( \dfrac{\partial g}{\partial x} - \dfrac{\partial f}{\partial y} \right) dx \wedge dy$.

**3.1.9**    $dw = 2z dx \wedge dy \wedge dz$.

## 3.2    Zero-Manifolds

**3.2.1**    $\int_M w = -12$.

**3.2.3**    $\int_M w = -44$.

**3.2.5**    $\int_M w = -14$.

**3.2.7**    $\int_M w = 4$.

## 3.3    One-Manifold

**3.3.1**    Let $L_1 : y = x + 1, L_2 : y = -x + 1$. Then, $\displaystyle\int_C x dx + y dy = 0$.

**3.3.3**    We put $x = \sin t, y = \cos t, t \in \left[ -\dfrac{\pi}{2}, \dfrac{\pi}{2} \right]$. Then $\displaystyle\int_C x dx + y dy = 0$.

**3.3.5**    Let $\Gamma_1$ denote the straight line segment path from $O$ to $A = (2\sqrt{3}, 2)$ and $\Gamma_2$ denote the arc of the circle centered at $(0,0)$ and radius 4 going counterclockwise from $\theta = \dfrac{\pi}{6}$ to $\theta = \dfrac{\pi}{5}$. Observe that the Cartesian equation of the line $\overrightarrow{OA}$ is $y = \dfrac{x}{\sqrt{3}}$. Then on $\Gamma_1$,

$x\,dx + y\,dy = \dfrac{4}{3}x\,dx$. Hence, $\displaystyle\int_{\Gamma_1} x\,dx + y\,dy = 8$. On the arc of the circle,

we may put $x = 4\cos\theta$ , $y = 4\sin\theta$ and integrate from $\theta = \dfrac{\pi}{6}$ to

$\theta = \dfrac{\pi}{5}$. Observe that there $x\,dx + y\,dy = 0$, and since the integrand is

0, the integral will be zero. Assembling these two pieces,
$\displaystyle\int_{\Gamma} x\,dx + y\,dy = \int_{\Gamma_1} x\,dx + y\,dy + \int_{\Gamma_2} x\,dx + y\,dy = 8$.

**3.3.7**   Let $\Gamma_1$ denote the straight line segment path from $O$ to
$A = \left(2\sqrt{3}, 2\right)$ and $\Gamma_2$ denote the arc of the circle centered at $(0,0)$

and radius 4 going counterclockwise from $\theta = \dfrac{\pi}{6}$ to $\theta = \dfrac{\pi}{5}$. Observe

that the Cartesian equation of the line $\overrightarrow{OA}$ is $y = \dfrac{x}{\sqrt{3}}$.

We find on $\Gamma_1$ that $x\|dx\| = \dfrac{2}{\sqrt{3}}x\,dx$, hence $\displaystyle\int_{\Gamma_1} x\|dx\| = 4\sqrt{3}$.

On $\Gamma_2$ that $x\|dx\| = 16\cos\theta\,d\theta$ , hence $\displaystyle\int_{\Gamma_2} x\|dx\| = 4\sin\dfrac{\pi}{5} - 8$.

Assembling these we gather that $\displaystyle\int_{\Gamma} x\|dx\| = 4\sin\dfrac{\pi}{5} - 8 + 16\sin\dfrac{\pi}{5}$.

**3.3.9**
> *with( Student[ VectorCalculus ]) :*
> *PathInt( x, [x, y] = Line( ⟨0, 0⟩, ⟨2 \* sqrt(3), 2⟩ )) + PathInt( x, [x, y]*
> *= Arc( Circle( ⟨0, 0⟩, 4), Pi/6, Pi/5 ) );*

$$4\sqrt{3} - 8 + 16\cos\left(\dfrac{3}{10}\,\pi\right)$$

> *is( 16 \* cos(3 \* Pi/10) = 16 \* sin(Pi/5) );*

*true*

## 3.4    Closed and Exact Forms

**3.4.1**    **1.**  True.

**3.4.3**    Since $x^2 dx + y dy = d\left(\dfrac{1}{3}x^3 + \dfrac{1}{2}y^2\right)$, the differential form is exact.

**3.4.5**    Let

$$w = a_1 + Pdxdw + Qdydw + Rdzdw,$$

on $\Pi$, we can find the functions $p, q, r$ so that $p_w = P$, $q_w = Q$, $r_w = R$. Let $\beta_1 = pdx + qdy + rdz$ and let $w_1 = w + d\beta_1$. Now, let $w_1 = Adxdy + Bdxdz + Cdydz$. Let $B = b_z, C = c_z$ and $\beta_2 = bdx + cdy$. Also, let $f_y = F$, where $f$ and $F$ depend only on $x$ and $y$. Then, let $\beta_3 = fdx$ and notice that $w_2 + d\beta_3 = 0$. Therefore, $w = d\left(-\beta_1 - \beta_2 - \beta_3\right)$, completing the proof.

## 3.5    Two-Manifolds

**3.5.1**    2.

**3.5.3**    $\dfrac{15\pi}{16}$.

**3.5.5**    $4\pi$.

**3.5.7**    $\dfrac{21}{8}$.

**3.5.9**    $e - 1$.

**3.5.11**    18.

**3.5.13**    $\dfrac{1}{4}$.

**3.5.15**    4.

**3.5.17**  $7$.

**3.5.19**  $\dfrac{4(\pi+2)}{\pi^3}$.

**3.5.21**  $\dfrac{8}{5}+4\sin^{-1}\left(\dfrac{\sqrt{5}}{5}\right)$.

**3.5.23**  $\dfrac{2}{3}\log_e 2+\dfrac{8}{9}-\dfrac{\pi}{3}$.

**3.5.25**  $\dfrac{8}{3}$.

**3.5.27**  $\displaystyle\int_0^1\int_0^1 f(x)f(y)(y-x)\big(f(x)-f(y)\big)\,dxdy\geq 0$.

**3.5.29**  $\displaystyle\int_0^1\int_0^1 (f(x)\,dx)\,dy\ \geq\int_0^1 (f\circ g)(y)\,dy-\int_0^1 g(y)\,dy$.

**3.5.31**  $\dfrac{e^{a^2b^2}-1}{ab}$.

**3.5.33**  $\displaystyle\int_R \sin 2\pi x\sin 2\pi y\ dxdy=\sum_{k=1}^{N}\int_{R_k}\sin 2\pi x\sin 2\pi y\ dxdy$.

**3.5.35**  $\dfrac{n}{2}$.

**3.5.37**  $\dfrac{275}{54}$.

**3.6**   **Change of Variables in Two Dimensions**

**3.6.1**   $dx\wedge dy=\dfrac{1}{2}du\wedge dv$.

**3.6.3**   $\dfrac{b-a}{4}$.

**3.6.5**   $\dfrac{13}{6}(2-\sin 2)$.

**3.6.7**   $0$.

**3.6.9**   $\dfrac{255}{4}$.

**3.6.11**   $\dfrac{1}{3} + \dfrac{1}{12}\sin 6 - \dfrac{1}{12}\sin 2$.

## 3.7   Change to Polar Coordinates

**3.7.1**   $\dfrac{49}{2}$.

**3.7.3**   $\dfrac{\pi\sqrt{2}}{12}$.

**3.7.5**   $\dfrac{\pi}{18} - \dfrac{16}{9} + \sqrt{3}$.

**3.7.7**   $\dfrac{\pi}{8}$.

**3.7.9**   $\dfrac{\pi\sqrt{2}}{4}$.

**3.7.11**   **1.**   $\pi\left(1 - e^{-a^2}\right)$.

   **3.**   Since both $I_a$ and $I_{a\sqrt{2}}$ tend to $\pi$ as $a \to +\infty$, we deduce that $J_a \to \pi$. This gives the result.

**3.7.13**   If no two points are farther than 2 units, their squares are no farther than 4 units, and so the area $< \dfrac{1}{2}\displaystyle\int_0^{\pi/2} 4d\theta = \pi$, a contradiction.

## 3.8   Three-Manifolds

**3.8.1**   $\dfrac{1}{6}$

**3.8.3**   $\dfrac{27}{8}$

**3.8.5** $\displaystyle\int_0^1\int_0^3\int_0^{\frac{(12-3x-2y)}{6}} f(x,y,z)\,dzdydx$.

**3.8.7** 2

**3.8.9** $3-e$

**3.8.11** $\dfrac{1}{6}$

**3.8.13** $\dfrac{16}{3}$

## 3.9 Change of Variables in Three Dimensions

**3.9.1** Cartesian:

$$\int\limits_{-1}^{1}\int\limits_{-\sqrt{1-y^2}}^{\sqrt{1-y^2}}\int\limits_{x^2+y^2}^{\sqrt{x^2+y^2}} dzdxdy$$

Cylindrical:

$$\int\limits_{0}^{1}\int\limits_{0}^{2\pi}\int\limits_{r^2}^{r} r\,dzd\theta\,dr$$

Spherical:

$$\int\limits_{\pi/4}^{\pi/2}\int\limits_{0}^{2\pi}\int\limits_{0}^{(\cos\phi)/(\sin\phi)^2} r^2\sin\phi\,drd\theta d\phi.$$

The volume is $\dfrac{\pi}{3}$.

**3.9.3** Cartesian:

$$\int\limits_{-\sqrt{3}}^{\sqrt{3}}\int\limits_{-\sqrt{3-y^2}}^{\sqrt{3-y^2}}\int\limits_{1}^{\sqrt{4-x^2-y^2}} dzdxdy.$$

Cylindrical:

$$\int_0^{\sqrt{3}}\int_0^{2\pi}\int_1^{\sqrt{4-r^2}} r\,\mathrm{d}z\,\mathrm{d}\theta\,\mathrm{d}r.$$

Spherical:

$$\int_0^{\pi/3}\int_0^{2\pi}\int_{1/\cos\phi}^{2} r^2\sin\phi\,\mathrm{d}r\,\mathrm{d}\theta\,\mathrm{d}\phi.$$

The volume is $\dfrac{5\pi}{3}$.

**3.9.5**   $\dfrac{\pi}{96}$.

**3.9.7**   $\pi^2$.

**3.9.9**   $\dfrac{4\pi}{5}$.

**3.9.11**   $\dfrac{81\pi}{2}$.

**3.9.13**   $\dfrac{2\pi}{9}$.

## 3.10   Surface Integrals

**3.10.1**   $\dfrac{13\sqrt{2}}{3}$.

**3.10.3**   $\dfrac{4\pi}{3}$.

**3.10.5**   $\dfrac{4\pi R^2}{n}$.

**3.10.7**   $8\sqrt{2}$.

**3.10.9** $\dfrac{3\pi}{4}$.

**3.10.11** $\dfrac{\pi}{48}\left(17^{3/2}-1\right)$.

## 3.11   Green's, Stokes', and Gauss' Theorems

**3.11.1**   $-4$.

**3.11.3**   $-6$.

**3.11.5**   $16\pi$.

**3.11.7**   $\dfrac{4R^3}{3}$.

**3.11.9**   $-\dfrac{2}{3}$.

**3.11.11**   $0$.

**3.11.13**   $\dfrac{1}{30}$.

**3.11.15**   $-\dfrac{96}{5}$.

**3.11.17**   $-\dfrac{1}{20}$.

**3.11.19**   $240$.

**3.11.21**   $96\pi$.

**3.11.23**   $\dfrac{3}{2}$.

**3.11.25**   $-2\pi$.

# D

# *FORMULAS*

## D.1 TRIGONOMETRIC IDENTITIES

Right triangle definitions, where $0 < \theta < \dfrac{\pi}{2}$.

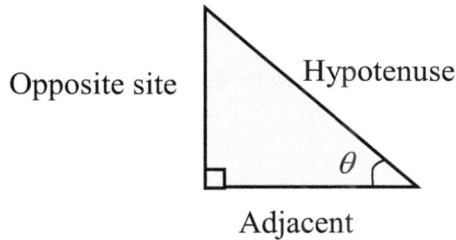

$$\sin\theta = \frac{opp}{hyp}, \cos\theta = \frac{adj}{hyp}, \tan\theta = \frac{opp}{adj}$$

$$\csc\theta = \frac{hyp}{opp}, \sec\theta = \frac{hyp}{adj}, \cot\theta = \frac{adj}{opp}$$

$$\cot\theta = \frac{1}{\tan\theta}, \ \sec\theta = \frac{1}{\cos\theta}, \ \csc\theta = \frac{1}{\sin\theta}$$

$$\tan\theta = \frac{\sin\theta}{\cos\theta}, \ \cot\theta = \frac{\cos\theta}{\sin\theta}$$

$$\sin^2\theta + \cos^2\theta = 1, \ \tan^2\theta + 1 = \sec^2\theta, \ \cot^2\theta + 1 = \csc^2\theta$$

$$\sin(-\theta) = -\sin\theta \,,\; \cos(-\theta) = \cos\theta \,,\; \tan(-\theta) = -\tan\theta$$

$$\csc(-\theta) = -\csc\theta \,,\; \sec(-\theta) = \sec\theta \,,\; \cot(-\theta) = -\cot\theta$$

$$\cos(\theta_1 \pm \theta_2) = \cos\theta_1 \cos\theta_2 \mp \sin\theta_1 \sin\theta_2$$

$$\sin(\theta_1 \pm \theta_2) = \sin\theta_1 \cos\theta_2 \pm \cos\theta_1 \sin\theta_2$$

$$\tan(\theta_1 \pm \theta_2) = \frac{\tan\theta_1 \pm \tan\theta_2}{1 \mp \tan\theta_1 \tan\theta_2}$$

$$\cos\theta_1 \cos\theta_2 = \frac{1}{2}\left[\cos(\theta_1 + \theta_2) + \cos(\theta_1 - \theta_2)\right]$$

$$\sin\theta_1 \sin\theta_2 = \frac{1}{2}\left[\cos(\theta_1 - \theta_2) - \cos(\theta_1 + \theta_2)\right]$$

$$\sin\theta_1 \cos\theta_2 = \frac{1}{2}\left[\sin(\theta_1 + \theta_2) + \sin(\theta_1 - \theta_2)\right]$$

$$\cos\theta_1 \sin\theta_2 = \frac{1}{2}\left[\sin(\theta_1 + \theta_2) - \sin(\theta_1 - \theta_2)\right]$$

$$\sin\theta_1 + \sin\theta_2 = 2\sin\left(\frac{\theta_1 + \theta_2}{2}\right)\cos\left(\frac{\theta_1 - \theta_2}{2}\right)$$

$$\sin\theta_1 - \sin\theta_2 = 2\cos\left(\frac{\theta_1 + \theta_2}{2}\right)\sin\left(\frac{\theta_1 - \theta_2}{2}\right)$$

$$\cos\theta_1 + \cos\theta_2 = 2\cos\left(\frac{\theta_1 + \theta_2}{2}\right)\cos\left(\frac{\theta_1 - \theta_2}{2}\right)$$

$$\cos\theta_1 - \cos\theta_2 = -2\sin\left(\frac{\theta_1 + \theta_2}{2}\right)\sin\left(\frac{\theta_1 - \theta_2}{2}\right)$$

$$a\cos\theta - b\sin\theta = \sqrt{a^2 + b^2}\,\cos(\theta + \phi) \,,\; \text{where } \phi = \tan^{-1}\left(\frac{b}{a}\right)$$

$$a\sin\theta + b\cos\theta = \sqrt{a^2 + b^2}\,\sin(\theta + \phi) \,,\; \text{where } \phi = \tan^{-1}\left(\frac{b}{a}\right)$$

$$\cos(90° - \theta) = \sin\theta \,,\; \sin(90° - \theta) = \cos\theta \,,\; \tan(90° - \theta) = \cot\theta$$

$$\cot(90° - \theta) = \tan\theta \,,\; \sec(90° - \theta) = \csc\theta \,,\; \csc(90° - \theta) = \sec\theta$$

$$\cos(\theta \pm 90°) = \mp\sin\theta \,,\; \sin(\theta \pm 90°) = \pm\sin\theta \,,\; \tan(\theta \pm 90°) = -\cot\theta$$

$$\cos(\theta \pm 180°) = -\cos\theta, \ \sin(\theta \pm 180°) = -\sin\theta, \ \tan(\theta \pm 180°) = \tan\theta$$

$$\cos 2\theta = \cos^2\theta - \sin^2\theta, \ \cos 2\theta = 1 - 2\sin^2\theta, \ \cos 2\theta = 2\cos^2\theta - 1$$

$$\sin 2\theta = 2\sin\theta\cos\theta, \ \tan 2\theta = \frac{2\tan\theta}{1 - \tan^2\theta}$$

$$\cos 3\theta = 4\cos^3\theta - 3\sin\theta$$

$$\sin 3\theta = 3\sin\theta - 4\sin^3\theta$$

$$\sin\frac{\theta}{2} = \pm\sqrt{\frac{1-\cos\theta}{2}}, \ \cos\frac{\theta}{2} = \pm\sqrt{\frac{1+\cos\theta}{2}},$$

$$\tan\frac{\theta}{2} = \pm\sqrt{\frac{1-\cos\theta}{1+\cos\theta}}, \ \tan\frac{\theta}{2} = \frac{\sin\theta}{1+\cos\theta}, \ \tan\frac{\theta}{2} = \frac{1-\cos\theta}{\sin\theta}$$

$$\sin\theta = \frac{e^{j\theta} - e^{-j\theta}}{2j}, \ \cos\theta = \frac{e^{j\theta} + e^{-j\theta}}{2} \ (j = \sqrt{-1}), \ \tan\theta = \frac{e^{j\theta} - e^{-j\theta}}{j\left(e^{j\theta} + e^{-j\theta}\right)}$$

$$e^{\pm j\theta} = \cos\theta \pm j\sin\theta \ \text{(Euler's identity)}$$

$$1\,rad = 57.296°$$

$$\pi = 3.1416$$

## D.2 HYPERBOLIC FUNCTIONS

$$\cosh x = \frac{e^x + e^{-x}}{2}, \ \sinh x = \frac{e^x - e^{-x}}{2}, \ \tanh x = \frac{\sinh x}{\cosh x},$$

$$\coth x = \frac{1}{\tanh x}, \ \text{sech}\,x = \frac{1}{\cosh x}, \ \text{csch}\,x = \frac{1}{\sinh x}$$

$$\sin jx = j\sinh x, \ \cos jx = \cosh x$$

$$\sinh jx = j\sin x, \ \cosh jx = \cos x$$

$$\sin(x \pm jy) = \sin x\cosh y \pm j\cos x\sinh y$$

$$\cos(x \pm jy) = \cos x\cosh y \mp j\sin x\sinh y$$

$$\sinh(x \pm y) = \sinh x \cosh y \pm \cosh x \sinh y$$

$$\cosh(x \pm y) = \cosh x \cosh y \pm \sinh x \sinh y$$

$$\sinh(x \pm jy) = \sinh x \cos y \pm j \cosh x \sin y$$

$$\cosh(x \pm jy) = \cosh x \cos y \pm j \sinh x \sin y$$

$$\tanh(x \pm jy) = \frac{\sinh 2x}{\cosh 2x + \cos 2y} \pm j \frac{\sin 2y}{\cosh 2x + \cos 2y}$$

$$\cosh^2 - \sinh^2 x = 1$$

$$\operatorname{sech}^2 + an \operatorname{h}^2 x = 1$$

## D.3   TABLE OF DERIVATIVES

| $y =$ | $\dfrac{dy}{dx} =$ |
|---|---|
| $c$  (constant) | $0$ |
| $cx^n$  (n any constant) | $cnx^{n-1}$ |
| $e^{ax}$ | $ae^{ax}$ |
| $a^x$  $(a > 0)$ | $a^x \ln a$ |
| $\ln x$  $(x > 0)$ | $\dfrac{1}{x}$ |
| $\dfrac{c}{x^a}$ | $\dfrac{-ca}{x^{a+1}}$ |
| $\log_a x$ | $\dfrac{\log_a e}{x}$ |
| $\sin ax$ | $a \cos ax$ |
| $\cos ax$ | $-a \sin ax$ |
| $\tan ax$ | $a \sec^2 ax = \dfrac{a}{\cos^2 ax}$ |
| $\cot ax$ | $-a \csc^2 ax = \dfrac{-a}{\sin^2 ax}$ |
| $\sec ax$ | $\dfrac{a \sin ax}{\cos^2 ax}$ |

| $y =$ | $\dfrac{dy}{dx} =$ |
|---|---|
| $\csc ax$ | $\dfrac{-a \cos ax}{\sin^2 ax}$ |
| $\arcsin ax = \sin^{-1} ax$ | $\dfrac{a}{\sqrt{1 - a^2 x^2}}$ |
| $\arccos ax = \cos^{-1} ax$ | $\dfrac{-a}{\sqrt{1 - a^2 x^2}}$ |
| $\arctan ax = \tan^{-1} ax$ | $\dfrac{a}{1 + a^2 x^2}$ |
| $arc \cot ax = \cot^{-1} ax$ | $\dfrac{-a}{1 + a^2 x^2}$ |
| $\sinh ax$ | $a \cosh ax$ |
| $\cosh ax$ | $a \sinh ax$ |
| $\tanh ax$ | $\dfrac{a}{\cosh^2 ax}$ |
| $\sinh^{-1} ax$ | $\dfrac{a}{\sqrt{1 + a^2 x^2}}$ |
| $\cosh^{-1} ax$ | $\dfrac{a}{\sqrt{a^2 x^2 - 1}}$ |
| $\tanh^{-1} ax$ | $\dfrac{a}{1 - a^2 x^2}$ |
| $u(x) + v(x)$ | $\dfrac{du}{dx} + \dfrac{dv}{dx}$ |
| $u(x)v(x)$ | $u\dfrac{dv}{dx} + v\dfrac{du}{dx}$ |
| $\dfrac{u(x)}{v(x)}$ | $\dfrac{1}{v^2}\left( v\dfrac{du}{dx} - u\dfrac{dv}{dx} \right)$ |
| $\dfrac{1}{v(x)}$ | $\dfrac{-1}{v^2}\dfrac{dv}{dx}$ |
| $y(v(x))$ | $\dfrac{dy}{dv}\dfrac{dv}{dx}$ |
| $y(v(u(x)))$ | $\dfrac{dy}{dv}\dfrac{dv}{du}\dfrac{du}{dx}$ |

## D.4 TABLE OF INTEGRALS

$$\int a\,dx = ax + c\,(c \text{ is an arbitrary constant})$$

$$\int \left[ f(x) \pm g(x) \right]\,dx = \int f(x)\,dx \pm \int g(x)\,dx$$

$$\int x\,dy = xy - \int y\,dx$$

$$\int x^n dx = \frac{x^{n+1}}{n+1} + c,(n \neq -1)$$

$$\int \frac{1}{x}\,dx = \ln|x| + c$$

$$\int e^{ax} dx = \frac{e^{ax}}{a} + c$$

$$\int a^x dx = \frac{a^x}{\ln a} + c \qquad \text{for } (a > 0)$$

$$\int \ln x\,dx = x\ln x - x + c \text{ for } (x > 0)$$

$$\int \sin ax\,dx = \frac{-\cos ax}{a} + c$$

$$\int \cos ax\,dx = \frac{\sin ax}{a} + c$$

$$\int \tan ax\,dx = \frac{-\ln|\cos ax|}{a} + c$$

$$\int \cot ax\,dx = \frac{\ln|\sin ax|}{a} + c$$

$$\int \sec ax\,dx = \frac{-\ln\left(\dfrac{1-\sin ax}{1+\sin ax}\right)}{2a} + c$$

$$\int \csc ax\,dx = \frac{\ln\left(\dfrac{1-\cos ax}{1+\cos ax}\right)}{2a} + c$$

$$\int \frac{1}{x^2 + a^2}\,dx = \frac{\tan^{-1}\left(\dfrac{x}{a}\right)}{a} + c$$

$$\int \frac{1}{x^2 - a^2}\,dx = \frac{\ln\left(\dfrac{x-a}{x+a}\right)}{2a} + c \text{ or } \frac{\tanh^{-1}\left(\dfrac{x}{a}\right)}{a} + c$$

$$\int \frac{1}{a^2 - x^2}\, dx = \frac{\ln\left(\frac{x+a}{x-a}\right)}{2a} + c$$

$$\int \frac{1}{\sqrt{a^2 - x^2}}\, dx = \sin^{-1}\left(\frac{x}{a}\right) + c$$

$$\int \frac{1}{\sqrt{a^2 + x^2}}\, dx = \frac{\sinh^{-1}\left(\frac{x}{a}\right)}{a} + c \quad \text{or} \quad \ln\left(x + \sqrt{x^2 + a^2}\right) + c$$

$$\int \frac{1}{\sqrt{x^2 - a^2}}\, dx = \ln\left(x + \sqrt{x^2 - a^2}\right) + c$$

$$\int \frac{1}{x\sqrt{x^2 - a^2}}\, dx = \frac{\sec^{-1}\left(\frac{x}{a}\right)}{a} + c$$

$$\int x e^{ax}\, dx = \frac{(ax - 1)e^{ax}}{a^2} + c$$

$$\int x \cos ax\, dx = \frac{\cos ax + ax \sin ax}{a^2} + c$$

$$\int x \sin ax\, dx = \frac{\sin ax - ax \cos ax}{a^2} + c$$

$$\int x \ln x\, dx = \frac{x^2}{2} \ln x - \frac{x^2}{4} + c$$

$$\int x e^{ax}\, dx = \frac{e^{ax}(ax - 1)}{a^2} + c$$

$$\int e^{ax} \cos bx\, dx = \frac{e^{ax}(a \cos bx + b \sin bx)}{a^2 + b^2} + c$$

$$\int e^{ax} \sin bx\, dx = \frac{e^{ax}(-b \cos bx + a \sin bx)}{a^2 + b^2} + c$$

$$\int \sin^2 x\, dx = \frac{x}{2} - \frac{\sin 2x}{4} + c$$

$$\int \cos^2 x\, dx = \frac{x}{2} + \frac{\sin 2x}{4} + c$$

$$\int \tan^2 x\, dx = \tan x - x + c$$

$$\int \cot^2 x\, dx = -\cot x - x + c$$

$$\int \sec^2 x \, dx = \tan x + c$$

$$\int \csc^2 x \, dx = -\cot x + c$$

$$\int \sec x \, \tan x \, dx = \sec x + c$$

$$\int \csc x \, \cot x \, dx = -\csc x + c$$

## D.5  SUMMATIONS (SERIES)

### D.5.1  Finite Element of Terms

$$\sum_{n=0}^{N} a^n = \frac{1 - a^{N+1}}{1 - a}; \quad \sum_{n=0}^{N} na^n = a\left(\frac{1 - (N+1)a^N + Na^{N+1}}{(1-a)^2}\right)$$

$$\sum_{n=0}^{N} n = \frac{N(N+1)}{2}; \quad \sum_{n=0}^{N} n^2 = \frac{N(N+1)(2N+1)}{6}$$

$$\sum_{n=0}^{N} n(n+1) = \frac{N(N+1)(N+2)}{3};$$

$$(a+b)^N = \sum_{n=0}^{N} NC_n a^{N-n} b^n, \text{ where } NC_n = NC_{N-n} = \frac{NP_n}{n!} = \frac{N!}{(N-n)!n!}$$

### D.5.2  Infinite Element of Terms

$$\sum_{n=0}^{\infty} x^n = \frac{1}{1-x}, \quad (|x| < 1); \quad \sum_{n=0}^{\infty} nx^n = \frac{1}{(1-x)^2}, \quad (|x| < 1)$$

$$\sum_{n=0}^{\infty} n^k x^n = \lim_{a \to 0} (-1)^k \frac{\partial^k}{\partial a^k}\left(\frac{x}{x - e^{-a}}\right), \quad (|x| < 1); \quad \sum_{n=0}^{\infty} \frac{(-1)^n}{2n+1} = 1 - \frac{1}{3} + \frac{1}{5} - \frac{1}{7} + \dots = \frac{1}{4}\pi$$

$$\sum_{n=0}^{\infty} \frac{1}{n^2} = 1 + \frac{1}{2^2} + \frac{1}{3^2} + \frac{1}{4^2} + \dots = \frac{1}{6}\pi^2$$

$$e^x = \sum_{n=0}^{\infty} \frac{x^n}{n!} = 1 + \frac{1}{1!}x + \frac{1}{2!}x^2 + \frac{1}{3!}x^3 + \dots$$

$$a^x = \sum_{n=0}^{\infty} \frac{(\ln a)^n x^n}{n!} = 1 + \frac{(\ln a)x}{1!} + \frac{(\ln a)^2 x^2}{2!} + \frac{(\ln a)^3 x^3}{3!} + \dots$$

$$\ln(1 \pm x) = -\sum_{n=1}^{\infty} \frac{(\pm 1)^n x^n}{n} = \pm x - \frac{x^2}{2} \pm \frac{x^3}{3} - \dots, \quad (|x| < 1)$$

$$\sin x = \sum_{n=0}^{\infty} \frac{(-1)^n x^{2n+1}}{(2n+1)!} = x - \frac{x^3}{3!} + \frac{x^5}{5!} - \frac{x^7}{7!} + \dots$$

$$\cos x = \sum_{n=0}^{\infty} \frac{(-1)^n x^{2n}}{(2n)!} = 1 - \frac{x^2}{2!} + \frac{x^4}{4!} - \frac{x^6}{6!} + \dots$$

$$\tan x = x + \frac{x^3}{3} + \frac{2x^5}{15} + \dots, \quad (|x| < 1)$$

$$\tan^{-1} x = \sum_{n=0}^{\infty} \frac{(-1)^n x^{2n+1}}{2n+1} = x - \frac{x^3}{3} + \frac{x^5}{5} - \frac{x^7}{7} + \dots, \quad (|x| < 1)$$

## D.6  LOGARITHMIC IDENTITIES

$$\log_e a = \ln a \text{ (natural logarithm)}$$

$$\log_{10} a = \log a \text{ (common logarithm)}$$

$$\log ab = \log a + \log b$$

$$\log \frac{a}{b} = \log a - \log b$$

$$\log a^n = n \log a$$

## D.7  EXPONENTIAL IDENTITIES

$$e^x = 1 + x + \frac{x^2}{2!} + \frac{x^3}{3!} + \frac{x^4}{4!} + \dots, \text{ where } e \simeq 2.7182$$

$$e^x e^y = e^{x+y}$$

$$\left(e^x\right)^n = e^{nx}$$

$$\ln e^x = x$$

## D.8  APPROXIMATIONS FOR SMALL QUANTITIES

If $|a| \ll 1$, then

$$\ln(1+a) \simeq a$$

$$e^a \simeq 1 + a$$

$$\sin a \simeq a$$

$$\cos a \simeq 1$$

$$\tan a \simeq a$$

$$(1 \pm a)^n \simeq 1 \pm na$$

# D.9  VECTORS

## D.9.1  Vector Derivatives

### 1.  Cartesian Coordinates

| Coordinates | $(x, y, z)$ |
|---|---|
| Vector | $\mathbf{A} = A_x \mathbf{a}_x + A_y \mathbf{a}_y + A_z \mathbf{a}_z$ |
| Gradient | $\nabla \mathbf{A} = \dfrac{\partial A}{\partial x} \mathbf{a}_x + \dfrac{\partial A}{\partial y} \mathbf{a}_y + \dfrac{\partial A}{\partial z} \mathbf{a}_z$ |
| Divergence | $\nabla \cdot \mathbf{A} = \dfrac{\partial A_x}{\partial x} + \dfrac{\partial A_y}{\partial y} + \dfrac{\partial A_z}{\partial z}$ |
| Curl | $\nabla \times \mathbf{A} = \begin{vmatrix} \mathbf{a}_x & \mathbf{a}_y & \mathbf{a}_z \\ \dfrac{\partial}{\partial x} & \dfrac{\partial}{\partial y} & \dfrac{\partial}{\partial z} \\ A_x & A_y & A_z \end{vmatrix} =$ $= \left( \dfrac{\partial A_z}{\partial y} - \dfrac{\partial A_y}{\partial z} \right) \mathbf{a}_x + \left( \dfrac{\partial A_x}{\partial z} - \dfrac{\partial A_z}{\partial x} \right) \mathbf{a}_y + \left( \dfrac{\partial A_y}{\partial x} - \dfrac{\partial A_x}{\partial y} \right) \mathbf{a}_z$ |
| Laplacian | $\nabla^2 \mathbf{A} = \dfrac{\partial^2 A}{\partial x^2} + \dfrac{\partial^2 A}{\partial y^2} + \dfrac{\partial^2 A}{\partial z^2}$ |

### 2.  Cylindrical Coordinates

| Coordinates | $(\rho, \phi, z)$ |
|---|---|
| Vector | $\mathbf{A} = A_\rho \mathbf{a}_\rho + A_\phi \mathbf{a}_\phi + A_z \mathbf{a}_z$ |
| Gradient | $\nabla \mathbf{A} = \dfrac{\partial A}{\partial \rho} \mathbf{a}_\rho + \dfrac{1}{\rho} \dfrac{\partial A}{\partial \phi} \mathbf{a}_\phi + \dfrac{\partial A}{\partial z} \mathbf{a}_z$ |

| Divergence | $\nabla \cdot \mathbf{A} = \dfrac{1}{\rho}\dfrac{\partial}{\partial\rho}(\rho A_\rho) + \dfrac{1}{\rho}\dfrac{\partial A_\phi}{\partial\phi} + \dfrac{\partial A_z}{\partial z}$ |
|---|---|
| Curl | $\nabla \times \mathbf{A} = \dfrac{1}{\rho}\begin{vmatrix} \mathbf{a}_\rho & \rho\,\mathbf{a}_\phi & \mathbf{a}_z \\ \dfrac{\partial}{\partial\rho} & \dfrac{\partial}{\partial\phi} & \dfrac{\partial}{\partial z} \\ A_\rho & \rho A_\phi & A_z \end{vmatrix} =$ <br><br> $= \left( \dfrac{1}{\rho}\dfrac{\partial A_z}{\partial\phi} - \dfrac{\partial A_\phi}{\partial z} \right)\mathbf{a}_\rho + \left( \dfrac{\partial A_\rho}{\partial z} - \dfrac{\partial A_z}{\partial\rho} \right)\mathbf{a}_\phi + \dfrac{1}{\rho}\left( \dfrac{\partial}{\partial x}(\rho A_\phi) - \dfrac{\partial A_\rho}{\partial\phi} \right)\mathbf{a}_z$ |
| Laplacian | $\nabla^2 \mathbf{A} = \dfrac{1}{\rho}\dfrac{\partial}{\partial\rho}\left( \rho\dfrac{\partial A}{\partial\rho} \right) + \dfrac{1}{\rho^2}\dfrac{\partial^2 A}{\partial\phi^2} + \dfrac{\partial^2 A}{\partial z^2}$ |

## 3. Spherical Coordinates

| Coordinates | $(r,\theta,\phi)$ |
|---|---|
| Vector | $\mathbf{A} = A_r\mathbf{a}_r + A_\theta\mathbf{a}_\theta + A_\phi\mathbf{a}_\phi$ |
| Gradient | $\nabla A = \dfrac{\partial A}{\partial r}\mathbf{a}_r + \dfrac{1}{r}\dfrac{\partial A}{\partial\theta}\mathbf{a}_\theta + \dfrac{1}{r\sin\theta}\dfrac{\partial A}{\partial\phi}\mathbf{a}_\phi$ |
| Divergence | $\nabla \cdot \mathbf{A} = \dfrac{1}{r^2}\dfrac{\partial}{\partial r}(r^2 A_r) + \dfrac{1}{r\sin\theta}\dfrac{\partial}{\partial\theta}(A_\theta\sin\theta) + \dfrac{1}{r\sin\theta}\dfrac{\partial A_\phi}{\partial\phi}$ |
| Curl | $\nabla \times \mathbf{A} = \dfrac{1}{r^2\sin\theta}\begin{vmatrix} \mathbf{a}_r & r\mathbf{a}_\theta & (r\sin\theta)\mathbf{a}_\phi \\ \dfrac{\partial}{\partial r} & \dfrac{\partial}{\partial\theta} & \dfrac{\partial}{\partial\phi} \\ A_r & rA_\theta & (r\sin\theta)A_\phi \end{vmatrix} =$ <br><br> $= \dfrac{1}{r\sin\theta}\left( \dfrac{\partial}{\partial\theta}(A_\phi\sin\theta) - \dfrac{\partial A_\theta}{\partial\phi} \right)\mathbf{a}_r + \dfrac{1}{r}\left( \dfrac{1}{\sin\theta}\dfrac{\partial A_r}{\partial\phi} - \dfrac{\partial}{\partial r}(rA_\phi) \right)\mathbf{a}_\theta$ <br><br> $+ \dfrac{1}{r}\left( \dfrac{\partial}{\partial r}(rA_\theta) - \dfrac{\partial A_r}{\partial\theta} \right)\mathbf{a}_\phi$ |
| Laplacian | $\nabla^2 \mathbf{A} = \dfrac{1}{r^2}\dfrac{\partial}{\partial r}\left( r^2\dfrac{\partial A}{\partial r} \right) + \dfrac{1}{r^2\sin\theta}\dfrac{\partial}{\partial\theta}\left( \sin\theta\dfrac{\partial A}{\partial\theta} \right) + \dfrac{1}{r^2\sin^2\theta}\dfrac{\partial^2 A}{\partial\phi^2}$ |

## D.9.2 Vector Identity

### 1. Triple Products

$$\mathbf{A}\cdot(\mathbf{B}\times\mathbf{C}) = \mathbf{B}\cdot(\mathbf{C}\times\mathbf{A}) = \mathbf{C}\cdot(\mathbf{A}\times\mathbf{B})$$
$$\mathbf{A}\times(\mathbf{B}\times\mathbf{C}) = \mathbf{B}(\mathbf{A}\cdot\mathbf{C}) - \mathbf{C}(\mathbf{A}\cdot\mathbf{B})$$

**2.** Product Rules

$$\nabla(fg) = f(\nabla g) + g(\nabla f)$$

$$\nabla(\mathbf{A} \cdot \mathbf{B}) = \mathbf{A} \times (\nabla \times \mathbf{B}) + \mathbf{B} \times (\nabla \times \mathbf{A}) + (\mathbf{A} \cdot \nabla)\mathbf{B} + (\mathbf{B} \cdot \nabla)\mathbf{A}$$

$$\nabla \cdot (f\mathbf{A}) = f(\nabla \cdot \mathbf{A}) + \mathbf{A} \cdot (\nabla f)$$

$$\nabla \cdot (\mathbf{A} \times \mathbf{B}) = \mathbf{B} \cdot (\nabla \times \mathbf{A}) - \mathbf{A} \cdot (\nabla \times \mathbf{B})$$

$$\nabla \times (f\mathbf{A}) = f(\nabla \times \mathbf{A}) - \mathbf{A} \times (\nabla f) = \nabla \times (f\mathbf{A}) = f(\nabla \times \mathbf{A}) + (\nabla f) \times \mathbf{A}$$

$$\nabla \times (\mathbf{A} \times \mathbf{B}) = (\mathbf{B} \cdot \nabla)\mathbf{A} - (\mathbf{A} \cdot \nabla)\mathbf{B} + \mathbf{A}(\nabla \cdot \mathbf{B}) - \mathbf{B}(\nabla \cdot \mathbf{A})$$

**3.** Second Derivative

$$\nabla \cdot (\nabla \times \mathbf{A}) = 0$$

$$\nabla \times (\nabla f) = 0$$

$$\nabla \cdot (\nabla f) = \nabla^2 f$$

$$\nabla \times (\nabla \times \mathbf{A}) = \nabla(\nabla \cdot \mathbf{A}) - \nabla^2 \mathbf{A}$$

**4.** Addition, Division, and Power Rules

$$\nabla(f + g) = \nabla f + \nabla g$$

$$\nabla \cdot (\mathbf{A} + \mathbf{B}) = \nabla \cdot \mathbf{A} + \nabla \cdot \mathbf{B}$$

$$\nabla \times (\mathbf{A} \times \mathbf{B}) = \nabla \times \mathbf{A} + \nabla \times \mathbf{B}$$

$$\nabla\left(\frac{f}{g}\right) = \frac{g(\nabla f) - f(\nabla g)}{g^2}$$

$$\nabla f^n = nf^{n-1}\nabla f \quad (\text{n = integer})$$

### D.9.3  Fundamental Theorems

**1.** Gradient Theorem

$$\int_a^b (\nabla f) \cdot d\mathbf{l} = f(b) - f(a)$$

**2.** Divergence Theorem

$$\int_{volume} (\nabla \cdot \mathbf{A})dv = \oint_{surface} \mathbf{A} \cdot d\mathbf{s}$$

**3.** Curl (Stokes) Theorem

$$\int\limits_{surface} (\nabla \times \mathbf{A}) \cdot d\mathbf{s} = \oint\limits_{line} \mathbf{A} \cdot d\mathbf{l}$$

**4.** $\oint\limits_{line} f d\mathbf{l} = - \int\limits_{surface} \nabla f \times d\mathbf{s}$

**5.** $\oint\limits_{surface} f d\mathbf{s} = \int\limits_{volume} \nabla f dv$

**6.** $\oint\limits_{surface} \mathbf{A} \times d\mathbf{s} = - \int\limits_{volume} \nabla \times \mathbf{A} dv$

# BIBLIOGRAPHY

1. H. Anton and C. Rorres, *Elementary Linear Algebra: Applications Version*, 8th Edition. John Wiley& Sons, 2000.

2. B. O'Neill, *Elementary Differential Geometry*, Academic Press, 1966.

3. A. V. Pogorelov, *Analytical Geometry*, Mir Publishers, 1980.

4. H. Cartan, *Differential Forms*, Hermann Publishers, 1970.

5. D. Lovelock and H. Rund, *Tensors, Differential Forms, and Variational Principles*, Dover Publications, Inc., 1989.

6. R. Larson, R. P. Hostetler, B. H. Edwards, and D. E. Heyd, *Calculus with Analytic Geometry*, Houghton Mifflin Company, 2002.

7. R. T. Seeley, *Calculus of Several Variables*, Scott, Foresman and Company, 1970.

8. W. Kaplan, *Advanced Calculus*, 2nd Edition, Addison-Wesley Publishing Company, 1973.

9. L. J. Goldstein, D. C. Lay, and D. I. Schneider, *Calculus and Its Applications*, 7th Edition, Prentice Hall, 1996.

10. G. James, *Advanced Modern Engineering Mathematics*, 3rd Edition, Pearson-Prentice Hall, 2004.

11. M. R. Spiegel, *Schaum's Outline Series Theory and Problems of Advanced Calculus*, McGraw-Hill, 1963.

12. P. M. Fitzpatrick, *Advanced Calculus: A Course in Mathematical Analysis*, PWS Publishing Company, 1996.

13. J. Stewart, *Calculus*, 5th Edition, Thomson Learning Inc., 2003.

14. E. J. Purcell and D. Varberg, *Calculus with Analytic Geometry*, 5th Edition, Prentice-Hall, Inc., 1987.

15. D. G. Zill and M. R. Cullen, *Advanced Engineering Mathematics*, 2nd Edition, Jones and Bartlett Publishers, 2000.

16. E. Kreyszig, *Advanced Engineering Mathematics*, 9th Edition, John Wiley & Sons, Inc., 2006.

17. H. E. Newell, Jr., *Vector Analysis*, McGraw-Hill Book Company, Inc., 1955.

18. P. V. O'Neil, *Advanced Engineering Mathematics*, 5th Edition, Brooks/Cole, 2003.

19. D. G. Duffy, *Advanced Engineering Mathematics with MATLAB*, 2nd Edition, Chapman& Hall/CRC, 2003.

20. R. Osserman, *Two-Dimensional Calculus*, Harcourt, Brace & World, Inc., 1968.

21. S. Lang, *Calculus of Several Variables*, 3rd Edition, Springer-Verlag, 1987.

22. S. Dineen, *Multivariate Calculus and Geometry*, 2nd Edition, Springer-Verlag, 2001.

23. P. S. Clarke, Jr., *Calculus with Analytic Geometry*, D.C. Health and Company, 1974.

24. H. S. Wilf, *Calculus and Linear Algebra*, Harcourt, Brace & World, Inc., 1966.

25. J. H. Hubbard and B. B. Hubbard, *Vector Calculus, Linear Algebra, and Differential Forms: A Unified Approach*, Prentice Hall, 1999.

26. B. Willcox, *Calculus of Several Variables*, Houghton Mifflin Company, 1971.

27. M. R. Spiegel, *Schaum's Outline Series Theory and Problems of Vector Analysis and An Introduction to Tensor Analysis*, McGraw-Hill, 1959.

28. K. A. Stroud and D. J. Booth, *Engineering Mathematics*, 6th Edition, Industrial Press, Inc., 2007.

29. S. Abbott, *Understanding Analysis*, Springer-Verlag, 2001.

**30.** M. N. O. Sadiku, *Elements of Electromagnetics*, 3rd Edition, Oxford University Press, 2001.

**31.** F. Ayres, Jr., *Schaum's Outline Series Theory and Problems of Modern Algebra*, McGraw-Hill, 1959.

**32.** R. C. Buck and E. F. Buck, *Advanced Calculus*, McGraw-Hill, 1956.

**33.** E. W. Swokowski, *Calculus with Analytic Geometry*, 2nd Edition, Prindle, Weber & Schmidt, 1979.

**34.** S. I. Grossman, *Multivariable Calculus, Linear algebra, and Differential Equations*, 2nd Edition, Academic Press, Inc., 1986.

**35.** W. G. May, *Linear Algebra*, Scott, Foresman, and Company, 1970.

**36.** H. Schneider and G. P. Barker, *Matrices and Linear Algebra*, 2nd Edition, Holt, Rinehart and Winston, Inc., 1968.

**37.** M. O'nan, *Linear Algebra*, 2nd Edition, Harcourt Brace Jovanovich, Inc.

**38.** T. Blyth and E. Robertson, *Basic Linear Algebra*, 2nd Edition, Springer, 2002.

**39.** J. R. Munkres, *Analysis on Manifolds*, Westview Press, 1997.

**40.** J. E. Marsden and A. Tromba, *Vector Calculus*, 6th Edition, W. H. Freeman, 2001.

**41.** J. H. Hubbard and B. B. Hubbard, *Vector Calculus, Linear Algebra, and Differential Forms: A Unified Approach*, 2nd Edition, Prentice Hall, 2001.

**42.** M. Spivak, *Calculus on Manifolds: A Modern Approach to Classical Theorems of Advanced Calculus*, Westview Press, 1971.

**43.** H. M. Edwards, *Advanced Calculus: A Differential Forms Approach*, Birkhäuser, 1994.

**44.** B. Kolman, *Elementary Linear Algebra*, Macmillan Publishing Co., Inc., 1970.

**45.** L. Smith, *Linear Algebra*, Springer-Verlag, 1978.

**46.** G. Strang, *Linear Algebra and Its Applications*, Academic Press, 1976.

**47.** M. Rahman, *Advanced Vector analysis for Scientist and Engineers*, WIT Press, 2007.

48. M. Rahman and I. Mulolani, *Applied Vector Analysis*, 2nd Edition, CRC Press, 2008.

49. G. Williams, *Linear Algebra with Applications*, 5th Edition, Jones and Bartlett Publishers, 2005.

50. H. Anton, *Elementary Linear Algebra*, 10th Edition, John Wiley & Sons, 2010.

# INDEX

## A

Approximations, 425–426
Arithmetic Mean-Geometric Mean
  Inequality, 157
Arithmetic operations
  in Maple™, 321–322
  in MATLAB, 329

## C

Canonical surfaces, 133–146
Cauchy-Bunyakovsky-Schwarz
  Inequality, 21, 154
Cavalieri's principle, 117–119
Chasles' rule, 7, 24
Cross product, 87–98
Cylindrical helix, 147

## D

Determinants
  in three dimensions, 104–111
  in two dimensions, 44–53
Differentiation
  derivatives
    definition of, 189–193
    of function, 190, 193
    gradients and directional, 207–218

extrema, 222–228
Jacobi matrix, 193–207
Lagrange multipliers, 229–234
Levi-Civitta and Einste,
  218–222
limits and continuity, 176–189
multivariable functions, 169–176
topology, 163–169
Dihedral angle, 121
  *vs.* polyhedral angles, 122
  rectilinear angle of, 121
  *vs.* trihedral angles, 122

## E

Einstein's Summation Convention,
  218–219
Ellipsoid, 144–145
Exponential identities, 425

## F

Formulas
  approximations, 425–426
  exponential identities, 425
  hyperbolic functions, 419–420
  logarithmic identities, 425
  summations (series), 424–425

table of derivatives, 420–421
table of integrals, 422–424
trigonometric identities,
    417–419
vectors, 426–429
Fubini's Theorem, 257
Fundamental parallelogram, 30

### G

Gauss' theorem, 307–313
Geometric transformations in two
    dimensions, 34–44
Green's theorem, 307–313

### H

Harmonic Mean-Arithmetic Mean
    Inequality, 159
Harmonic Mean-Geometric Mean
    Inequality, 159
Hessian matrix, 222
Hyperbolic functions, 419–420

### I

Identity transformation, 34
Integration
    change of variables
        in double integrals, 273–280
        in triple integrals, 297–301
    closed and exact forms, 250–256
    differential forms, concept
        of, 235–240
    Green's, Stokes', and Gauss'
        theorems, 307–313
    one-manifold, 243–248
    polar coordinates, 280–287
    surface integrals, 303–306
    three-manifolds, 290–295
    two-manifolds, 257–267
    zero-manifold, 241–242

### J

Jacobi's Identity, 90
Jordan curve, 53

### K

Kroenecker's Delta, 219–220

### L

Lagrange's Identity, 92
Levi-Civitta's Alternating
    Tensor, 220–221
Linear independence, 28–33
Logarithmic identities, 425

### M

Maple™, 50
    arithmetic operations, 321–322
    assignments, 323
    in document mode, 320
    help system, 321
    int command, 326
    limit and diff command, 326
    Matrix command, 326
    percent sign, 323
    plot command, 324–325
    solve command, 323–324
    symbolic calculations, 322
    worksheet mode, 320
MATLAB
    arithmetic operators, 329
        on arrays, 335
    considerations, 360
    default environment, 328
    math functions, 330
    Matrix operations, 331
    named constants, 334
    programming
        if, else, and elseif statements,
            357–359

for loops, 355
rational and logical operators,
354–355
switch statement, 359–360
while loops, 356–357
symbolic expressions, 361–368
three-dimensional plotting, 347–352
two-dimensional plotting
axis command, 343
fplot command, 338
Greek characters, 344–345
hold on and hold off commands,
341
label command, 342
legend command, 343
lines, points, and color types, 337
logarithm scaling, 345
pos value, 343
with special graphics, 346
text and gtext command, 342–343
title command, 342
windows, 328–329
Matrices in three dimensions, 98–104
Matrix operations
in Maple™, 326
in MATLAB, 331
Multidimensional vectors, 152–162

**O**

One-sheet hyperboloid, 140, 144
Orthocentre of the triangle, 25

**P**

Pappus-Guldin Rule, 119–121
Paraboloid
elliptic, 142
hyperbolic, 142–143
Parametric curves
on the plane, 53–70

in space, 146–152
Platonic solids, 125–126
Plotting, MATLAB
three-dimensional plotting, 347–352
two-dimensional plotting
axis command, 343
fplot command, 338
Greek characters, 344–345
hold on and hold off commands,
341
label command, 342
legend command, 343
lines, points, and color types, 337
logarithm scaling, 345
pos value, 343
with special graphics, 346
text and gtext command, 342–343
title command, 342
Points and vectors on the plane, 1–17
Polyhedral angles, 122, 123
Programming in MATLAB
if, else, and elseif statements,
357–359
for loops, 355
rational and logical operators,
354–355
switch statement, 359–360
while loops, 356–357
Pythagorean Theorem, 22

**R**

Repeated partial derivatives, 198

**S**

Scalar product on the plane, 17–28
Spherical trigonometry, 127–132
Stokes' theorem, 307–313
Summations (series), 424–425
Surveyor's Theorem, 49

**T**

Table of derivatives, 420–421
Table of integrals, 422–424
Thales' Theorem, 23, 113
Three-dimensional geometry, 111–117
Triangle Inequality, 22, 154
Trigonometric identities, 417–419
Trihedral angles, 122
Two-sheet hyperboloid, 143

**V**

Vector normalization, 6
Vectors, 426–429
    addition of, 7
    difference of, 7
    and parametric curves
        canonical surfaces, 133–146
        Cavalieri's principle, 117–119
        cross product, 87–98
        determinants, 44–53, 104–111
        geometric transformations, 34–44
        linear independence, 28–33
        matrices in three dimensions,
            98–104
    multidimensional vectors,
        152–162
    Pappus-Guldin Rule, 119–121
    on the plane, 53–70
    points and vectors on the
        plane, 1–17
    scalar product on the
        plane, 17–28
    in space, 70–86, 146–152
    spherical trigonometry,
        127–132
    three-dimensional geometry,
        111–117
    scalar multiplication of, 8
Vector spaces, 1, 70–86

**Y**

Young's Inequality, 155

**Z**

Zero-manifolds, 235, 241, 408
zero matrix, 38
zero vector, 3, 6, 27, 28, 77, 92, 156,
    162, 386